P. Sawamura

INVERSE METHODS FOR ATMOSPHERIC SOUNDING
Theory and Practice

SERIES ON ATMOSPHERIC, OCEANIC AND PLANETARY PHYSICS

Series Editor: F. W. Taylor *(Oxford Univ., UK)*

- Vol. 1: TITAN: The Earth-Like Moon
  Athena Coustenis & Fredric W. Taylor

- Vol. 2: Inverse Methods for Atmospheric Sounding: Theory and Practice
  Rodgers Clive D.

- Vol. 3: Non-LTE Radiative Transfer in the Atmosphere
  M. López-Puertas & F. W. Taylor

- Vol. 4: TITAN: Exploring an Earthlike World (2nd Edition)
  Athena Coustenis & Fredric W. Taylor

Series on Atmospheric, Oceanic and Planetary Physics — Vol. 2

INVERSE METHODS FOR ATMOSPHERIC SOUNDING
Theory and Practice

Clive D. Rodgers
University of Oxford

Singapore • New Jersey • London • Hong Kong

Published by

World Scientific Publishing Co. Pte. Ltd.
5 Toh Tuck Link, Singapore 596224
USA office: 27 Warren Street, Suite 401–402, Hackensack, NJ 07601
UK office: 57 Shelton Street, Covent Garden, London WC2H 9HE

British Library Cataloguing-in-Publication Data
A catalogue record for this book is available from the British Library.

First published 2000
Reprinted 2004, 2008

INVERSE METHODS FOR ATMOSPHERIC SOUNDING: THEORY AND PRACTICE
Copyright © 2000 by World Scientific Publishing Co. Pte. Ltd.

All rights reserved. This book, or parts thereof, may not be reproduced in any form or by any means, electronic or mechanical, including photocopying, recording or any information storage and retrieval system now known or to be invented, without written permission from the Publisher.

For photocopying of material in this volume, please pay a copying fee through the Copyright Clearance Center, Inc., 222 Rosewood Drive, Danvers, MA 01923, USA. In this case permission to photocopy is not required from the publisher.

ISBN-13 978-981-02-2740-1
ISBN-10 981-02-2740-X

Printed in Singapore by B & JO Enterprise

To my wife Terry

Preface

This book is aimed at graduate students and at post-doctoral and other researchers who need to solve inverse problems for atmospheric measurements. The intention is both to provide a well-founded background to the inverse problem and its solution, with which the student will develop a good intuition about the nature of the problem, and to give practical recipes for solving real problems. It has developed from various courses that I have given over many years, including informal seminars to research groups at the National Center for Atmospheric Research in Boulder, Colorado, the Subdepartment of Atmospheric, Oceanic and Atmospheric Physics at Oxford, the National Institute for Water and Air at Lauder, New Zealand, and formal courses given at the 1995 NATO Advanced Study Institute at Val Morin, Canada (Brasseur, 1997) and the 1999 Oxford/RAL Spring Schools on Quantitative Earth Observation, which is to be repeated in 2000 and future years.

The usual approach to solving an atmospheric retrieval problem will consist of several stages: design a forward model to describe the instrument and the physics of the measurement; determine the criterion by which a solution is acceptable as valid; construct a numerical method to find a solution which satisfies the criterion; carry out an error analysis; validate the process by reference to internal diagnostics and independent measurements; and finally attempt to understand how the result obtained is related to reality and examine how much information has been obtained. In this book I go through these stages more or less in the reverse order to try to develop the reader's intuition about the nature of the inverse problem. I first discuss the information content of remote measurements and the corresponding retrievals, introducing the Bayesian approach as a conceptual background for the problem, then carry out a formal characterisation and error analysis of a generic retrieval to show how retrievals are related to reality, and how their errors can be described. The characterisation and error analysis provides a basis from which we can understand what is meant by 'optimality' in designing an inverse method. Next I discuss numerical methods for solving both linear and nonlinear problems, optimally and otherwise. After a detour into related issues of Kalman filtering and assimilation of meteorological observations, both of which are concerned with the time variation of measured quantities, I discuss forward modelling, prior information and systematic approaches to the design and validation of remote sounding data analysis systems.

A set of exercises for the student are given within the text. Many of these are part of the algebraic or conceptual development of the topic being described, so they should be attempted. Full answers are given in Appendix B.

There is a range of mathematical tools and concepts which are of particular value for the understanding of inverse methods. These include

Linear algebra: matrix operations; eigenvector and singular vector decomposition; vector spaces; quadratic forms; calculus applied to matrices.

Probability: multivariate probability density functions; Gaussian and other distributions; conditional probabilities; Bayes' theorem.

Estimation theory: Maximum likelihood and maximum *a posteriori* estimates;

Information theory: Shannon information content.

Numerical methods: Minimisation methods; numerical integration; interpolation; least squares fitting; eigenvector and singular vector decomposition;

Mostly some knowledge of these concepts will be assumed, but some will be introduced in the text. Basic mathematical background is given in appendix A.

I use the notation developed for Rodgers (1976) with a few minor changes, the most noticeable being the use of **G** for gain matrix (after Kalman gain) rather than **D** for contribution function, which was not mnemonic. Ide *et al.* (1997) have proposed a standard notation for data assimilation and retrieval, but I have found it insufficiently flexible for the purposes of this book. So many different variants of state and measurement vectors and covariance matrices are required that the use of subscripts and other diacritical marks on **x**, **y** and **S** are preferred to different letters for each variant. Appendix C lists the main notation used.

There are several recent books that treat remote atmospheric measurement techniques, including Hanel, Conrath, Jennings and Samuelson (1992), Kidder and Vonder Haar (1995) and Stephens (1994), but few that deal in detail with inverse theory for atmospheric problems since Twomey's classic 1977 book, recently reprinted by Dover (Twomey, 1996). In other areas, Menke (1989), Tarantola (1987) and Parker (1994) in solid Earth studies and Daley (1991) in atmospheric data assimilation are useful, as are the more general books of Deutsch (1965) on estimation theory, Jazwinski (1970) on Kalman filtering and Gelb (1974) on both.

Acknowledgements

It was fun to have been in almost at the start of the subject, at least in the context of satellite remote sounding, and to have known most of the major players in the early days, particularly Lewis Kaplan and Jean King who started it all by realising that satellite remote sounding could be done, Dave Wark and Rudi Hanel who made it happen with the earliest satellite spectrometers, Carl Mateer, Henry Fleming, Neil Strand, Ed Westwater, Sean Twomey, Barney Conrath, Dave Staelin and Bill Smith who were among the pioneers of using inverse methods based on estimation theory, Moustafa Chahine, with whom I had many pleasant arguments, Bob Parker who introduced us to the Solid Earth community who were developing similar techniques, and many, many others. I would like to thank everyone who has been responsible for moving this subject forward, including all those whom I have not found the opportunity to reference in this book.

Many people have made helpful comments on early drafts, and I would particularly like to thank Kevin Bowman, Edwin Sarkissian, Randy Van Valkenberg, Alan O'Neill, Alan Lambert, Andrew Collard, Arun Gopalan, Tilman Steck, Brian Connor, Anu Dudhia and Victoria Jay for their contributions. However there are no doubt still many errors and deficiencies. I will welcome comments addressed to me at c.rodgers@physics.ox.ac.uk, and will publicise errata on my web site at www.atm.ox.ac.uk/user/rodgers.

My especial thanks go to John Gille and National Center for Atmospheric Research for hosting me for three sabbaticals, during which I developed some of the ideas presented here, and in the last of which wrote the bulk of this book.

C. D. Rodgers   Oxford, January 2000

The Reprinting of 2008

It is gratifying that this book is still of sufficient interest to require reprinting again. For this printing I have corrected all of the errors, deficiencies and typos that have been brought to my attention. I would especially like to thank Randy Van Valkenberg and Daniela Wurl for their particularly plentiful harvests of mistakes, as well as Justus Notholt, Yasjka Meijer, Bojan Bojkov, Susan Sund Kukawik, Anu Dudhia, Javier Martin-Torres and Robert Watson. For details of changes, see the errata to the first printing on my web site.

C. D. Rodgers   Oxford, February 2008

Contents

Preface   vii

Chapter 1   Introduction   1
1.1 The Beginnings   2
1.2 Atmospheric Remote Sounding Methods   3
     1.2.1 Thermal emission nadir and limb sounders   3
     1.2.2 Scattered solar radiation   4
     1.2.3 Absorption of solar radiation   6
     1.2.4 Active techniques   6
1.3 Simple Solutions to the Inverse Problem   7

Chapter 2   Information Aspects   13
2.1 Formal Statement of the Problem   13
     2.1.1 State and measurement vectors   13
     2.1.2 The forward model   14
     2.1.3 Weighting function matrix   15
     2.1.4 Vector spaces   15
2.2 Linear Problems without Measurement Error   17
     2.2.1 Subspaces of state space   17
     2.2.2 Identifying the null space and the row space   18
2.3 Linear Problems with Measurement Error   20
     2.3.1 Describing experimental error   20
     2.3.2 The Bayesian approach to inverse problems   21
            2.3.2.1 Bayes' theorem   22
            2.3.2.2 Example: The Linear problem with Gaussian statistics   24
2.4 Degrees of Freedom   27
     2.4.1 How many independent quantities can be measured?   27
     2.4.2 Degrees of freedom for signal   29
2.5 Information Content of a Measurement   32
     2.5.1 The Fisher information matrix   32

	2.5.2	Shannon information content	33
		2.5.2.1 Entropy of a probability density function	33
		2.5.2.2 Entropy of a Gaussian distribution	34
		2.5.2.3 Information content in the linear Gaussian case	36
2.6	The Standard Example: Information Content and Degrees of Freedom		37
2.7	Probability Density Functions and the Maximum Entropy Principle		40

Chapter 3 Error Analysis and Characterisation 43

3.1	Characterisation	43
	3.1.1 The forward model	43
	3.1.2 The retrieval method	44
	3.1.3 The transfer function	45
	3.1.4 Linearisation of the transfer function	45
	3.1.5 Interpretation	46
	3.1.6 Retrieval method parameters	47
3.2	Error Analysis	48
	3.2.1 Smoothing error	48
	3.2.2 Forward model parameter error	49
	3.2.3 Forward model error	50
	3.2.4 Retrieval noise	50
	3.2.5 Random and systematic error	50
	3.2.6 Representing covariances	51
3.3	Resolution	52
3.4	The Standard Example: Linear Gaussian Case	55
	3.4.1 Averaging kernels	56
	3.4.2 Error components	58
	3.4.3 Modelling error	60
	3.4.4 Resolution	61

Chapter 4 Optimal Linear Inverse Methods 65

4.1	The Maximum a Posteriori Solution	66
	4.1.1 Several independent measurements	68
	4.1.2 Independent components of the state vector	69
4.2	Minimum Variance Solutions	71
4.3	Best Estimate of a Function of the State Vector	73
4.4	Separately Minimising Error Components	73
4.5	Optimising Resolution	74

Chapter 5 Optimal Methods for Non-linear Inverse Problems 81

5.1	Determination of the Degree of Nonlinearity	82
5.2	Formulation of the Inverse Problem	83
5.3	Newton and Gauss–Newton Methods	85
5.4	An Alternative Linearisation	86

5.5	Error Analysis and Characterisation		86
5.6	Convergence		87
	5.6.1	Expected convergence rate	87
	5.6.2	A popular mistake	88
	5.6.3	Testing for convergence	89
	5.6.4	Testing for correct convergence	90
	5.6.5	Recognising and dealing with slow convergence	91
5.7	Levenberg–Marquardt Method		92
5.8	Numerical Efficiency		93
	5.8.1	Which formulation for the linear algebra?	93
		5.8.1.1 The n-form	94
		5.8.1.2 The m-form	97
		5.8.1.3 Sequential updating	97
	5.8.2	Computation of derivatives	98
	5.8.3	Optimising representations	99

Chapter 6   Approximations, Short Cuts and Ad-hoc Methods   101

6.1	The Constrained Exact Solution	101
6.2	Least Squares Solutions	105
	6.2.1 The overconstrained case	105
	6.2.2 The underconstrained case	106
6.3	Truncated Singular Vector Decomposition	107
6.4	Twomey–Tikhonov	108
6.5	Approximations for Optimal Methods	110
	6.5.1 Approximate *a priori* and its covariance	110
	6.5.2 Approximate measurement error covariance	111
	6.5.3 Approximate weighting functions	111
6.6	Direct Multiple Regression	113
6.7	Linear Relaxation	114
6.8	Nonlinear Relaxation	116
6.9	Maximum Entropy	118
6.10	Onion Peeling	119

Chapter 7   The Kalman Filter   121

7.1	The Basic Linear Filter	122
7.2	The Kalman Smoother	124
7.3	The Extended Filter	125
7.4	Characterisation and Error Analysis	126
7.5	Validation	127

Chapter 8   Global Data Assimilation   129

8.1	Assimilation as a Inverse Problem	129
8.2	Methods for Data Assimilation	130

	8.2.1	Successive correction methods 130
	8.2.2	Optimal interpolation 131
	8.2.3	Adjoint methods 132
	8.2.4	Kalman filtering 134
8.3	Preparation of Indirect Measurements for Assimilation 135
	8.3.1	Choice of profile representation 137
	8.3.2	Linearised measurements 137
	8.3.3	Systematic errors 138
	8.3.4	Transformation of a characterised retrieval 139

Chapter 9 Numerical Methods for Forward Models and Jacobians 141

9.1 The Equation of Radiative Transfer 141
9.2 The Radiative Transfer Integration 143
9.3 Derivatives of Forward Models: Analytic Jacobians 145
9.4 Ray Tracing 147
    9.4.1 Choosing a coordinate system 148
    9.4.2 Ray tracing in radial coordinates 149
    9.4.3 Horizontally homogeneous case 149
    9.4.4 The general case 151
9.5 Transmittance Modelling 152
    9.5.1 Line-by-line modelling 153
    9.5.2 Band transmittance 154
    9.5.3 Inhomogeneous paths 155
        9.5.3.1 Curtis–Godson approximation 155
        9.5.3.2 Emissivity growth approximation 156
        9.5.3.3 McMillin–Fleming method 156
        9.5.3.4 Multiple absorbers 157

Chapter 10 Construction and Use of Prior Constraints 159

10.1 Nature of *a Priori* 159
10.2 Effect of Prior Constraints on a Retrieval 161
10.3 Choice of Prior Constraints 162
    10.3.1 Retrieval grid 162
        10.3.1.1 Transformation between grids 162
        10.3.1.2 Choice of grid for maximum likelihood retrieval 163
        10.3.1.3 Choice of grid for maximum *a priori* retrieval 164
    10.3.2 *Ad hoc* Soft constraints 165
        10.3.2.1 Smoothness constraints 165
        10.3.2.2 Markov process 165
    10.3.3 Estimating *a priori* from real data 166
        10.3.3.1 Estimating *a priori* from independent sources 166
        10.3.3.2 Maximum entropy and the estimation of *a priori* . . . 166
    10.3.4 Validating and improving *a priori* with indirect measurements . 168

	10.3.4.1 The nearly linear case	169
	10.3.4.2 The moderately non-linear case	170
10.4	Using Retrievals Which Contain *a Priori*	171
	10.4.1 Taking averages of sets of retrievals	172
	10.4.2 Removing *a priori*	172

Chapter 11 Designing an Observing System — 175

11.1 Design and Optimisation of Instruments 175
    11.1.1 Forward model construction 176
    11.1.2 Retrieval method and diagnostics 177
    11.1.3 Optimisation 178
    11.1.4 Specifying requirements for the accuracy of parameters 179
11.2 Operational Retrieval Design 179
    11.2.1 Forward model construction 180
    11.2.2 State vector choice 180
    11.2.3 Choice of vertical grid coordinate 181
        11.2.3.1 Choice of parameters describing constituents 182
    11.2.4 *A priori* information 183
    11.2.5 Retrieval method 183
    11.2.6 Diagnostics 183

Chapter 12 Testing and Validating an Observing System — 185

12.1 Error Analysis and Characterisation 186
12.2 The χ^2 Test 187
12.3 Quantities to be Compared and Tested 188
    12.3.1 Internal consistency 188
    12.3.2 Does the retrieval agree with the measurement? 189
    12.3.3 Consistency with the *a priori* 190
        12.3.3.1 Measured signal and *a priori* 190
        12.3.3.2 Retrieval and *a priori* 191
        12.3.3.3 Comparison of the retrieved signal and the *a priori* . . 191
12.4 Intercomparison of Different Instruments 192
    12.4.1 Basic requirements for intercomparison 192
    12.4.2 Direct comparison of indirect measurements 193
    12.4.3 Comparison of linear functions of measurements 194

Appendix A Algebra of Matrices and Vectors — 197

A.1 Vector Spaces 197
A.2 Eigenvectors and Eigenvalues 199
A.3 Principal Axes of a Quadratic Form 201
A.4 Singular Vector Decomposition 201
A.5 Determinant and Trace 203
A.6 Calculus with Matrices and Vectors 204

Appendix B   Answers to Exercises **207**

Appendix C   Terminology and Notation **225**
C.1   Summary of Terminology 225
C.2   List of Symbols Used 227

Bibliography 231

Index 237

Chapter 1

Introduction

Remote or indirect measurements are used for a wide variety of purposes, from examining the structure of the nucleus to the structure of stars, from oil exploration to medical tomography. Whenever direct measurements are difficult or expensive, remote measurements are used, although they often bring in their wake complex problems of interpretation generally known as *Inverse Problems*. The atmosphere is a typical subject for which remote measurements are cost effective, and for which inverse methods are required. Much of the conceptual and mathematical approach is common to inverse problems in all subjects, and has been developed independently many times using many different terminologies. This book will treat the subject from the point of view of atmospheric problems, incorporating material such as atmospheric radiative transfer which is not part of general inverse theory.

When we make a remote measurement it is usually one in which the quantity actually measured is some more or less complicated function of the parameter that is actually required. The distinguishing characteristic of these measurements is not the 'remoteness', rather it that the measurements are indirect. *Inverse Theory* refers to the inversion of complicated functions regardless of whether the measurement is physically remote, and the theory to be discussed here applies to inverse problem, rather than remote measurements as such. The typical atmospheric example is one in which electromagnetic radiation emerging from an atmosphere is measured, when the quantity required is the distribution of temperature or constituents. In another area, solid earth studies, the quantity measured might be the travel time of seismic waves from earthquakes or variations in the acceleration due to gravity, when the quantity required is the internal structure of the earth.

A physically remote measurement can have many advantages. Medical tomography is not physically invasive, an important consideration in the case of living subjects. In atmospheric studies global data can be obtained from a single instrument on a satellite, or vertical profile data can be obtained from a single instrument on the ground. In astrophysics or in solid earth physics, data can be obtained from places where it is virtually impossible to send an in-situ instrument to make measurements.

The inverse problem is the question of finding the best representation of the required parameter given the measurements made, together with any appropriate prior information that may be available about the system and the measuring device. Associated with the inverse problem there are also questions of understanding and describing the information content of the measurement, the relationship between the true state of the system and that retrieved using inverse methods, the error analysis of the overall measuring system, optimising observing systems, and validating results.

1.1 The Beginnings

Atmospheric inverse problems began in the 1920's when Dobson started to make measurements of ozone in the stratosphere using an ultraviolet spectrometer on the ground. The initial measurements were of total ozone, based on the absorption of solar ultraviolet light, but it was realised by Götz in 1930 that by measuring the Rayleigh scattered sunlight from the zenith sky as the sun sets it is possible to obtain information about the vertical distribution of ozone (Dobson 1968). This is the so-called Umkehr Method. The word umkehr means 'reversal', and refers to the time variation of the ratio of the scattered intensity at two different wavelengths.

Methods for deriving the vertical distribution from the umkehr curve were developed by Götz, Dobson and Meetham (1934). Method A, a graphical method, expressed the distribution in terms of the ozone amounts in two layers, together with an overall constraint provided by a separate total ozone measurement. It used pretabulated nonlinear relationships between the two unknowns and a quantity derived nonlinearly from the measurement, Method B was more complicated in that it divided the distribution into nine layers and used a relaxation method (by hand!) based on a linear expansion of the data about one of a set of standard profiles. The equations used were underconstrained, so the solutions found depended on the solver. It was realised that the inverse problem was not well-posed, and a series of *ad hoc* methods based on method B were developed. It was not until Mateer (1965, 1966) carried out his seminal work on the information content of umkehr observations that the subject was put on a firm footing.

With the advent of meteorological satellites, it became clear that remote measurements from space would be very important in determining the state of the atmosphere globally. Kaplan (1959) proposed that the vertical distribution of temperature in the atmosphere could be determined globally by measuring from satellites, as a function of wavelength, the thermal emission from the 15 μm band of carbon dioxide, and also pointed out that ozone and water vapour should also be amenable to remote measurement. At about the same time King (1956, 1959), showed that it should be possible to determine the temperature profile by measuring the angular distribution of emitted radiation in a single spectral interval. However this proved to be of less practical value than Kaplan's spectral approach.

Kaplan only gave a qualitative suggestion about how the profiles might be retrieved; detailed methods were developed by Wark (1961) and Yamamoto (1961) in two of the earliest papers on the subject of inverse theory for atmospheric remote sounding from satellites. These methods were in the same category as the early umkehr methods, Wark considered measurements in three spectral intervals, and retrieved the temperatures and mean lapse rates in two layers, i.e. linear interpolation between three levels, while Yamamoto took four spectral intervals, and used three different kinds of polynomials in a height coordinate proportional to $p^{1/5}$ to represent the unknown profile. Both carried out simulations, without considering the effects of measurement error on the solution.

The subject did not take long to develop after the stimulus provided by the possibility of satellite measurements, and the first sounders, the Infrared Interferometer Spectrometer (IRIS; Hanel et al., 1970; Conrath et al., 1970) and the Satellite Infrared Spectrometer (SIRS; Wark and Hilleary, 1969; Wark, 1970) were launched on the Nimbus III satellite in 1969.

1.2 Atmospheric Remote Sounding Methods

Remote sounding of the atmosphere has been carried out by a wide variety of instruments, using many different principles of measurement. Almost all techniques involve the measurement of electromagnetic radiation, although sound propagation has also been used. The physical effects exploited may involve refraction, transmittance, scattering, thermal emission and non-thermal emission of radiation at all wavelengths from radio to the ultraviolet, and the measurement techniques include grating spectrometers, Michelson interferometers, Fabry-Perot etalons, filter radiometers, gas correlation methods, microwave and radio receivers. Most methods are passive, measuring naturally generated radiation, but some, notably lidar (using lasers) and GPS (Global Positioning System) occultation, are active in that they use man-made sources. The following descriptions are not intended to provide an exhaustive review of instruments and techniques, they are rather a brief introduction to illustrate the possibilities, and to indicate the main features of each kind of measurement. There are several books which describe instrumental techniques and their physical basis in more detail, for example Houghton, Taylor and Rodgers (1984), now somewhat out of date, Hanel, Conrath, Jennings and Samuelson (1992), Kidder and Vonder Haar (1995) and Stephens (1994).

1.2.1 *Thermal emission nadir and limb sounders*

Thermal emission in the infrared and microwave is perhaps the most useful of the measurable quantities, because every part of the atmosphere emits, and the details of the emission spectrum depend on the molecules present in very distinctive ways. Almost every molecule has its own characteristic spectral signature in the infrared and/or microwave.

Nadir sounders are those that measure thermal emission from some relatively small field of view in a direction that is generally downwards, but may be scanned away from the nadir, or may use an array of detectors to record images. In this geometry there is the possibility of measuring at good horizontal spatial resolution. However the equation of radiative transfer limits the vertical resolution to several kilometers in most cases. Near-nadir thermal emission can be used for making measurements of temperature and constituent concentrations, as both of these quantities are important in determining the thermal emission.

Thermal emission limb sounders look through the edge of the atmosphere with cold space as a background, looking at a range of elevations above the surface. They have the same kind features, problems and flexibility as nadir sounders, and are useful for measuring temperature and constituent concentrations, but different observing geometry means that horizontal resolution is limited, and cannot be better than a few hundred kilometers along the line of sight. However the vertical resolution can be much better, and one or two kilometers is feasible. Furthermore, because the path through the limb is geometrically much longer than that in the nadir, sensitivity to small amounts of trace gases can be much better.

Thermal emission signals are not so large as those that depend on the sun as a source, because the source temperature is much lower, so signal-to-noise ratio has always been a limitation, and the technology has been stretched to its limits to obtain good measurements. In the early days, spectral resolution was relatively low, but as improvements have been made in detectors, and as it has become possible to fly larger instruments with larger optics, the greater energy grasp has allowed spectral resolution to improve, and more information to be obtained.

A large number of instruments measuring thermal emission have been flown. Spectral selection has been obtained in a wide variety of ways, including grating spectrometers, Michelson interferometers, Fabry-Perot etalons, filter radiometers, gas correlation radiometers and microwave receivers.

1.2.2 *Scattered solar radiation*

Scattered solar radiation has the advantage that it is more intense and more easily measured, but it is of course only available during the daytime. There are fewer candidate molecules that can be measured this way, as not everything has a suitable spectral feature in the appropriate part of the spectrum, primarily the ultraviolet and visible, but the near infrared has also been used. The main gases measured this way are ozone and nitrogen dioxide, although sulphur dioxide has been measured after major volcanic eruptions.

Solar radiation which has been Rayleigh scattered, or scattered by aerosols, and partially absorbed by the target gas can be measured in various geometries, specifically in the nadir or limb from satellites, or in the zenith from the surface.

The relative absorption across the spectrum is used to determine the amount of absorber. The physics of the radiative transfer imposes a vertical resolution of about 8 or 10 km on nadir and zenith measurements, but as with thermal emission the limb view has better vertical and poorer horizontal resolution.

The most well known instrument which work on this principle in the near nadir view are the Solar Backscatter Ultraviolet and Total Ozone Mapping Spectrometer (SBUV and TOMS: Heath, Krueger and Park, 1978), versions of which have been flown several times to measure ozone distribution and total ozone respectively. Both are grating spectrometers operating in the ultraviolet and visible. It is possible, as with TOMS, to obtain good horizontal resolution and hence images of total ozone. SBUV, which obtains the vertical distribution of ozone, does not scan across the sub-satellite track.

Scattered solar radiation has also been used to measure wind from the doppler shift of spectral lines. This has been done in the near nadir and limb views for the stratosphere and mesosphere by HRDI (High Resolution Doppler Imager, Hays et al., 1993), a Fabry-Perot instrument, and in the limb view for the mesosphere and lower thermosphere by WINDII (Wind Imaging Interferometer, Shepherd, et al., 1993), a field-widened Michelson interferometer.

The limb view has not otherwise been much used for scattered solar radiation, even though it has the advantage of better vertical resolution and sensitivity. An UV spectrometer flew on the Solar Mesosphere Explorer in 1981, which measured ozone in this view, and a more complex instrument, SCIAMACHY (Bovensmann et al., 1999), with a fully steerable view, is due for launch on the European Space Agency ENVISAT in 2001.

The zenith view of scattered solar radiation has been widely used for measuring ozone, since the umkehr method was invented by Götz. It has the advantage that measurements can be made from the surface, with relatively inexpensive instruments (relative to spacecraft!), but in order to obtain the geographical distribution of ozone, a network with a large number of instruments is needed. Because the measurement is of scattered radiation as a function of solar zenith angle, only one, or possibly two, measurement can be made each day as the sun rises or sets, and variations of ozone through the day can cause problems of interpretation. As with the nadir view from space, the physics of the measurement limits the vertical resolution to about 8 or 10 km, The Umkehr method has allowed us to have a reasonable understanding of the ozone distribution stretching back to the International Geophysical Year in 1957, and beyond in a few places.

The primary instrument used has been the Dobson spectrophotometer, a grating spectrometer, but other instruments, including filter radiometers have been used.

1.2.3 *Absorption of solar radiation*

The first remote measurement of the atmospheric composition was of total ozone using absorption of solar radiation. The same principle can be used from the surface for a wide variety of gases, particularly in the infrared and microwave where many constituents have absorption lines and bands, but also in the ultraviolet and visible. Spectrometers of various kinds are the most popular technique at present.

Surprisingly, information about profiles can be obtained if pressure-broadened spectral lines are well resolved, leading to measurements of the vertical distribution of constituents with a resolution of around 10 km, up to the altitude where Doppler broadening becomes important. This is most effective in the microwave where good spectral resolution is straightforwardly obtained, and Doppler broadening is small. Vertical resolution can also be obtained with more difficulty in the infrared using high resolution instruments such as Fourier transform spectrometers.

Another geometry where absorption of solar radiation is useful is in occultation, i.e. measurement through the limb from a spacecraft as the sun sets or rises. This has the same advantage of high vertical resolution as other limb measurements, but its horizontal resolution is poor - not only is the line-of-sight averaging over the same large distance, but only two measurements can be made each orbit. It also has the advantage of very high precision, as the direct solar signal is extremely large. Filter radiometers and grating spectrometers have been used for this kind of measurement, and one instrument, HALOE (Halogen Occultation Experiment, Russell *et al.*, 1993), has used gas correlation.

Stellar occultation provides more opportunities for making sounding than solar occultation, because there are more targets. However the signal is much weaker, so precision is not so good. A stellar occultation instrument, GOMOS (Popescu and Ingmann, 1993), is included in the ENVISAT payload.

1.2.4 *Active techniques*

A few active techniques have been developed, in which the source of the quantity sensed is artificial rather than natural.

Radio occultation. The GPS-MET instrument (Ware *et al.*, 1996 is basically a GPS receiver in low earth orbit, but as well as locating itself using the GPS system it measures the phase delay of the signal from GPS transmitters as they are occulted by the atmosphere. This provides a measure of the refraction of the radio signal, which depends on the refractivity of the atmosphere, itself dependent on air density, water vapour density, and ionospheric electron density. The retrieval can provide the temperature distribution at good vertical resolution (around 100 m), but the problem disentangling the effect of water vapour has not yet been fully solved without the use of external information. There are enough opportunities to observe occultations that this should provide a very useful source of meteorological data at a relatively low cost.

Lidar. 'Light detection and ranging' involves measuring the backscatter from a pulsed laser pointed upwards from the surface. It has also been used from aircraft and experimentally from space. The backscatter at time t after the pulse depends on Rayleigh and aerosol scattering at a distance $ct/2$ above the instrument, and if aerosol scattering is small and there is no absorption by constituents then it is proportional to air density, from which temperature can be derived with the aid of the hydrostatic equation. Alternatively, if there is absorption by constituents the use of two (or more) wavelengths can provide a measurement of the constituent distribution as well as the temperature. At altitudes from which there is an adequate return signal, the vertical resolution depends mainly on the length of the lidar pulse.

Sound propagation. A method which was significant in the 1960's and 70's was the grenade sonde, for which a sequence of grenades were released from a sounding rocket and exploded at different altitudes. The time of flight of the sound pulse to an array of microphones on the surface was measured, which depends on the speed of sound as a function of altitude.

1.3 Simple Solutions to the Inverse Problem

One of the earliest inverse problems to be attempted for actual atmospheric remote sounding from satellites was that for the nadir sounding infrared spectrometer, SIRS, measuring thermal emission from carbon dioxide in order to obtain the temperature profile. We will use an idealised version of this problem to show how easy it is to find a solution which is completely useless, and to illustrate the nature of inverse problems and the pitfalls that can appear.

If the question of clouds and other absorbers is ignored, the equation of radiative transfer can easily be solved (for example see Goody and Yung, 1989). The result for the radiance $L(\nu)$ emerging vertically at the top of the atmosphere at wavenumber ν is

$$L(\nu) = \int_0^\infty B[\nu, T(z)] \frac{\mathrm{d}\tau(\nu, z)}{\mathrm{d}z} \mathrm{d}z, \qquad (1.1)$$

where $B[\nu, T(z)]$ is the Planck radiance at temperature T at height z, and $\tau(\nu, z)$ is the atmospheric transmittance from height z to the measuring instrument above the atmosphere. For simplicity we assume that the absorption of the whole atmosphere is so great that the transmittance from the surface to the instrument is zero, and emission from the surface can be ignored. In the case of carbon dioxide, the mixing ratio is known, and constant, so that atmospheric temperature is the only unknown on the right hand side, and a measurement of the spectrum of radiance should contain information to determine the temperature profile. In practice there is also an integral over the finite spectral bandwidth of the spectrometer, but that makes no qualitative difference to the analysis of the simple kind of solution discussed here. Consider making a set of measurements of radiance, $L(\nu_i)$, $i = 1 \ldots m$, at a set of

m closely spaced wavenumbers ν_i, so that the frequency dependence of the Planck function can be ignored, but that of the transmittance varies considerably:

$$L_i = L(\nu_i) = \int_0^\infty B(\bar{\nu}, T(z)) K_i(z) \, \mathrm{d}z, \qquad (1.2)$$

where $\bar{\nu}$ is some representative wavenumber and $K_i(z) = \mathrm{d}\tau(\nu, z)/\mathrm{d}z$ is a function of only z and i. The equation is now linear in $B(\bar{\nu}, T(z))$, which we may take to be the unknown. If $B(\bar{\nu}, T(z))$ can be found, then $T(z)$ follows immediately, as the Planck function can be inverted algebraically. The radiance is thus a weighted mean of the Planck function profile, with $K_i(z)$ as the weighting function. It is a true mean, because the assumption of zero transmittance from the surface to space ensures that $\int_0^\infty K_i(z) \, \mathrm{d}z = 1$. Because of this, any quantity that takes the part of $K_i(z)$ in any inverse problem tends to be called the *weighting function* in the atmospheric literature, regardless of whether it is normalised.

Solving Eq. (1.2) is clearly going to cause some difficulties, not least because it is underconstrained, or ill-posed, as there are only a finite number of measurements, and the unknown is a continuous function. The obvious approach is to express the unknown as a function of a finite number of parameters, such as a polynomial or a sum of sines and cosines. The general linear form, of which these are but two examples can be written

$$B(\bar{\nu}, T(z)) = \sum_{j=1}^m w_j W_j(z), \qquad (1.3)$$

where w_j is a set of coefficients to be found, and $W_j(z)$ is a set of functions, such as z^{j-1} or $\sin(2\pi j z/Z)$ and $\cos(2\pi j z/Z)$ for a finite height range $(0, Z)$, in terms of which the profile is to be represented. Substituting (1.3) into (1.2) gives

$$L_i = L(\nu_i) = \sum_{j=1}^m w_j \int_0^\infty W_j(z) K_i(z) \, \mathrm{d}z = \sum_{j=1}^m C_{ij} w_j, \qquad (1.4)$$

thus defining the square matrix $\mathbf{C}$ whose elements $C_{ij} = \int_0^\infty W_j(z) K_i(z) \, \mathrm{d}z$ can easily be calculated.* We now have a set of m equations for m unknowns, which can in principle be solved exactly. This was the approach explored in the early papers of Wark (1961) and Yamamoto (1961).

Unfortunately this type of solution is ill-conditioned in many practical situations, as will be illustrated. This means that any experimental error in the measurements can be greatly amplified, and the solution can be meaningless, even though it agrees with the measurements. Eq. (1.2) is a Fredholm integral equation of the first kind, which has long been recognised in the mathematical and numerical literature as likely to be ill-conditioned. Nevertheless, we will proceed with the formal solution,

---

*The notation used for matrices is described in Appendix A. Bold face upper case is be used for matrices, bold face lower case for column vectors.

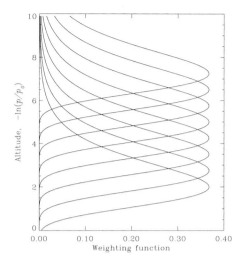

Fig. 1.1 A set of synthetic weighting functions representing a typical nadir sounder measuring thermal emission. The vertical spacing is approximately 5 km.

and carry out an elementary error analysis. Solve Eq. (1.4) by inverting the matrix, so that the vector of coefficients **b** is $\mathbf{C}^{-1}\mathbf{l}$, where **l** is the vector of radiances, and substitute the result back in Eq. (1.3):

$$B(\bar{\nu}, T(z)) = \sum_{i,j} W_j(z) C_{ji}^{-1} L_i = \sum_i G_i(z) L_i, \qquad (1.5)$$

where C_{ji}^{-1} is the jith component of the inverse matrix $\mathbf{C}^{-1}$. This equation also defines the set of functions $G_i(z)$. These have various names in various branches of inverse theory. In atmospheric literature they are often called *contribution functions*, because $G_i(z)L_i$ is the contribution to the solution profile due to the measured radiance L_i. This solution gives back exactly the measured radiances when inserted in the measurement Eq. (1.2), and can therefore be termed *exact*. If there is an error ϵ_i in the measurement of L_i, then there will obviously be a corresponding error $G_i(z)\epsilon_i$ in the profile. Thus the size of the functions $G_i(z)$ gives an indication of the ill-conditioning of the solution method.

A simple illustration of this can be provided by a simulation. To this end, we will set up an example case which will be used to illustrate a range of solutions throughout this book. The set of synthetic weighting functions shown in Fig. 1.1 will be used; these are typical of nadir sounders measuring thermal emission. The vertical coordinate is $\zeta = -\ln(p/p_0)$, where p is pressure and p_0 is the surface pressure. The transmittance from level p to space in channel n is modelled by $\tau(p) = \exp(-p/p_n)$, corresponding to a uniformly mixed absorber with an absorption coefficient which does not vary with temperature and pressure. The resulting weighting function is proportional to $p \exp(-p/p_n)$ which has a maximum at $p = p_n$. The spacing chosen

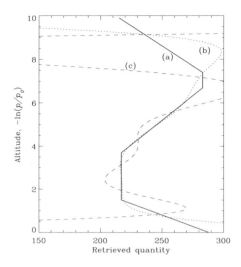

Fig. 1.2 Simulated retrievals using the weighting functions of Fig. 1.1 and a polynomial representation. (a) The original profile; (b) An exact retrieval with no simulated experimental error; (c) An exact retrieval with a 0.5 K simulated experimental error.

is $\delta \ln p_n = 0.75$, corresponding to about 5 km.

The results of a simulated retrieval with and without noise are shown in Fig. 1.2. The solid line is the simulated 'true' profile, based on a U.S. standard atmosphere. The nonlinearity of the Planck function is ignored, i.e. $B(\zeta) = T(\zeta)$, so the 'radiances' are simply weighted means of the temperature profile using the weighting function of Fig. 1.1, using Eq. (1.2). The calculated radiances have been used to retrieve the temperature function profile according to Eq. (1.5), with a polynomial representation $W_j(\zeta) = \zeta^{j-1}$ for $j = 0...m$. The dotted line is the result of this noise-free retrieval. Noise has then been added to the simulated measurements, with an r.m.s. value of 0.5 K, and the retrieval repeated. The result of the noisy retrieval is shown as the dashed line, showing that even with this small amount of noise, the retrieval is very different.

To carry out an error analysis, the contribution functions $G_i(\zeta)$ have been calculated and are shown in Fig. 1.3. The sensitivity to noise is shown by the large values of $G_j(\zeta)$, even in the mid range. The profile has the same units as the measurement, Kelvin, so noise is multiplied by a factor of up to about 10 between about $\zeta = 2$ and $\zeta = 5.5$ for each radiance measured. Outside the middle range it has been necessary to change the scale by a factor of 50 to show the large values of the contribution function!

From Eq. (1.5) it can be seen that if the measurement error in L_i has a variance σ^2 then the solution variance will be given by $\sigma_B^2 = \sum_i G_i(\zeta)^2 \sigma^2$, so that $[\sum_i G_i(\zeta)^2]^{\frac{1}{2}}$ may be regarded as an error amplification factor. This is plotted on the right in Fig. 1.3, showing large values even in the altitude range covered by the

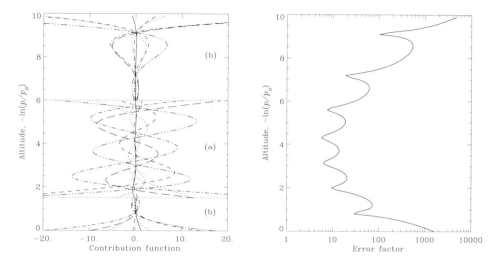

Fig. 1.3 Left: Contribution functions for the retrieval of Fig. 1.2 (a) on the scale give in the abscissa (b) multiplied by 0.02. Right: The error amplification factor for the simulated retrieval.

weighting functions, and a very large error amplification outside that range.

It is clear that difficulties can easily arise if we try to solve inverse problems by simplified methods without much thought. The purpose of this book is to show how we can solve inverse problems successfully, and understand what we are doing at the same time.

Chapter 2
Information Aspects

As the simple illustration in Chapter 1 shows, there can be more to an inverse problem than meets the eye. The problem may be ill-conditioned, as that example is, or even ill-posed. To make things more complicated, most practical inverse problems are nonlinear too. To gain some understanding of the nature of the problem we will first examine various aspects of the information contained in the measurement. For this purpose we will use only the linear case. Nonlinear problems may be linearised, and the information content and behaviour of the resulting linear equations examined. Further implications of nonlinearity will be dealt with in chapter 5.

2.1 Formal Statement of the Problem

The general inverse problem can be regarded as a question of setting up and solving a set of simultaneous linear or non-linear equations, in the presence of experimental error in some of the parameters (the 'measurements'), and quite possibly in the presence of approximations in the formulation of the equations. To examine the information content of the indirect measurement, we will consider the measurements assembled into a vector $\mathbf{y}$, the *measurement vector*, and the unknowns into a *state vector* $\mathbf{x}$, describing the state of the atmosphere. Some aspects of the state of the atmosphere, for example the temperature distribution, are properly described by continuous functions rather discrete values, but ways can always be found of approximating continuous functions by discrete values to any desired accuracy. The process of measurement will be described by a *forward model*, which describes the physics of the measurement process.

2.1.1 *State and measurement vectors*

The quantities to be retrieved can be represented by a state vector, $\mathbf{x}$, with n elements, $x_1, x_2, \ldots, x_n$. Often it will represent a profile of some quantity given at a finite number of levels, enough to adequately represent the possible atmospheric variations. However, it may in principle comprise any set of relevant variables, such as coefficients for some other kind of representation of the profile, such as a fourier series, or it may include a range of different types of parameters, e. g. the air density and ozone mixing ratio at a set of pressure levels together with the altitude of one of those pressure levels.

The quantities actually measured in order to retrieve $\mathbf{x}$ can be represented by the measurement vector $\mathbf{y}$, with m elements, $y_1, y_2, \ldots, y_m$. This vector should include all of the quantities measured that are functions of the state vector. It is quite possible for the same physical quantity to appear in both the measurement and state vector if a direct measurement is made of it. Measurements are made to a finite accuracy; random error or *measurement noise* will be denoted by the vector ϵ.

In some parts of the literature of inverse problems the state vector is called the *model*. This term will not be used here as it has many other connotations in atmospheric applications. The measurement vector is also called the *data vector*, another term which will not be used here.

2.1.2 The forward model

For each state vector there is a corresponding ideal measurement vector, $\mathbf{y}_I$, determined by the physics of the measurement. We can describe the physics formally as the *forward function* $\mathbf{f}(\mathbf{x})$:

$$\mathbf{y}_I = \mathbf{f}(\mathbf{x}). \tag{2.1}$$

However, in practice not only is there always experimental error, but it is often necessary to approximate the detailed physics by some *forward model* $\mathbf{F}(\mathbf{x})$. Therefore we will write the relationship between the measurement vector and the state vector as

$$\mathbf{y} = \mathbf{F}(\mathbf{x}) + \epsilon, \tag{2.2}$$

where the vector $\mathbf{y}$ is the measurement with error ϵ, and $\mathbf{F}(\mathbf{x})$ is a vector valued function of the state, which encapsulates our understanding of the physics of the measurement. To construct a forward model we must of course know how the measuring device works, and understand how the quantity measured, such as the thermal emission from CO_2 or the backscattered solar ultraviolet radiation, is related to the quantity that is really wanted, e.g., the temperature distribution or the ozone distribution. The word *understanding* is used deliberately, and $\mathbf{F}$ is described as a *model*, because there may be underlying physics which is not fully understood, or the real physics may be so complicated that approximations are necessary. The treatment of modelling errors will be discussed in Chapter 3.

The quantities to be retrieved in most inverse problems are continuous functions, while the measurements are always of discrete quantities. Thus most inverse problems are formally ill-posed or underconstrained in this trivial sense. This is simply dealt with by replacing the truly continuous state function, corresponding to an infinite number of variables, with a representation in terms of a finite number of parameters. This can be done to whatever spatial resolution or degree of accuracy is required for scientific use of the retrieval, e.g., the temperature profile could be represented on a finite grid of points with a spacing appropriate to the application. After discretisation the problem may or may not be underconstrained, depending on the grid spacing required and the information content of the measurement. Discretising the problem before attempting to solve it allows us to use the algebra of vectors and matrices, rather than the more general algebra of Hilbert space, and in any case numerical solutions must always be represented in some discrete form.

2.1.3 Weighting function matrix

For the purpose of examining the information content of a measurement it is most convenient to consider a linear problem. A linearisation of the forward model about some reference state $\mathbf{x}_0$ will be adequate for this purpose, provided that $\mathbf{F}(\mathbf{x})$ is linear within the error bounds of the retrieval. Write

$$\mathbf{y} - \mathbf{F}(\mathbf{x}_0) = \frac{\partial \mathbf{F}(\mathbf{x})}{\partial \mathbf{x}}(\mathbf{x} - \mathbf{x}_0) + \boldsymbol{\epsilon} = \mathbf{K}(\mathbf{x} - \mathbf{x}_0) + \boldsymbol{\epsilon}, \qquad (2.3)$$

which defines the $m \times n$ *weighting function matrix* $\mathbf{K}$, not necessarily square, in which each element is the partial derivative of a forward model element with respect to a state vector element, i.e. $K_{ij} = \partial F_i(\mathbf{x})/\partial x_j$. Derivatives of this type are known as Fréchet derivatives, see Appendix A.6. If $m < n$ the equations are described as underconstrained (or ill-posed or under-determined) because there are fewer measurements than unknowns. Similarly if $m > n$ the equations are often described as overconstrained or over-determined. Unfortunately this description is over-simplified, and can easily be wrong; it is possible for a set of equations to be simultaneously over- and under-determined, as will be shown.

The term *weighting function* is peculiar to the atmospheric remote sounding literature, as mentioned in section 1.3 it arises because in the early application of nadir sounding for temperature the forward model takes the form of a weighted mean of the vertical profile of the Planck function. It may also be called the Jacobian (it is a matrix of derivatives), the kernel (hence $\mathbf{K}$), the sensitivity kernel, the tangent linear model or the adjoint, amongst other terms.

2.1.4 Vector spaces

The concept of a *linear vector space* is very useful in considering linear equations. We will give two such spaces special names: The *state space* (or model space) is a vector space of dimension n, within which each conceivable state is represented by a point, or equivalently by a vector from the origin to the point. (Consider the

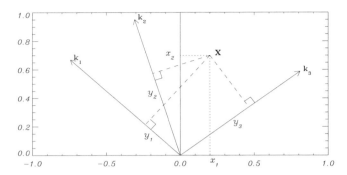

Fig. 2.1 Illustrating a state space in two dimensions, and a measurement space in three dimensions.

origin to be at $\mathbf{x}_0$ for the moment). The *measurement space* (or data space), is a vector space of dimension m, in which each conceivable measurement is likewise represented by a point or a vector (with the origin at $\mathbf{F}(\mathbf{x}_0)$). The act of measurement is then equivalent to a mapping from the state space into the measurement space, and the inverse problem is that of finding an appropriate inverse mapping from the measurement space back into the state space. The weighting function matrix $\mathbf{K}$ represents the forward mapping, equivalent to the measurement apart from measurement error $\boldsymbol{\epsilon}$. Each row of $\mathbf{K}$, of dimension n, can be thought of as a vector $\mathbf{k}_i$ in state space, even though it is not a state. It corresponds to the i-th measurement y_i in the sense that the value of the i-th coordinate of measurement space for a given state vector $\mathbf{x}$ is the vector product of $\mathbf{x}$ and $\mathbf{k}_i$, plus measurement error. Thus each of the m rows of $\mathbf{K}$ corresponds to a coordinate in measurement space, providing the mapping from state space to measurement space. The mapping is fuzzy, to an extent determined by the measurement error statistics.

Fig. 2.1 illustrates a state space in two dimensions, in which the state is represented by the point $\mathbf{x}$, with components x_1 and x_2. Three measurements are made, with weighting functions corresponding to the three vectors $\mathbf{k}_1$, $\mathbf{k}_2$ and $\mathbf{k}_3$, here chosen to be unit vectors, and therefore of unit length. The measurements y_1, y_2 and y_3 correspond to the orthogonal projections of $\mathbf{x}$ onto the $\mathbf{k}$'s, i.e. the distances from the origin to the intersections of the perpendiculars with the $\mathbf{k}$-vectors. These three numbers then define a point in 3-dimensional measurement space. This figure also illustrates that such a measurement is overconstrained. Given only the weighting function vectors and the measurements, the state can be found from the intersections of the perpendiculars from the weighting function vectors. However, only two are needed to determine $\mathbf{x}$, the third is superfluous, and could be inconsistent with the other two if measurement error prevents all three perpendiculars from intersecting in one point. The measurement space is three dimensional, but the state space maps onto a two-dimensional subspace of it because one measurement is always a function of the other two. In the absence of measurement error, points in measurement space but not in this subspace of it do not correspond to possible states.

2.2 Linear Problems without Measurement Error

Consider first a linear problem with arbitrary numbers of dimensions in the absence of measurement error. In this case the problem reduces to the exact solution of linear simultaneous equations,

$$\mathbf{y} = \mathbf{Kx}, \tag{2.4}$$

and involves determining whether there are no solutions, one solution, or an infinite number. More generally, it is a matter of investigating what information can be extracted from the measurements $\mathbf{y}$ about the state $\mathbf{x}$ particularly when there is no solution or no unique solution.

2.2.1 *Subspaces of state space*

The m weighting function vectors $\mathbf{k}_j$ will span some subspace of state space which will be of dimension not greater then m, and may be less than m if the vectors are not linearly independent. The dimension of this subspace is known as the *rank* of the matrix $\mathbf{K}$, denoted by p, and is equal to the number of linearly independent rows (or columns). Even when $m > n$, more measurements than unknowns, the rank cannot be greater than n, and may even be less. This subspace, spanned by the vectors forming the rows of $\mathbf{K}$, is called the *row space* or *range* of $\mathbf{K}$ and may or may not comprise the whole of state space. $\mathbf{K}$ also has a *column space* of dimension p which is a subspace of measurement space.

We can imagine an orthogonal *coordinate system* or *basis* for state space which has p orthogonal *base vectors* (coordinates) in row space, and $n - p$ base vectors outside which are orthogonal to row space and therefore orthogonal to all of the weighting function vectors. Only components of the state vector which lie in row space will contribute to the measurement vector, all other components being orthogonal to it, will give a zero contribution to the measurement, i.e. are unmeasurable. This undetermined part of state space is called the *null space* of $\mathbf{K}$.

The problem will be under-determined if $p < n$, i.e. if a null space exists. In this case the solution is non-unique because there are components of the state which are not determined by the measurements, and which could therefore take any value. Their sizes (whether they are taken to be zero or something else) must be determined from other arguments, or they must be explicitly ignored:

> If a retrieved state has components which lie in the null space, their values cannot have been obtained from the measurements.

Consider just the components of the state vector in the row space. They will be over-determined if $m > p$ and there are more measurements than the rank of **K**, or well-determined if $m = p$. Thus it is possible for a problem to be simultaneously over-determined (in row space) and under-determined (if there is a null space), a condition called *mixed-determined*. It is even possible for there to be more measurements than unknowns, $m > n$, and for the problem to be under-determined, if $p < n$. A problem is well-determined only if $m = n = p$.

> ⇒ *Exercise* 2.1: Construct a simple example of a set of linear equations which is simultaneously over- and under-determined.

If the problem is well determined then a unique solution can be found by solving a set of $p \times p$ equations. If the problem is overdetermined in the row space and there is no measurement error, then either the measurements must be linearly related in the same way as the $\mathbf{k}_j$-vectors, or they are inconsistent, and there is no exact solution. The latter is the normal situation when measurement error is present, and will be dealt with in section 2.4.1.

In summary, the measurement represented by **K** provides not more than p independent quantities or pieces of information with which to describe the state.

2.2.2 *Identifying the null space and the row space*

The row space of a given **K** can be identified by finding a basis, or coordinate system, for state space of the kind described above, i.e. an orthogonal set of vectors that every $\mathbf{k}_j$ can be expressed in terms of, which therefore must be an orthogonal set of linear combinations of the $\mathbf{k}_j$. There are many ways of doing this, and there are clearly an infinite number of bases, as any rotation of an orthogonal basis within its own space is also an orthogonal basis for the space. The Gram–Schmidt orthogonalisation is probably the simplest method, but singular vector decomposition* (Appendix A.4) has some properties which make it rather more useful. We can express **K** in the form

$$\mathbf{K} = \mathbf{U}\mathbf{\Lambda}\mathbf{V}^T, \tag{2.5}$$

where **U** ($m \times p$) and **V** ($n \times p$) are matrices of its left and right singular vectors, and $\mathbf{\Lambda}$ is a $p \times p$ diagonal matrix of non-zero singular values, where p is the rank of **K**. The p columns of **V** form the desired orthogonal basis for the row space, and the columns of **U** form a corresponding basis in measurement space for the column space. Inserting Eq. (2.5) in the linearised forward model, $\mathbf{y} = \mathbf{K}\mathbf{x}$, and multiplying

---

*Eigenvector and singular vector decompositions are fundamental to understanding the information content of indirect measurements. If you are not familiar with these concepts, you should read Appendix A or relevant textbooks before proceeding.

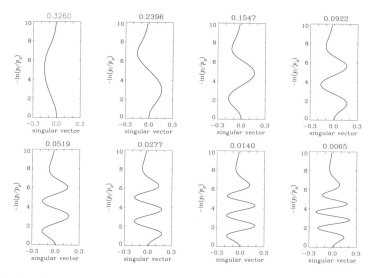

Fig. 2.2 The singular vectors of the weighting functions of Figure 1.1. The corresponding singular values are given above each panel.

by $\mathbf{U}^T$ gives

$$\mathbf{U}^T\mathbf{y} = \mathbf{\Lambda}\mathbf{V}^T\mathbf{x}, \tag{2.6}$$

so that, putting $\mathbf{y}' = \mathbf{U}^T\mathbf{y}$ and $\mathbf{x}' = \mathbf{V}^T\mathbf{x}$ we obtain

$$\mathbf{y}' = \mathbf{\Lambda}\mathbf{x}' \tag{2.7}$$

indicating that the p transformed measurements $\mathbf{y}'$ in column space, are each proportional to a component of a transformed state $\mathbf{x}'$ in p-dimensional row space. The $\mathbf{V}$ matrix forms a natural basis for row space, closely related to the $\mathbf{U}$ matrix basis for column space.

Fig. 2.2 shows the eight singular vectors of the weighting functions of Fig. 1.1. The corresponding singular values are given above each frame. This representation can be seen to be qualitatively similar to a Fourier representation, in that the functions are orthogonally, oscillatory, and with an increasing number of oscillations. Structures in the profile corresponding to the singular vectors can be thought of as being measured independently, with a sensitivity given by the singular value. If there were measurement error, then the finer scale structures with smaller singular values would be measured to a lower precision, for example vector number five is measured with a precision about 6.3 times smaller (i.e. 0.3260/0.0519) than vector number one.

2.3 Linear Problems with Measurement Error

2.3.1 *Describing experimental error*

All real measurements are subject to experimental error or 'noise', so that any practical retrieval must allow for this. The proper treatment of experimental error is the dominant consideration in designing retrieval methods. Thus we need a formalism in which to express uncertainty in measurements and the resulting uncertainty in retrievals, and with which to ensure that the latter is as small as possible.

A description of experimental error in terms of probability density functions (*pdf*'s), using a Bayesian approach to probability, gives useful insight. The statement that a scalar measurement has a value $\bar{y}$ and an error σ is a shorthand way of saying that our knowledge of the true value of the measured parameter is described by a *pdf* $P(y)$ with a mean $\bar{y}$ and variance σ^2:

$$\bar{y} = \int y P(y)\,\mathrm{d}y$$
$$\sigma^2 = \int (y-\bar{y})^2 P(y)\,\mathrm{d}y \qquad (2.8)$$

where $P(y)\,\mathrm{d}y$ is the probability that y lies in $(y, y+\mathrm{d}y)$. Probability interpreted this way is a measure of knowledge about y, rather than a distribution function for repeated trials of a measurement of y, though repeated trials are a useful way of estimating probability when appropriate. The form of $P(y)$ is almost always taken to be Gaussian or 'normal':

$$P(y) = N(y-\bar{y},\sigma) = \frac{1}{(2\pi)^{1/2}\sigma}\exp\left\{-\frac{(y-\bar{y})^2}{2\sigma^2}\right\}. \qquad (2.9)$$

This is usually a good approximation for experimental error, and is very convenient for algebraic manipulations. As we shall see in section 2.7, if the only information available about a *pdf* is its mean and variance, then the Gaussian distribution is the one which contains the least information about the measured quantity, in the sense used in information theory, to be discussed in section 2.5.

When the measured quantity is a vector, a probability density can still be defined over measurement space, $P(\mathbf{y})$, with the interpretation that $P(\mathbf{y})\,\mathrm{d}\mathbf{y}$ (where $\mathrm{d}\mathbf{y}$ is shorthand for $\mathrm{d}y_1 \ldots \mathrm{d}y_m$) is the probability that the true value of the measurement lies in a multidimensional interval $(\mathbf{y}, \mathbf{y}+\mathrm{d}\mathbf{y})$ in measurement space. Different elements of a vector may be correlated, in the sense that

$$S_{ij} = \mathcal{E}\{(y_i - \bar{y}_i)(y_j - \bar{y}_j)\} \neq 0, \qquad (2.10)$$

where S_{ij} is called the *covariance* of y_i and y_j, and $\mathcal{E}$ is the expected value operator. These covariances can be assembled into a matrix, which we will denote by $\mathbf{S}_y$ for the covariance matrix of $\mathbf{y}$. Its diagonal elements are clearly the variances of the individual elements of $\mathbf{y}$. A covariance matrix is symmetric and non-negative definite, and is almost always positive definite. I use the notation $\mathbf{S}_y$ is used for covariance in preference to $\mathbf{\Sigma}_y$, which might be thought more logical by analogy to σ^2 for variance, because of the possible confusion with summation.

The Gaussian distribution for a vector is of the form:

$$P(\mathbf{y}) = \frac{1}{(2\pi)^{n/2}|\mathbf{S}_y|^{1/2}} \exp\left\{-\tfrac{1}{2}(\mathbf{y}-\bar{\mathbf{y}})^T\mathbf{S}_y^{-1}(\mathbf{y}-\bar{\mathbf{y}})\right\} \qquad (2.11)$$

where $\mathbf{S}_y$ must be nonsingular. If it is singular then there are components of $\mathbf{y}$ that are known exactly, equivalent to $\sigma = 0$ in the scalar case. Eq. (2.11) can be seen to be related to the scalar Gaussian distribution by transforming to a basis in which $\mathbf{S}_y$ is diagonal, using the eigenvector decomposition $\mathbf{S}_y = \mathbf{L}\mathbf{\Lambda}\mathbf{L}^T$:

$$\begin{aligned} P(\mathbf{y}) &= \frac{1}{(2\pi)^{n/2}|\mathbf{L}\mathbf{\Lambda}\mathbf{L}^T|^{1/2}} \exp\left\{-\tfrac{1}{2}(\mathbf{y}-\bar{\mathbf{y}})^T\mathbf{L}\mathbf{\Lambda}^{-1}\mathbf{L}^T(\mathbf{y}-\bar{\mathbf{y}})\right\} & (2.12)\\ &= \frac{1}{(2\pi)^{n/2}|\mathbf{\Lambda}|^{1/2}} \exp\left\{-\tfrac{1}{2}\mathbf{z}^T\mathbf{\Lambda}^{-1}\mathbf{z}\right\}, & (2.13)\end{aligned}$$

where $\mathbf{z} = \mathbf{L}^T(\mathbf{y}-\bar{\mathbf{y}})$. [†] Thus the *pdf* can be written as a product of the independent *pdf*'s of each element of $\mathbf{z}$:

$$P(\mathbf{z}) = \prod_i \frac{1}{(2\pi\lambda_i)^{1/2}} \exp\left\{-\frac{z_i^2}{2\lambda_i}\right\}, \qquad (2.14)$$

where the eigenvalue λ_i is the variance of z_i. The eigenvector transformation provides as basis for measurement space in which the transformed measurements are statistically independent. A singular covariance matrix would have one or more zero eigenvalues, corresponding to elements of $\mathbf{z}$ that are known without error. Such components presumably would not correspond to real physical measurements, and can be eliminated or ignored as appropriate.

Notice that surfaces of constant probability of the *pdf* are of the form

$$(\mathbf{y}-\bar{\mathbf{y}})^T\mathbf{S}_y^{-1}(\mathbf{y}-\bar{\mathbf{y}}) = \sum_i z_i^2/\lambda_i = \text{constant} \qquad (2.15)$$

and are ellipsoids in measurement space, with principal axes corresponding to the eigenvectors of $\mathbf{S}_y$, and the lengths of these principal axes proportional to $\lambda_i^{1/2}$. These ellipsoids can be thought of as the multivariate equivalent of error bars.

2.3.2 The Bayesian approach to inverse problems

We now turn to the question of relating the *pdf* of the measurement to the *pdf* of the state. The act of measurement maps the state into the measurement space according to the forward model, Eq. (2.2). The measurement error $\boldsymbol{\epsilon}$ is known only statistically, so that even though $\mathbf{F}(\mathbf{x})$ is a deterministic mapping, in the presence

---
[†]See Appendix A.2 for the algebra of eigenvectors and eigenvalues.

of measurement error a point in state space maps into a region in measurement space determined by the probability density function of ϵ. Conversely, if $\mathbf{y}$ is a given measurement, it could be the result of a mapping from anywhere in a region of state space described by some *pdf*, rather than from a single point, even in the absence of a null space. Furthermore, we may have some prior information about the state, for example a climatology, which can also be conveniently described by a *pdf*, and which can be used to constrain the solution. Such prior knowledge can be thought of as a *virtual measurement*, as, like a real measurement, it provides us with an estimate of some function (often the identity) of the state, together with a measure of the accuracy of the estimate, albeit usually rather a poor one.

We saw in Chapter 1 that the simple-minded approach of solving equations can lead to disaster, so we need to find a more subtle approach. The Bayesian approach is a very helpful way of looking at the noisy inverse problem, in which we have some prior understanding or expectation about some quantity, and want to update the understanding in the light of new information. Imperfect prior knowledge can be quantified as a probability density function over the state space. A measurement, also imperfect because of experimental error, can be quantified as a *pdf* over measurement space. We would like to know how the measurement *pdf* maps into state space, and combines with prior knowledge. Bayes' theorem tells us how.

2.3.2.1 *Bayes' theorem*

Probability density is a scalar-valued function, in our case a function of the state vector or the measurement vector. Let us define:

$P(\mathbf{x})$ as the prior *pdf* the state $\mathbf{x}$. This means that the quantity $P(\mathbf{x})\,\mathrm{d}\mathbf{x}$ is the probability before the measurement that $\mathbf{x}$ lies in the multidimensional volume $(\mathbf{x}, \mathbf{x} + \mathrm{d}\mathbf{x})$, expressing quantitatively our knowledge of $\mathbf{x}$ before the measurement is made. It is normalised so that $\int P(\mathbf{x})\,\mathrm{d}\mathbf{x} = 1$.

$P(\mathbf{y})$ as the prior *pdf* of the measurement, with a similar meaning. This is the *pdf* of the measurement *before it is made*.

$P(\mathbf{x}, \mathbf{y})$ as the joint prior *pdf* of $\mathbf{x}$ and $\mathbf{y}$, meaning that $P(\mathbf{x}, \mathbf{y})\,\mathrm{d}\mathbf{x}\,\mathrm{d}\mathbf{y}$ is the probability that $\mathbf{x}$ lies in $(\mathbf{x}, \mathbf{x} + \mathrm{d}\mathbf{x})$ *and* $\mathbf{y}$ lies in $(\mathbf{y}, \mathbf{y} + \mathrm{d}\mathbf{y})$.

$P(\mathbf{y}|\mathbf{x})$ as the conditional *pdf* of $\mathbf{y}$ given $\mathbf{x}$, meaning that $P(\mathbf{y}|\mathbf{x})\,\mathrm{d}\mathbf{y}$ is the probability that $\mathbf{y}$ lies in $(\mathbf{y}, \mathbf{y} + \mathrm{d}\mathbf{y})$ when $\mathbf{x}$ has a given value.

$P(\mathbf{x}|\mathbf{y})$ as the conditional *pdf* of $\mathbf{x}$ given $\mathbf{y}$, meaning that $P(\mathbf{x}|\mathbf{y})\,\mathrm{d}\mathbf{x}$ is the probability that $\mathbf{x}$ lies in $(\mathbf{x}, \mathbf{x} + \mathrm{d}\mathbf{x})$ when $\mathbf{y}$ has a given value. This is the quantity that is of interest for solving the inverse problem.

It may seem slightly odd to use the same symbol P for all of these different functions, but in each case the argument makes it clear which *pdf* is referred to, and the convention eliminates a forest of subscripts.

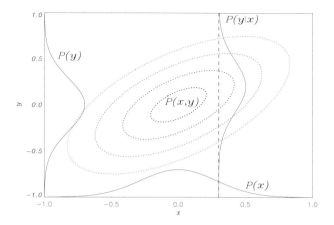

Fig. 2.3 Illustrating Bayes' theorem for a two-dimensional case.

Fig. 2.3 illustrates these concepts when **x** and **y** are scalars. The contours are of $P(x, y)$, and $P(x)$ is given by the integral over all values of y:

$$P(x) = \int_{-\infty}^{\infty} P(x, y) \, dy, \qquad (2.16)$$

and likewise $P(y)$ is found by integrating over all values of x. The conditional *pdf* $P(y|x)$ is proportional to the values of $P(y, x)$ as a function of y for a given value of x, e.g., along the cut marked by a dotted line. The constant of proportionality is such that $\int P(y|x) \, dy = 1$, so $P(y, x)$ should be divided by the integral along the line:

$$P(y|x) = \frac{P(x, y)}{\int P(x, y) \, dy}. \qquad (2.17)$$

Inserting Eq. (2.16) for the integral we obtain

$$P(y|x) = P(x, y)/P(x). \qquad (2.18)$$

We can use equivalent arguments to show that $P(x|y) = P(x, y)/P(y)$, and by eliminating $P(x, y)$ between the two equations we obtain Bayes' theorem as the relationship between the two different conditional *pdf*'s. When generalised for the vector case it states that

$$\boxed{P(\mathbf{x}|\mathbf{y}) = \frac{P(\mathbf{y}|\mathbf{x})P(\mathbf{x})}{P(\mathbf{y})}.} \qquad (2.19)$$

The left hand side of Eq. (2.19), $P(\mathbf{x}|\mathbf{y})$, is the posterior *pdf* of the state when the measurement is given. This equation is what we need in order to update the prior knowledge $P(\mathbf{x})$ of the state with the measurement. $P(\mathbf{y}|\mathbf{x})$ describes the knowledge of **y** that would be obtained if the state were **x**. To write it down explicitly requires only the forward model and the statistical description of the measurement error. The only remaining quantity required is the denominator, $P(\mathbf{y})$. Formally this can be obtained by integrating the expression for $P(\mathbf{x}, \mathbf{y})$, namely $P(\mathbf{y}|\mathbf{x})P(\mathbf{x})$, over all $\mathbf{x}$, but in practice it is only a normalising factor, and is often not needed.

We now have a conceptual approach to the inverse problem:

- Before we make a measurement we have prior knowledge expressed as a prior *pdf*;
- The measurement process is expressed as a forward model which maps the state space into measurement space;
- Bayes' theorem provides a formalism to invert this mapping and calculate a posterior *pdf* by updating the prior *pdf* with a measurement *pdf*.

Note that the Bayesian view is general. It is not just an inverse method which produces a solution which may be compared with those produced by other methods, rather it encompasses all inverse methods by providing a way of characterising the class of possible solutions, considering all possible states, and assigning a probability density to each.

The forward model is never explicitly inverted in this approach, but then an explicit state is not produced as an 'answer'. It provides us with some intuition about how the measurement improves knowledge of the state, but to obtain an explicit retrieval we must choose one state from the ensemble described by the posterior *pdf*, perhaps the expected value or most probable value of the state, together with some measure of the width of the *pdf* to give the accuracy of the retrieval. Thus further work is needed, resulting in expressions mathematically equivalent to inverting the forward model.

2.3.2.2 *Example: The Linear problem with Gaussian statistics*

As a simple illustration of the Bayesian approach consider a linear problem in which all of the *pdf*'s are Gaussian. The Gaussian distribution is commonly used to model probability density functions because many processes are well described by it and because it is algebraically convenient. We use it here for just those reasons.

The multivariate Gaussian distribution for a random vector $\mathbf{x}$ is obtained from Eq. (2.11). The maximum probability value for $\mathbf{x}$ is clearly equal to the expected value $\bar{\mathbf{x}}$, because the *pdf* is symmetric about $\mathbf{x} = \bar{\mathbf{x}}$, but the reader may like to try:

⇒ **Exercise** 2.2: Solve $\partial P/\partial \mathbf{x} = 0$ for the maximum probability value, and integrate $\int \mathbf{x} P(\mathbf{x}) \, d\mathbf{x}$ to confirm that both are equal to $\bar{\mathbf{x}}$

A linear problem is one for which the forward model is linear, the simplest being:

$$\mathbf{y} = \mathbf{F}(\mathbf{x}) + \boldsymbol{\epsilon} = \mathbf{K}\mathbf{x} + \boldsymbol{\epsilon}. \tag{2.20}$$

Gaussian statistics are usually a good approximation for the errors in real measurements, so express $P(\mathbf{y}|\mathbf{x})$ as

$$-2\ln P(\mathbf{y}|\mathbf{x}) = (\mathbf{y} - \mathbf{Kx})^T \mathbf{S}_\epsilon^{-1}(\mathbf{y} - \mathbf{Kx}) + c_1, \qquad (2.21)$$

where c_1 is independent of $\mathbf{x}$, and $\mathbf{S}_\epsilon$ is the measurement error covariance. Less realistic, but convenient, is to describe prior knowledge of $\mathbf{x}$ by a Gaussian *pdf*:

$$-2\ln P(\mathbf{x}) = (\mathbf{x} - \mathbf{x}_a)^T \mathbf{S}_a^{-1}(\mathbf{x} - \mathbf{x}_a) + c_2, \qquad (2.22)$$

where $\mathbf{x}_a$ is the *a priori* value of $\mathbf{x}$, and $\mathbf{S}_a$ is the associated covariance matrix

$$\mathbf{S}_a = \mathcal{E}\{(\mathbf{x} - \mathbf{x}_a)(\mathbf{x} - \mathbf{x}_a)^T\}. \qquad (2.23)$$

Substituting Eqs. (2.21) and (2.22) in Eq. (2.19) we obtain for the posterior *pdf*:

$$-2\ln P(\mathbf{x}|\mathbf{y}) = (\mathbf{y} - \mathbf{Kx})^T \mathbf{S}_\epsilon^{-1}(\mathbf{y} - \mathbf{Kx}) + (\mathbf{x} - \mathbf{x}_a)^T \mathbf{S}_a^{-1}(\mathbf{x} - \mathbf{x}_a) + c_3, \qquad (2.24)$$

where c_3 is independent of $\mathbf{x}$. This is a quadratic form in $\mathbf{x}$, so it must be possible to write it as

$$-2\ln P(\mathbf{x}|\mathbf{y}) = (\mathbf{x} - \hat{\mathbf{x}})^T \hat{\mathbf{S}}^{-1}(\mathbf{x} - \hat{\mathbf{x}}) + c_4, \qquad (2.25)$$

i.e. the posterior *pdf* is also a Gaussian distribution with expected value $\hat{\mathbf{x}}$ and covariance $\hat{\mathbf{S}}$. We can relate Eq. (2.25) to Eq. (2.24) by equating like terms. Equating terms that are quadratic in $\mathbf{x}$:

$$\mathbf{x}^T \mathbf{K}^T \mathbf{S}_\epsilon^{-1} \mathbf{Kx} + \mathbf{x}^T \mathbf{S}_a^{-1} \mathbf{x} = \mathbf{x}^T \hat{\mathbf{S}}^{-1} \mathbf{x} \qquad (2.26)$$

gives

$$\hat{\mathbf{S}}^{-1} = \mathbf{K}^T \mathbf{S}_\epsilon^{-1} \mathbf{K} + \mathbf{S}_a^{-1}. \qquad (2.27)$$

Likewise equating the terms linear in $\mathbf{x}^T$ gives:

$$(-\mathbf{Kx})^T \mathbf{S}_\epsilon^{-1}(\mathbf{y}) + (\mathbf{x})^T \mathbf{S}_a^{-1}(-\mathbf{x}_a) = \mathbf{x}^T \hat{\mathbf{S}}^{-1}(-\hat{\mathbf{x}}). \qquad (2.28)$$

I have used $\mathbf{x}^T$ for convenience only, as equating terms linear in $\mathbf{x}$ simply gives the transpose of this equation. Cancelling the $\mathbf{x}^T$'s, because this must be valid for any value of $\mathbf{x}$, and substituting for $\hat{\mathbf{S}}^{-1}$ from Eq. (2.27) gives

$$\mathbf{K}^T \mathbf{S}_\epsilon^{-1} \mathbf{y} + \mathbf{S}_a^{-1} \mathbf{x}_a = (\mathbf{K}^T \mathbf{S}_\epsilon^{-1} \mathbf{K} + \mathbf{S}_a^{-1})\hat{\mathbf{x}} \qquad (2.29)$$

and hence

$$\begin{aligned}\hat{\mathbf{x}} &= (\mathbf{K}^T \mathbf{S}_\epsilon^{-1} \mathbf{K} + \mathbf{S}_a^{-1})^{-1}(\mathbf{K}^T \mathbf{S}_\epsilon^{-1} \mathbf{y} + \mathbf{S}_a^{-1} \mathbf{x}_a) \\ &= \mathbf{x}_a + (\mathbf{K}^T \mathbf{S}_\epsilon^{-1} \mathbf{K} + \mathbf{S}_a^{-1})^{-1} \mathbf{K}^T \mathbf{S}_\epsilon^{-1}(\mathbf{y} - \mathbf{Kx}_a).\end{aligned} \qquad (2.30)$$

We could equate the constant terms, but they add no further information. From Eq. (2.11) c_1, c_2 and c_4 must all be of the form $\ln[(2\pi)^n |\mathbf{S}|]$ in order to correctly normalise the *pdf*'s. An alternate form for $\hat{\mathbf{x}}$ can be obtained from Eq. (2.30):

$$\hat{\mathbf{x}} = \mathbf{x}_a + \mathbf{S}_a \mathbf{K}^T (\mathbf{K} \mathbf{S}_a \mathbf{K}^T + \mathbf{S}_\epsilon)^{-1}(\mathbf{y} - \mathbf{Kx}_a). \qquad (2.31)$$

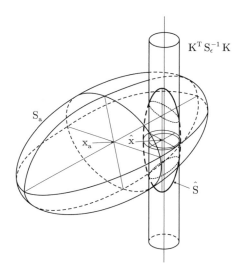

Fig. 2.4 Illustrating the relationship between the prior state estimate, the measurement mapped into state space, and the posterior estimate, for a three-dimensional state space and a two-dimensional measurement space. The large ellipsoid is a contour of the prior *pdf*, the cylinder is a contour of the *pdf* of the state given only the measurement, and the small ellipsoid is a contour of the posterior *pdf*.

A derivation of this will be given in section 4.1. I will describe Eq. (2.30) as the 'n-form', and Eq. (2.31) as the 'm-form' from the order of the matrix to be inverted.

Remember the Bayesian solution to the inverse problem is not $\hat{\mathbf{x}}$, but is the Gaussian *pdf* $P(\mathbf{x}|\mathbf{y})$, of which $\hat{\mathbf{x}}$ is the expected value and $\hat{\mathbf{S}}$ is the covariance. Note $\hat{\mathbf{x}}$ is a linear function of the prior expected value and the measurement, as might be expected for a linear problem, and that the *inverse* covariance matrix $\hat{\mathbf{S}}^{-1}$ is a linear function of the inverse prior covariance matrix and the inverse measurement error covariance. The terms in Eq. (2.27) are known as Fisher information matrices, and will be discussed in more detail in section 2.5.1.

Fig. 2.4 provides a geometric illustration of the relationship between the prior covariance, the measurement and the posterior covariance when state space has three dimensions and measurement space has two. The large ellipsoid centred on $\mathbf{x}_a$ represents a contour of the prior covariance, outlining the region of state space within which the state is likely to lie. The cylinder represents the region in which states consistent with the measurement might lie, where the axis of the cylinder is the set of states corresponding exactly to the measurement, and the cylinder encloses experimental error. The direction of the axis represents the null space. The weighting functions for the two measurement elements are orthogonal to the axis. The small ellipsoid outlines the region which is consistent with both the prior information and the measurement. Note that its centre, $\hat{\mathbf{x}}$, does not lie on the axis of the cylinder, i.e. the expected value does not fit the measurement exactly.

2.4 Degrees of Freedom

2.4.1 *How many independent quantities can be measured?*

When experimental error is not present there are p independent pieces of information that can in principle be determined from the measurement, but with error, these p values will now be uncertain to some extent. It is conceivable that the error in some of the values is large enough that the component is worthless, so that the number of useful independent pieces of information may be less than p. It is possible even in the exact problem that numerical rounding error causes some components to become useless, a phenomenon called 'ill-conditioning'. The number of useful independent pieces of information can be regarded as the 'effective' rank of the problem, and the subspace in which they lie as the 'effective row space'.

As a contrived example, consider a state vector with two elements, (x_1, x_2), and an instrument which can measure two rather similar quantities, y_1 and y_2 with the forward model:

$$\begin{pmatrix} y_1 \\ y_2 \end{pmatrix} = \begin{pmatrix} 1.01 & 0.99 \\ 0.99 & 1.01 \end{pmatrix} \begin{pmatrix} x_1 \\ x_2 \end{pmatrix} + \begin{pmatrix} \epsilon_1 \\ \epsilon_2 \end{pmatrix} \qquad (2.32)$$

where the errors ϵ_i are independent with variance σ^2. This is equivalent to measuring the orthogonal combinations z_1 and z_2:

$$\begin{aligned} z_1 &= y_1 + y_2 &= 2(x_1 + x_2) + \epsilon_1 + \epsilon_2 \\ z_2 &= y_1 - y_2 &= 0.02(x_1 - x_2) + \epsilon_1 - \epsilon_2 \end{aligned} \qquad (2.33)$$

where the errors are still independent, with variances $2\sigma^2$. However z_2 is likely to be a much smaller quantity than z_1 for arbitrary values of x_1 and x_2, so that there may be no useful information about the difference between x_1 and x_2.

To identify the effective row space, we compare the measurement error covariance with the natural variability of the measurement vector as expressed by its prior covariance. Any component whose natural variability is smaller than the measurement error is in effect not measurable, and is not part of the effective row space. Unfortunately, covariance matrices generally have non-zero off-diagonal elements indicating correlations between variability of different elements of the vectors, so it is not immediately obvious how to do this. To compare them properly we transform the basis so that the off-diagonal elements disappear. The simplest way to do this is to transform from $\mathbf{K}$ to $\tilde{\mathbf{K}}$ by

$$\tilde{\mathbf{x}} = \mathbf{S}_a^{-\frac{1}{2}}(\mathbf{x} - \mathbf{x}_a) \text{ and } \tilde{\mathbf{y}} = \mathbf{S}_\epsilon^{-\frac{1}{2}}(\mathbf{y} - \mathbf{y}_a) \qquad (2.34)$$

where $\mathbf{y}_a = \mathbf{K}\mathbf{x}_a$, so that the transformed forward model is

$$\tilde{\mathbf{y}} = \mathbf{S}_\epsilon^{-\frac{1}{2}} \mathbf{K} \mathbf{S}_a^{\frac{1}{2}} \tilde{\mathbf{x}} + \mathbf{S}_\epsilon^{-\frac{1}{2}} \boldsymbol{\epsilon} = \tilde{\mathbf{K}} \tilde{\mathbf{x}} + \tilde{\boldsymbol{\epsilon}}. \qquad (2.35)$$

where $\tilde{\mathbf{K}} = \mathbf{S}_\epsilon^{-\frac{1}{2}} \mathbf{K} \mathbf{S}_a^{\frac{1}{2}}$. The covariances of $\tilde{\mathbf{x}}$ and $\tilde{\boldsymbol{\epsilon}}$ are both unit matrices in this transformed system, for example

$$\mathbf{S}_{\tilde{\epsilon}} = \mathcal{E}\{\tilde{\boldsymbol{\epsilon}}\tilde{\boldsymbol{\epsilon}}^T\} = \mathbf{S}_\epsilon^{-\frac{1}{2}} \mathcal{E}\{\boldsymbol{\epsilon}\boldsymbol{\epsilon}^T\} \mathbf{S}_\epsilon^{-\frac{1}{2}} = \mathbf{S}_\epsilon^{-\frac{1}{2}} \mathbf{S}_\epsilon \mathbf{S}_\epsilon^{-\frac{1}{2}} = \mathbf{I}_m. \qquad (2.36)$$

A covariance matrix, or its inverse, has an infinite number of square roots, see Appendix A.2. Two are particularly useful, namely the positive definite symmetric square root constructed from its eigenvector decomposition,

$$\mathbf{S} = \mathbf{L}\boldsymbol{\Lambda}\mathbf{L}^T \;\Rightarrow\; \mathbf{S}^{\frac{1}{2}} = \mathbf{L}\boldsymbol{\Lambda}^{\frac{1}{2}}\mathbf{L}^T \;\text{ and }\; \mathbf{S}^{-\frac{1}{2}} = \mathbf{L}\boldsymbol{\Lambda}^{-\frac{1}{2}}\mathbf{L}^T. \qquad (2.37)$$

and the upper triangular Cholesky decomposition, of the form $\mathbf{S} = \mathbf{T}^T\mathbf{T}$, see section 5.8.1.1 and exercise 5.3. Algebraic derivations are clearer when written in terms of $\mathbf{S}^{\frac{1}{2}}$, but numerically the Cholesky decomposition is considerably more efficient, and should be used in preference. With this square root we can write

$$\tilde{\mathbf{x}} = \mathbf{T}_a^{-T}(\mathbf{x} - \mathbf{x}_a), \;\; \tilde{\mathbf{y}} = \mathbf{T}_\epsilon^{-T}(\mathbf{y} - \mathbf{y}_a) \;\text{ and }\; \tilde{\mathbf{K}} = \mathbf{T}_\epsilon^{-T}\mathbf{K}\mathbf{T}_a^T, \qquad (2.38)$$

where the superscript $-T$ means the transpose of the inverse. This transformation of $\mathbf{y}$ using $\mathbf{S}_\epsilon^{-\frac{1}{2}}$ or $\mathbf{T}^{-T}$ is called *prewhitening* as it transforms the noise into white noise.

⇒ **Exercise** 2.3: Why does it not matter which square root we use for prewhitening?

The measurement error covariance and the prior state covariance cannot be compared directly, as they are in different spaces. Rather we should compare the measurement error covariance with the prior covariance of $\tilde{\mathbf{y}}$:

$$\mathbf{S}_{\tilde{y}} = \mathcal{E}\{\tilde{\mathbf{y}}\tilde{\mathbf{y}}^T\} = \mathcal{E}\{(\tilde{\mathbf{K}}\tilde{\mathbf{x}} + \tilde{\boldsymbol{\epsilon}})(\tilde{\mathbf{K}}\tilde{\mathbf{x}} + \tilde{\boldsymbol{\epsilon}})^T\} = \tilde{\mathbf{K}}\tilde{\mathbf{K}}^T + \mathbf{I}_m. \qquad (2.39)$$

The component of this covariance due to the variability of the state is $\tilde{\mathbf{K}}\tilde{\mathbf{K}}^T$, and that due to measurement noise is $\mathbf{I}_m$. $\tilde{\mathbf{K}}\tilde{\mathbf{K}}^T$ is not normally diagonal, and we still cannot easily compare, so we carry out one more transformation of the forward model, the singular value decomposition $\tilde{\mathbf{K}} = \mathbf{U}\boldsymbol{\Lambda}\mathbf{V}^T$, as in section 2.2.2 but now with prewhitening, giving

$$\mathbf{y}' = \mathbf{U}^T\tilde{\mathbf{y}} = \mathbf{U}^T\tilde{\mathbf{K}}\tilde{\mathbf{x}} + \mathbf{U}^T\tilde{\boldsymbol{\epsilon}} = \boldsymbol{\Lambda}\mathbf{V}^T\tilde{\mathbf{x}} + \boldsymbol{\epsilon}' = \boldsymbol{\Lambda}\mathbf{x}' + \boldsymbol{\epsilon}' \qquad (2.40)$$

where $\mathbf{y}'$, $\mathbf{x}'$ and $\boldsymbol{\epsilon}'$ are of dimension p, the rank of $\mathbf{K}$. The covariance matrices of $\boldsymbol{\epsilon}' = \mathbf{U}^T\tilde{\boldsymbol{\epsilon}}$ and $\mathbf{x}' = \mathbf{V}^T\tilde{\mathbf{x}}$ are both still unity, because $\mathbf{U}^T\mathbf{U}$ and $\mathbf{V}^T\mathbf{V}$ are both unit matrices. The covariance of the variability of $\mathbf{y}'$ is therefore $\boldsymbol{\Lambda}^2 + \mathbf{I}_p$, a diagonal matrix, and the elements of $\mathbf{y}'$ which vary more than the noise are those for which $\lambda_i^2 \gtrsim 1$. Elements corresponding to zero singular values carry no information, only noise. Thus

> the number of independent measurements made to better than measurement error, the effective rank of the problem, is the number of singular values of $\mathbf{S}_\epsilon^{-\frac{1}{2}}\mathbf{K}\mathbf{S}_a^{\frac{1}{2}}$ which are greater than about unity.

$\tilde{\mathbf{K}}$ will depend on which square roots have been used for scaling, but its singular values will not. The corresponding singular vectors in row space, the columns of $\mathbf{V}$, comprise the effective row space. These are vectors in the state space transformed by $\mathbf{S}_a^{-\frac{1}{2}}$; to return them to the original untransformed state space, transform back with $\mathbf{S}_a^{\frac{1}{2}}$.

To use the standard example from Chapter 1 to illustrate this point, we must first define a state vector and choose some covariance matrices. The measurement vector is clearly a set of eight radiances, and we can take $\mathbf{S}_\epsilon$ to be diagonal with a variance of $\sigma_\epsilon^2 = 0.25\,\mathrm{K}^2$, as in chapter 1. Let us choose the state vector to be the temperature at levels spaced at intervals of 0.1 in $z = \ln(p/p_0)$, and, to make a fairly noncommittal assumption, take $\mathbf{S}_a$ to be diagonal with a variance of say $\sigma_a^2 = 100\,\mathrm{K}^2$. In this case the singular vectors of $\tilde{\mathbf{K}}$ are the same as those of $\mathbf{K}$, Fig. 2.2, and the singular values are a factor of $\sigma_a/\sigma_\epsilon = 20$ larger. The measurements might appear to have a signal-to-noise ratio of 20, but that would be an overestimate, because the variance of each element of radiance $\mathbf{Kx}$ is not σ_a^2, but $\sigma_a^2 \sum_j K_{ij}^2 \simeq 2.5\,\mathrm{K}^2$, giving a signal-to-noise ratio of 3.16 for a single radiance. Even that is an overestimate because of the correlations between radiance variability expressed by $\mathbf{KS}_a\mathbf{K}^T$. The real signal-to-noise ratio is expressed by the singular values of $\tilde{\mathbf{K}}$. The first five singular vectors would be measured with a signal-to-noise ratios of 6.5, 4.8, 3.1, 1.8 and 1.0 respectively, and the last three with signal to noise of less than unity, so the effective rank of $\mathbf{K}$ is about four.

2.4.2 Degrees of freedom for signal

The above discussion gives a qualitative description of the number of independent pieces of information in a measurement as the number of singular values of $\tilde{\mathbf{K}}$ which are greater than about unity, which is the same as the number of eigenvalues of $\tilde{\mathbf{K}}\tilde{\mathbf{K}}^T$ or of $\tilde{\mathbf{K}}^T\tilde{\mathbf{K}}$ greater than about unity.

We can formalise this concept and make it more precise by relating it to the concept of 'degrees of freedom', and enquiring how many of the degrees of freedom of a measurement are related to signal, and how many are related to noise. In the simplest case with one degree of freedom let us consider making a single direct measurement of a scalar, with noise:

$$y = x + \epsilon, \tag{2.41}$$

where x has prior variance σ_a^2 and ϵ has σ_ϵ^2. The prior variance of y will be $\sigma_y^2 =$

$\sigma_a^2 + \sigma_\epsilon^2$. The best estimate of x will be, from Eq. (2.30),

$$\hat{x} = \frac{\sigma_\epsilon^{-2} y + \sigma_a^{-2} x_a}{\sigma_\epsilon^{-2} + \sigma_a^{-2}} = \frac{\sigma_a^2 y + \sigma_\epsilon^2 x_a}{\sigma_a^2 + \sigma_\epsilon^2}. \tag{2.42}$$

If $\sigma_a^2 \gg \sigma_\epsilon^2$ then y will be providing information about x, but if $\sigma_a^2 \ll \sigma_\epsilon^2$ then it will be providing information about ϵ. In the first case we can say that the measurement provides a 'degree of freedom for signal', and in the second case a 'degree of freedom for noise'.

Now consider the general case of measuring a vector $\mathbf{y}$ with m degrees of freedom. The most probable state in the Gaussian linear case is the one which minimises

$$\chi^2 = (\mathbf{x} - \mathbf{x}_a)^T \mathbf{S}_a^{-1} (\mathbf{x} - \mathbf{x}_a) + \boldsymbol{\epsilon}^T \mathbf{S}_\epsilon^{-1} \boldsymbol{\epsilon}, \tag{2.43}$$

where $\boldsymbol{\epsilon} = \mathbf{y} - \mathbf{K}\mathbf{x}$. As we have seen, Eq. (2.30), the minimum is at

$$\hat{\mathbf{x}} - \mathbf{x}_a = \mathbf{G}(\mathbf{y} - \mathbf{K}\mathbf{x}_a) = \mathbf{G}[\mathbf{K}(\mathbf{x} - \mathbf{x}_a) + \boldsymbol{\epsilon}], \tag{2.44}$$

where the equivalent of the contribution function of Eq. (1.5) is now a matrix $\mathbf{G}$, given by

$$\mathbf{G} = (\mathbf{K}^T \mathbf{S}_\epsilon^{-1} \mathbf{K} + \mathbf{S}_a^{-1})^{-1} \mathbf{K}^T \mathbf{S}_\epsilon^{-1} = \mathbf{S}_a \mathbf{K}^T (\mathbf{K} \mathbf{S}_a \mathbf{K}^T + \mathbf{S}_\epsilon)^{-1}. \tag{2.45}$$

This matrix has various names in different fields, for example it may be described as a generalised inverse of $\mathbf{K}$, or as a *gain matrix*, the term I will normally use in this book in preference to contribution function matrix. At the minimum the expected value of χ^2 is equal to the number of degrees of freedom, or the number of measurements, m. This can be divided into two parts, corresponding to the two terms in Eq. (2.43):

$$d_s = \mathcal{E}\{(\hat{\mathbf{x}} - \mathbf{x}_a)^T \mathbf{S}_a^{-1} (\hat{\mathbf{x}} - \mathbf{x}_a)\} \tag{2.46}$$
$$d_n = \mathcal{E}\{\hat{\boldsymbol{\epsilon}}^T \mathbf{S}_\epsilon^{-1} \hat{\boldsymbol{\epsilon}}\}. \tag{2.47}$$

The first term measures the part of χ^2 attributable to the state vector, and the second term that attributable to noise. They may therefore described as the number of 'degrees of freedom for signal' and 'degrees of freedom for noise', respectively. Thus d_s describes the number of useful independent quantities there are in a measurement, and hence is a measure of information. It is not necessarily an integer—for example a component with signal-to-noise ratio of unity would provide 0.5 to each of d_s and d_n.

We can find explicit expressions for d_s and d_n as follows. In the derivation we use the relation for the trace of a product of two matrices $\text{tr}(\mathbf{CD}) = \text{tr}(\mathbf{DC})$ where $\mathbf{D}$ and $\mathbf{C}^T$ are the same shape. It is clear that

$$d_s = \mathcal{E}\{(\hat{\mathbf{x}} - \mathbf{x}_a)^T \mathbf{S}_a^{-1} (\hat{\mathbf{x}} - \mathbf{x}_a)\} \tag{2.48}$$
$$= \mathcal{E}\{\text{tr}[(\hat{\mathbf{x}} - \mathbf{x}_a)(\hat{\mathbf{x}} - \mathbf{x}_a)^T \mathbf{S}_a^{-1}]\} \tag{2.49}$$
$$= \text{tr}(\mathbf{S}_{\hat{\mathbf{x}}} \mathbf{S}_a^{-1}) \tag{2.50}$$

where, from Eqs. (2.44) and (2.45),

$$\begin{align}
\mathbf{S}_{\hat{\mathbf{x}}} &= \mathcal{E}\{(\hat{\mathbf{x}} - \mathbf{x}_a)(\hat{\mathbf{x}} - \mathbf{x}_a)^T\} \tag{2.51} \\
&= \mathbf{G}(\mathbf{K}\mathbf{S}_a\mathbf{K}^T + \mathbf{S}_\epsilon)\mathbf{G}^T \tag{2.52} \\
&= \mathbf{S}_a\mathbf{K}^T(\mathbf{K}\mathbf{S}_a\mathbf{K}^T + \mathbf{S}_\epsilon)^{-1}\mathbf{K}\mathbf{S}_a. \tag{2.53}
\end{align}$$

Therefore

$$\begin{align}
d_s &= \operatorname{tr}(\mathbf{S}_a\mathbf{K}^T[\mathbf{K}\mathbf{S}_a\mathbf{K}^T + \mathbf{S}_\epsilon]^{-1}\mathbf{K}) \tag{2.54} \\
&= \operatorname{tr}(\mathbf{K}\mathbf{S}_a\mathbf{K}^T[\mathbf{K}\mathbf{S}_a\mathbf{K}^T + \mathbf{S}_\epsilon]^{-1}) \tag{2.55}
\end{align}$$

and using Eq. (2.45) in the first of these two forms we also obtain

$$d_s = \operatorname{tr}([\mathbf{K}^T\mathbf{S}_\epsilon^{-1}\mathbf{K} + \mathbf{S}_a^{-1}]^{-1}\mathbf{K}^T\mathbf{S}_\epsilon^{-1}\mathbf{K}). \tag{2.56}$$

By following a similar derivation we can show that

$$d_n = \operatorname{tr}(\mathbf{S}_\epsilon[\mathbf{K}\mathbf{S}_a\mathbf{K}^T + \mathbf{S}_\epsilon]^{-1}) = \operatorname{tr}([\mathbf{K}^T\mathbf{S}_\epsilon^{-1}\mathbf{K} + \mathbf{S}_a^{-1}]^{-1}\mathbf{S}_a^{-1}) + m - n \tag{2.57}$$

⇒ ***Exercise*** 2.4: Derive these expressions.

and hence $d_s + d_n = \operatorname{tr}(\mathbf{I}_m) = m$ as expected. Note that one of the forms for d_s is $\operatorname{tr}(\mathbf{G}\mathbf{K})$. The matrix $\mathbf{A} = \mathbf{G}\mathbf{K}$ is a very useful quantity, as will be seen in more detail in Chapter 3. For the moment simply note that it relates the expected state to the true state through

$$\hat{\mathbf{x}} - \mathbf{x}_a = \mathbf{G}\mathbf{y} = \mathbf{G}[\mathbf{K}(\mathbf{x} - \mathbf{x}_a) + \epsilon] = \mathbf{A}(\mathbf{x} - \mathbf{x}_a) + \mathbf{G}\epsilon. \tag{2.58}$$

It is known as the *averaging kernel matrix* (Backus and Gilbert, 1970), the *model resolution matrix* (Menke, 1989), the *state resolution matrix*, or the *resolving kernel*, amongst other terms. It plays a significant rôle in various descriptions of the information content, as it describes the subspace of state space in which the retrieval must lie.

Degrees of freedom will be unchanged through linear transformations. Hence we can also consider d_s in the $\tilde{\mathbf{x}}, \tilde{\mathbf{y}}$ and $\mathbf{x}', \mathbf{y}'$ transformations. In the former, we can show

$$\begin{align}
d_s &= \operatorname{tr}(\tilde{\mathbf{K}}^T[\tilde{\mathbf{K}}\tilde{\mathbf{K}}^T + \mathbf{I}_m]^{-1}\tilde{\mathbf{K}}) \tag{2.59} \\
&= \operatorname{tr}(\tilde{\mathbf{K}}\tilde{\mathbf{K}}^T[\tilde{\mathbf{K}}\tilde{\mathbf{K}}^T + \mathbf{I}_m]^{-1}) \tag{2.60} \\
&= \operatorname{tr}([\tilde{\mathbf{K}}^T\tilde{\mathbf{K}} + \mathbf{I}_n]^{-1}\tilde{\mathbf{K}}^T\tilde{\mathbf{K}}) \tag{2.61}
\end{align}$$

and in the latter

$$d_s = \operatorname{tr}(\mathbf{\Lambda}^2(\mathbf{\Lambda}^2 + \mathbf{I}_m)^{-1}) = \sum_{i=1}^{m} \lambda_i^2/(1 + \lambda_i^2). \tag{2.62}$$

The corresponding number of degrees of freedom for noise is given by

$$d_\mathrm{n} = \mathrm{tr}((\mathbf{\Lambda}^2 + \mathbf{I}_m)^{-1}) = \sum_{i=1}^{m} 1/(1+\lambda_i^2). \tag{2.63}$$

To evaluate d_n it is necessary to regard $\mathbf{\Lambda}$ as an order m matrix, keeping any zero singular values it might have.

> ⇒ **Exercise** 2.5: Relate the eigenvectors and values of $\mathbf{A}$ to the singular vectors and values of $\tilde{\mathbf{K}}$. Remember $\mathbf{A}$ may be asymmetric.

2.5 Information Content of a Measurement

'Information' is a very general term that has been used in many different ways by different authors. Degrees of freedom for signal, for example, may be considered to be a measure of information. However in this section I want to discuss two widely used measures that carry the name 'information', namely the Fisher information matrix and the Shannon information content.

2.5.1 The Fisher information matrix

The Fisher information matrix arises in the theory of maximum likelihood estimation (e.g. see Stuart, Ord and Arnold (1999) chapters 17 and 18 for a detailed development). Likelihood, as defined by Fisher (1921) is the conditional *pdf* $P(\mathbf{y}|\mathbf{x})$, considered as a function of $\mathbf{x}$ for a given value of $\mathbf{y}$. In the context of the Bayesian approach used here, it is the same as the posterior *pdf* $P(\mathbf{x}|\mathbf{y})$ when there is no prior information, or equivalently when any prior information is considered as being of the same nature as a measurement—a virtual measurement. A maximum likelihood estimator finds the value of $\mathbf{x}$ which maximises $L(\mathbf{x}) = P(\mathbf{y}|\mathbf{x})$. It can be shown (the Cramer–Rao inequality) that the variance of the estimate of x satisfies, in the scalar case,

$$\mathrm{var}(x) \geq 1/\mathcal{E}\left\{\left(\frac{\partial \ln L}{\partial x}\right)^2\right\} \tag{2.64}$$

with the equality holding under certain conditions, which include the linear Gaussian case. The denominator is known as the Fisher information, and generalises to a matrix

$$\mathcal{F} = \mathcal{E}\left\{\left(\frac{\partial \ln L}{\partial \mathbf{x}}\right)\left(\frac{\partial \ln L}{\partial \mathbf{x}}\right)^T\right\} = \int L(\mathbf{x})\left(\frac{\partial \ln L(\mathbf{x})}{\partial \mathbf{x}}\right)\left(\frac{\partial \ln L(\mathbf{x})}{\partial \mathbf{x}}\right)^T d\mathbf{x} \tag{2.65}$$

in the multidimensional case. It can be shown that the information matrix of a product of two independent likelihoods, i.e. $P(\mathbf{y}_1|\mathbf{x})P(\mathbf{y}_2|\mathbf{x})$, is the sum of the individual matrices, so that information of two independent measurements is additive.

In the Gaussian linear case, the information matrix is equal to the inverse covariance matrix, so we may interpret the terms in the equation

$$\hat{\mathbf{S}}^{-1} = \mathbf{K}^T \mathbf{S}_\epsilon^{-1} \mathbf{K} + \mathbf{S}_a^{-1} \qquad (2.66)$$

as information matrices, indicating that the posterior information matrix is the sum of that of the prior and that of the measurement.

2.5.2 Shannon information content

The Shannon definition of information content arises from Information Theory, which he developed in the 1940's (Shannon and Weaver, 1949). Its original purpose was to describe the information carrying capacity of communication channels, but its applications are much wider. Information as he defined it depends on the *entropy* of probability density functions, which is very closely related to thermodynamic entropy. Information content is a scalar quantity, and as such is useful for optimising observing systems, as well as characterising and comparing them. We will discuss entropy and information first in terms of direct measurements, and then show how the concepts carry over to indirect measurements and to retrieval theory.

The concept of signal-to-noise ratio gives a hint about how a useful definition of information content might be constructed. Consider a simple direct measurement of a scalar x, with accuracy σ. The signal-to-noise ratio for the measurement is commonly taken to be x/σ. This is appropriate for measurements of quantities which have fixed values such as physical constants, but for quantities which vary, like most interesting atmospheric parameters, and for which it is the variation rather than the absolute value that is usually of interest, a more useful definition is σ_x/σ, where σ_x is the standard deviation of the variability of x. For example a measurement of a temperature of 300 K to an accuracy of 1 K might have an apparent signal-to-noise ratio of 300, but if it is of a tropical sea surface temperature whose variation through the year is known to be, say, 4 K then the useful signal-to-noise ratio is only 4.

The information content of a measurement can be defined qualitatively as the factor by which knowledge of a quantity is improved by making the measurement. In practice the logarithm of the factor is used, with the base of the logarithm depending on the application and determining the dimensionless 'units'. In information theory it is often convenient to use base two, when the units are 'bits'. In the above example, knowledge is improved by a factor of four, so the measurement provides 2 bits of information about the sea surface temperature. For simplicity of algebraic manipulations, however, natural logarithms are more convenient.

2.5.2.1 *Entropy of a probability density function*

Entropy in the thermodynamic sense is the logarithm of the number of distinct internal states of a thermodynamic system consistent with a measured macro-state (pressure, temperature, etc). Information is the change in the logarithm of the number of distinct possible internal states of the system being measured, consistent with the change in knowledge of the system resulting from a measurement.

To construct a formal definition of information, as distinct from the informal signal-to-noise ratio, we use the *pdf* as a measure of knowledge of a system. The Gibbs definition of thermodynamic entropy and the Shannon definition for discrete information systems are the same, apart from a numerical factor:

$$S(P) = -k \sum_i p_i \ln p_i, \qquad (2.67)$$

where p_i is the probability of the system being in state i. In thermodynamics, k is the Boltzmann constant, and in information theory $k = 1$ and the logarithm is usually taken to base two. For the detailed reason why this form is chosen, the reader is referred to Shannon and Weaver (1949), but as a simple example consider a discrete system with 2^n states. If all states are equally likely, and nothing is known about the state, then $p_i = 2^{-n}$, so that the prior entropy $S(P) = n$ using the logarithm to base two. If the system is examined and found to be in some definite state s, then $p_s = 1$ and $p_i = 0$ for $i \neq s$. In this case the posterior entropy is zero, and the information content is the difference n, in units of bits. This corresponds to the number of digits of a binary number required to identify the state.

For a continuous *pdf*, p_i corresponds to $P(x)\,dx$ so the $\ln p_i$ factor needs some attention. Entropy of a continuous *pdf* is defined as

$$S(P) = -\int P(x) \log_2[P(x)/M(x)]\, dx, \qquad (2.68)$$

where the place of dx is taken by a measure function, M, chosen so that P/M is dimensionless, and so that $P = M$ corresponds to a state of no knowledge of x. Choice of M is analogous to the third law of thermodynamics, in that $P = M$ determines the zero of entropy. It is often constant, when it may be safely omitted, but it may also be interpreted as a prior *pdf*, in which case Eq. (2.68) is a relative entropy.

If $P_1(x)$ describes knowledge before a measurement, and $P_2(x)$ describes it afterwards, then the information content of the measurement is the reduction in entropy:

$$H = S(P_1) - S(P_2). \qquad (2.69)$$

2.5.2.2 *Entropy of a Gaussian distribution*

To illustrate that the definition of Eq. (2.68) has the desired properties for continuous *pdf*'s, we will apply it to a scalar Gaussian case, and show that the information content of the measurement is the logarithm of the signal-to-noise ratio. Using the natural logarithm for algebraic convenience, the entropy of the Gaussian *pdf*

Eq. (2.9) is:

$$S = \frac{1}{(2\pi)^{\frac{1}{2}}\sigma} \int \exp\left\{-\frac{(x-\bar{x})^2}{2\sigma^2}\right\} \left(\ln[(2\pi)^{\frac{1}{2}}\sigma] + \frac{(x-\bar{x})^2}{2\sigma^2}\right) \, \mathrm{d}x. \qquad (2.70)$$

⇒ **Exercise** 2.6: Integrate Eq. (2.70) to obtain $S = \ln\sigma(2\pi\mathrm{e})^{1/2}$.

The information content of a measurement where the prior knowledge is Gaussian with variance σ_1^2, and the posterior knowledge is Gaussian with variance σ_2^2 is therefore $\ln(\sigma_1/\sigma_2)$, i.e. the logarithm of the signal-to-noise ratio, as anticipated.

In the case of a multivariate Gaussian distribution, we have already shown that the *pdf* is equivalent to a product of independent distributions with variances equal to the eigenvalues of $\mathbf{S}_y$. We would expect the entropy of a product of independent *pdf*s to be the sum of the entropies of individual *pdf*s.

⇒ **Exercise** 2.7: Prove that $S[P(x)P(y)] = S[P(x)] + S[P(y)]$ if x and y are independent

Hence the entropy of the multivariate Gaussian distribution for a vector with m elements is, using Eq. (2.14)

$$\begin{aligned} S[P(\mathbf{y})] &= \sum_{i=1}^{m} \ln(2\pi\mathrm{e}\lambda_i)^{\frac{1}{2}} \\ &= m\ln(2\pi\mathrm{e})^{\frac{1}{2}} + \tfrac{1}{2}\ln(\textstyle\prod_i \lambda_i) \\ &= m\ln(2\pi\mathrm{e})^{\frac{1}{2}} + \tfrac{1}{2}\ln|\mathbf{S}_y| \end{aligned} \qquad (2.71)$$

because the determinant is equal to the product of the eigenvalues. The square root of the product is also proportional to the volume of an ellipsoid describing a surface of constant probability, as each eigenvalue is proportional to the square of the principal axis of the ellipsoid (Appendix A.3). Therefore the entropy of the *pdf* is the logarithm of the volume inside a surface of constant probability, plus a constant which depends on the surface chosen. It is a measure of the volume of state space occupied by the *pdf* which describes knowledge of the state.

When we make a measurement, this 'volume of uncertainty' decreases; the information content of the measurement is a measure of the factor by which it decreases, a generalisation of the scalar concept of signal-to-noise ratio.

Zero eigenvalues, corresponding to a singular covariance matrix, or to quantities known exactly, will lead to negative infinite contributions to the entropy, and zero length principal axes. Such terms should cancel exactly when calculating information content, as knowledge afterwards of these quantities will be the same as knowledge before. However to avoid mathematical difficulties it is best to eliminate them first by not including those basis vectors in the measurement or state space. If a component has finite variance before a measurement and zero variance afterwards, however, then the measurement has provided infinite information! Such a measurement is probably non-physical.

The information content of a measurement when the prior covariance is $\mathbf{S}_1$ and the posterior covariance is $\mathbf{S}_2$ can be written as

$$H = \tfrac{1}{2}\ln|\mathbf{S}_1| - \tfrac{1}{2}\ln|\mathbf{S}_2| = \tfrac{1}{2}\ln|\mathbf{S}_1\mathbf{S}_2^{-1}| = -\tfrac{1}{2}\ln|\mathbf{S}_2\mathbf{S}_1^{-1}|. \tag{2.72}$$

2.5.2.3 Information content in the linear Gaussian case

The information content of a measurement can be evaluated either in state space (H_s) or in measurement space (H_m), and we would expect to obtain the same value in either case. In state space it depends on the entropies of the *pdf* of the state before and after the measurement:

$$\begin{aligned}
H_s &= S[P(\mathbf{x})] - S[P(\mathbf{x}|\mathbf{y})] \\
&= \tfrac{1}{2}\ln|\mathbf{S}_a| - \tfrac{1}{2}\ln|\hat{\mathbf{S}}| \\
&= \tfrac{1}{2}\ln|\hat{\mathbf{S}}^{-1}\mathbf{S}_a|.
\end{aligned} \tag{2.73}$$

Putting Eq. (2.27) for $\hat{\mathbf{S}}$ in the linear Gaussian case into the third expression in Eq. (2.73) gives the following, using the properties of determinants given in Appendix A.5:

$$\begin{aligned}
H_s &= \tfrac{1}{2}\ln|(\mathbf{K}^T\mathbf{S}_\epsilon^{-1}\mathbf{K} + \mathbf{S}_a^{-1})\mathbf{S}_a| \\
&= \tfrac{1}{2}\ln|\mathbf{S}_a^{\tfrac{1}{2}}\mathbf{K}^T\mathbf{S}_\epsilon^{-1}\mathbf{K}\mathbf{S}_a^{\tfrac{1}{2}} + \mathbf{I}_n| \\
&= \tfrac{1}{2}\ln|\tilde{\mathbf{K}}^T\tilde{\mathbf{K}} + \mathbf{I}_n|.
\end{aligned} \tag{2.74}$$

In measurement space, the information content is the difference between the entropy of the prior estimate of $\mathbf{y}$ and the posterior estimate:

$$H_m = S[P(\mathbf{y})] - S[P(\mathbf{y}|\mathbf{x})]. \tag{2.75}$$

The covariance for $P(\mathbf{y})$ before the measurement is

$$\begin{aligned}
\mathbf{S}_{y_a} &= \mathcal{E}\{(\mathbf{y}-\mathbf{y}_a)(\mathbf{y}-\mathbf{y}_a)^T\} \\
&= \mathcal{E}\{\mathbf{K}(\mathbf{x}-\mathbf{x}_a)(\mathbf{x}-\mathbf{x}_a)^T\mathbf{K}^T + \epsilon\epsilon^T\} \\
&= \mathbf{K}\mathbf{S}_a\mathbf{K}^T + \mathbf{S}_\epsilon
\end{aligned} \tag{2.76}$$

and the posterior covariance is $\mathbf{S}_\epsilon$, so the information content is, in various forms:

$$\begin{aligned}
H_m &= \tfrac{1}{2}\ln|\mathbf{S}_\epsilon^{-1}(\mathbf{K}\mathbf{S}_a\mathbf{K}^T + \mathbf{S}_\epsilon)| \\
&= \tfrac{1}{2}\ln|\mathbf{S}_\epsilon^{-\tfrac{1}{2}}\mathbf{K}\mathbf{S}_a\mathbf{K}^T\mathbf{S}_\epsilon^{-\tfrac{1}{2}} + \mathbf{I}_m| \\
&= \tfrac{1}{2}\ln|\tilde{\mathbf{K}}\tilde{\mathbf{K}}^T + \mathbf{I}_m|.
\end{aligned} \tag{2.77}$$

Note that $\tilde{\mathbf{K}}\tilde{\mathbf{K}}^T$ and $\tilde{\mathbf{K}}^T\tilde{\mathbf{K}}$ have the same non-zero eigenvalues, therefore both H_m and H_s are equal to $\sum_i \tfrac{1}{2}\ln(1+\lambda_i^2)$ where λ_i is a singular value of $\tilde{\mathbf{K}}$.

⇒ **Exercise** 2.8: With the aid of Bayes' theorem, show that for general *pdf*'s the information content is the same whether computed in measurement space or in state space.

The averaging kernel matrix for the linear Gaussian problem was introduced briefly in section 2.4.2 in connexion with degrees of freedom for signal. It can also be related to the information content as follows. From Eq. (2.45), we can write it as

$$\mathbf{A} = \mathbf{GK} = (\mathbf{K}^T \mathbf{S}_\epsilon^{-1} \mathbf{K} + \mathbf{S}_a^{-1})^{-1} \mathbf{K}^T \mathbf{S}_\epsilon^{-1} \mathbf{K} \qquad (2.78)$$

and hence

$$\mathbf{I} - \mathbf{A} = (\mathbf{K}^T \mathbf{S}_\epsilon^{-1} \mathbf{K} + \mathbf{S}_a^{-1})^{-1} \mathbf{S}_a^{-1} = \hat{\mathbf{S}} \mathbf{S}_a^{-1} \qquad (2.79)$$

which we may relate to information content through Eq. (2.73). This, together with Eq. (2.54) or (2.56) gives the following relations between information content, degrees of freedom for signal, the singular values of $\tilde{\mathbf{K}}$ and the averaging kernel matrix:

$$\begin{array}{ll} H = \frac{1}{2} \sum_i \ln(1 + \lambda_i^2) = & -\frac{1}{2} \ln |\mathbf{I}_n - \mathbf{A}| \\ d_s = \sum_i \lambda_i^2/(1 + \lambda_i^2) = & \operatorname{tr}(\mathbf{A}). \end{array} \qquad (2.80)$$

2.6 The Standard Example: Information Content and Degrees of Freedom

In order to illustrate the concepts of information content and degrees of freedom with the standard example, an *a priori* covariance matrix is needed. I will use two cases, firstly and most simply a diagonal matrix with a variance at each level of $\sigma_a^2 = 100 \, \mathrm{K}^2$, and secondly a matrix with the same values on the diagonal but with non-zero off-diagonal elements. The first case has reasonable values at each level, corresponding to knowledge of the temperature to about $\pm 10\,\mathrm{K}$ before we make the measurement, but it implicitly assumes that the temperature difference between adjacent levels (in this case about 700 m apart) is known to only about $\pm 14\,\mathrm{K}$. This is unrealistically large.

A more reasonable *a priori* would have adjacent levels correlated. Therefore for the second case I have constructed a simple covariance matrix model based on a first order auto-regressive model, or Markov process. This is a variant of a random walk. Take the departure of the temperature from the mean, $\delta T_{i+1} = T_{i+1} - \bar{T}_{i+1}$ at level $i+1$ to be related to that at level i by:

$$\delta T_{i+1} = \beta \delta T_i + \xi_i \qquad (2.81)$$

where the regression coefficient β is a constant between zero and unity and ξ_i is Gaussian random variable, uncorrelated with δT_i, and with constant variance σ_ξ^2.

In the limit of large i, it is easy to see that the variance of δT tends to $\sigma_\xi^2/(1-\beta^2)$, so if we choose the variance of ξ_i to be $\sigma_\xi^2 = \sigma_a^2(1-\beta^2)$, we get a covariance matrix with elements

$$S_{ij} = \sigma_a^2 \beta^{2|i-j|} \tag{2.82}$$

which has σ_a^2 on the diagonal, and non-zero off-diagonal elements. We can also write this as

$$S_{ij} = \sigma_a^2 \exp\left(-|i-j|\frac{\delta z}{h}\right) \tag{2.83}$$

where δz is the level spacing and $h = -\delta z/2\ln\beta$ is the length scale at which the interlevel correlation is $1/e$. I have chosen a value for β which gives a length scale $h = 1$ in units of $\ln p$, about 7 km, giving $\beta \simeq 0.95$ and $\sigma_\xi \simeq 3$ K. This provides some modelling of the inter-level correlations that undoubtedly exist in the atmosphere, and suppresses structures on an arbitrarily fine scale. The atmosphere is represented on a grid of levels with a spacing of $\delta z = 0.1$ in units of $\ln p$.

Table 2.1 Singular values of $\tilde{\mathbf{K}}$, together with contributions of each vector to the degrees of freedom d_s and information content H for both covariance matrices. Measurement noise variance is $0.25\,\mathrm{K}^2$.

i	Diagonal covariance			Full covariance		
	λ_i	d_{si}	H_i (bits)	λ_i	d_{si}	H_i (bits)
1	6.51929	0.97701	2.72149	27.81364	0.99871	4.79865
2	4.79231	0.95827	2.29147	18.07567	0.99695	4.17818
3	3.09445	0.90544	1.70134	9.94379	0.98999	3.32105
4	1.84370	0.77269	1.06862	5.00738	0.96165	2.35227
5	1.03787	0.51858	0.52731	2.39204	0.85123	1.37443
6	0.55497	0.23547	0.19368	1.09086	0.54337	0.56546
7	0.27941	0.07242	0.05423	0.46770	0.17948	0.14270
8	0.13011	0.01665	0.01211	0.17989	0.03135	0.02297
totals		4.45653	8.57024		5.55272	16.75571

Table 2.1 give the singular values of $\tilde{\mathbf{K}}$, and the contributions of each singular vector to the degrees of freedom d_s and information content H for both prior covariances, and for a measurement noise of $0.25\,\mathrm{K}^2$. The singular vectors for the diagonal case are as shown in Fig. 2.2. The vectors for the non-diagonal case are very similar and have not been plotted.

For the diagonal matrix, we obtain a total $d_s = 4.46$ and $H = 8.6$ bits, indicating that about four and a half quantities are measured, and about $2^{8.6} = 380$ different atmospheric states can be distinguished. This corresponds to the first four of five singular values being greater than about unity, the first three giving close to one degree of freedom, next two or three giving a fraction of a degree of freedom, and

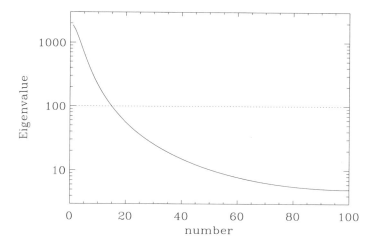

Fig. 2.5 Solid: eigenvalues of a non-diagonal model covariance matrix; dotted: variance of a diagonal covariance matrix with the same total variance.

the last two giving little. This is echoed in the information content of each of the singular vectors.

For the non-diagonal matrix, we would expect that as the *a priori* is more tightly constrained, both the information content and the degrees of freedom would be smaller. However this does not seem to be the case, the singular values and contributions to d_s and H all being larger. The entropy of the diagonal prior covariance is 332.2 bits (i.e. $100\log_2 10$), while that of the non-diagonal one is 210.2 bits. The nondiagonal covariance is restricted to a much smaller volume of state space, by a factor of $2^{122} \simeq 5 \times 10^{37}$.

The reason for this paradox is that the weighting functions see only large scale structure. The non-diagonal covariance has a larger variance at these scales, and smaller variance at small scales, in contrast to the diagonal covariance which has variance at all scales equally. The eigenvalues of the non-diagonal matrix are plotted as the solid line in Fig. 2.5, and those of the diagonal matrix are plotted as the dotted line. The first 14 values of the non-diagonal matrix are larger than 100. The corresponding vectors are quasi-sinusoidal, with the largest wavelength corresponding to the largest eigenvalue, see Fig. 3.2. Thus at the scale of the width and spacing of the weighting functions, the non-diagonal *a priori* is less constrained, while it is more constrained at scales which cannot be measured.

It is clear that the *a priori* covariance is important, and that an incorrect one can constrain the solution inappropriately. This will be discussed in more detail in Chapter 10.

2.7 Probability Density Functions and the Maximum Entropy Principle

It is necessary to be able to estimate probability density functions, and often on the basis of very little information. They are not quantities that are measured directly, and have to be related somehow to the various different kinds of information we have about a system. It is common to assume that if we make a measurement of some quantity, and make an error estimate, that the corresponding *pdf* is Gaussian, but why? The central limit theorem is on reason often cited, but there is another, more subtle, reason.

In the Bayesian view, probability density is a measure of knowledge. It is reasonable to take as a guiding principle in quantifying a *pdf* the 'maximum entropy principle' that the region of state space that we believe the state to lie in should be as large as possible consistent with what is known about the variability of the state. Entropy is an appropriate measure of the size of the region, so a set of parameters describing the system can be interpreted in terms of a *pdf* by finding the function for which the entropy is maximum, or equivalently, the information is minimum, consistent with the available parameters.

I will illustrate the ideas using three examples of *pdf*s of a single random variable x. Assume in the first example that we know nothing at all about its variability, except that x lies in some known range (a, b). What is an appropriate prior *pdf* with which to describe x? Intuition tells us that all values in (a, b) are equally likely, so that $P(x) = 1/(b - a)$ within the range, and zero outside. The principle of maximum entropy tells us that $P(x)$ must maximise:

$$S(P) = -\int_a^b P(x) \ln P(x)\, dx \tag{2.84}$$

subject to the condition that $\int_a^b P(x)\, dx = 1$. Maximising $S(P)$ with respect to P gives:

$$-\ln P(x) - 1 + \mu = 0 \tag{2.85}$$

where μ is the Lagrangian multiplier for the condition. $P(x)$ is clearly a constant, and its value is found from the unit area condition to be $1/(b - a)$, as expected. If we know nothing at all about x, i.e. $b - a \to \infty$, and $P(x) \to 0$ everywhere, a reasonable conclusion.

Now consider the case where we also know that the expected value of x is $\bar{x}$, but we know nothing about its variability. Now we have the extra condition:

$$\int_a^b xP(x)\, dx = \bar{x} \tag{2.86}$$

and the maximisation gives:

$$-\ln P(x) - 1 + \mu + \nu x = 0 \tag{2.87}$$

where ν is the Lagrangian multiplier for the extra condition. For the case where a and b are general, this leads to complicated but straightforward algebra which the reader may explore at leisure. Consider only the case where it is known that x is positive, i.e. $a = 0$ and $b = \infty$. On substituting $P(x) = \exp(-1 + \mu + \nu x)$ into the constraints, it is straightforward to show that $\nu = -1/\bar{x}$, and hence that:

$$P(x) = \begin{cases} (1/\bar{x})\exp(-x/\bar{x}) & \text{for } x \geq 0 \\ 0 & \text{for } x < 0. \end{cases} \qquad (2.88)$$

This is an example of a *pdf* where the most likely value, at $x = 0$, is not near the mean, $\bar{x}$, and is in fact at an extreme of the range of x. Without the positivity constraint the *pdf* would be zero everywhere, as in the case of no information at all.

The third example is when an expected value and a variance are both known, and there is no finite range limitation. Now we have three constraints on $P(x)$:

$$\int P(x)\,\mathrm{d}x = 1$$
$$\int xP(x)\,\mathrm{d}x = \bar{x}$$
$$\int (x - \bar{x})^2 P(x)\,\mathrm{d}x = \sigma^2. \qquad (2.89)$$

In this case it is possible to show that the maximum entropy *pdf* is the normal distribution with mean $\bar{x}$ and variance σ^2.

⇒ ***Exercise*** 2.9: Show this

This is the primary reason why the Gaussian distribution is an appropriate default when only the mean and variance (or covariance, in general) are known. Of all possible distributions with a given mean and variance, the Gaussian has maximum entropy, or minimum information.

The measure function $M(x)$, Eq. (2.68), has been ignored (or set to a constant) in the above examples, and a few words of explanation are in order. The first example above could be restated by describing the system in terms of a single variable $y = x^2$, where all that is known is that y lies in the range (a^2, b^2) where a and b are both positive for the purpose of this discussion. We could then follow the arguments above to determine that y must be uniformly distributed in (a^2, b^2), a different result from x being uniformly distributed in (a, b). The measure function is a means of making the maximum entropy *pdf* independent of the coordinate system, because it will transform with the coordinate system. The choice of measure function is equivalent to selecting a coordinate system in which a non-informative *pdf* would be uniform.

Construction of *a priori* for practical retrievals will be discussed in more detail in Chapter 10.

Chapter 3
Error Analysis and Characterisation

When an inverse problem is ill-posed, so that it does not have a unique solution, we may wish to select from the infinite number of possible solutions one which is the 'best' or the 'optimal' one in some sense. However there are many ways in which 'optimal' can be defined. Before optimal solutions can be developed, it is necessary to determine what can sensibly be optimised, and how the solution is to be related to the true state. As indicated in Chapter 2, the retrieved state could be the most likely, consistent with all of the available information. Alternatively it could be the solution method which minimises the error variance when applied to an ensemble of cases (I will show later that these two approaches lead to the same solution in some circumstances). A completely different kind of possible requirement is that the retrieval have the best spatial resolution.

Therefore before we attempt to design specific optimal retrieval methods, we should consider what the characteristics of retrievals in general might be, so that we can decide which properties would be the most appropriate to optimise. In this chapter we will carry out a general characterisation and error analysis that can be applied to any inverse method and which will show how a retrieval is related to the true state of the atmosphere, and how various sources of error propagate into the final product. The analysis will help identify characteristics of retrievals that can be optimised, and is in any case an essential part of the documentation of any retrieved data set. It can be argued that a retrieval method without an error analysis and characterisation is of little value.

The analysis used in this chapter follows the methods developed by Eyre (1987) and Rodgers (1990).

3.1 Characterisation

3.1.1 *The forward model*

For any remote measurement, or any indirect measurement of any kind, the quantity measured, $\mathbf{y}$, is some vector valued function $\mathbf{f}$ of the unknown state vector $\mathbf{x}$, and of some other set of parameters **b** that we have decided not to include in the state vector. There is also an experimental error term ϵ. Therefore we write, more generally than Eq. (2.2):

$$\mathbf{y} = \mathbf{f}(\mathbf{x}, \mathbf{b}) + \boldsymbol{\epsilon} \qquad (3.1)$$

where the *Forward Function* **f** describes the complete physics of the measurement, including for example the radiative transfer theory required to relate the state to the measured signal, as well as a full description of the measuring instrument. The vector of parameters **b** comprises those quantities which influence the measurement, are known to some accuracy, but are not intended as quantities to be retrieved. They will be termed the *forward function parameters* and will contribute to the total measurement accuracy. A typical example might include such things as spectral line strengths. The error term ϵ includes errors from sources such as detector noise, which are not related to the forward function parameters. We could in principle include ϵ as one of the forward function parameters, but it is clearer if left as an explicit term. Generally it is a purely random term, which we call *measurement noise*, reserving *(total) measurement error* for the sum of contributions from all sources, including systematic errors.

3.1.2 The retrieval method

The retrieval $\hat{\mathbf{x}}$ is the result of operating on the measurement with some *Inverse* or *Retrieval Method* **R**

$$\hat{\mathbf{x}} = \mathbf{R}(\mathbf{y}, \hat{\mathbf{b}}, \mathbf{x}_a, \mathbf{c}) \qquad (3.2)$$

where the circumflex indicates an estimated quantity, rather than a true state, $\hat{\mathbf{b}}$ is the our best estimate of the forward function parameters, as distinct from the true value **b** which is the value that the atmosphere and the instrument know about. The vectors $\mathbf{x}_a$ and **c** together comprise parameters that do not appear in the forward function, but do affect the retrieval, and may be subject to uncertainty; $\mathbf{x}_a$ is any *a priori* estimate of **x** that may be used, and **c** contains any other parameters of this nature. Some inverse methods use an explicit *a priori*, some do not, and some claim not to use it, but when examined closely are found to use it. It can be thought of as any kind of estimate of the state, or part of the state, used in the inverse method and which is unrelated to the actual measurement. It might for example be an independent state estimate such as a climatology that is explicitly combined with a measurement. The parameters comprising **c** will be called *retrieval method parameters* as a catch-all for anything else that may be used in the retrieval method, such as convergence criteria. We will find later that *a priori* is the *only* example of a retrieval method parameter that should matter in most reasonable circumstances.

3.1.3 *The transfer function*

We are now in a position to relate the retrieval to the true state by substituting Eq. (3.1) into Eq. (3.2):

$$\hat{\mathbf{x}} = \mathbf{R}(\mathbf{f}(\mathbf{x}, \mathbf{b}) + \boldsymbol{\epsilon}, \hat{\mathbf{b}}, \mathbf{x}_a, \mathbf{c}) \tag{3.3}$$

This may be regarded as a *Transfer Function* describing the operation of the whole observing system, including both the measuring instrument and the retrieval method. Understanding the properties of the transfer function is fundamental to both the error analysis and the characterisation of the observing system. By *characterisation* is meant the sensitivity of the retrieval to the true state, most simply expressed by the matrix of derivatives $\partial\hat{\mathbf{x}}/\partial\mathbf{x}$, and by *error analysis* is meant the sensitivity of the retrieval to all of the sources of error in the transfer function, including noise in the measurement, error in the non-retrieved parameters and in the retrieval method parameters, and the effect of modelling the true physics of the measurement by some forward *model*, if needed.

3.1.4 *Linearisation of the transfer function*

The forward function itself is often a source of difficulty, for various reasons. It may be that the real physics is far too complex to deal with explicitly, for example if both scattering and molecular band absorption are both involved in the radiative transfer. It may even be that the detailed physics is still uncertain, for example in the treatment of cloud, but a model can be constructed to adequate accuracy. Whatever the reason, a *Forward Model*, $\mathbf{F}$, with associated errors, may be used:

$$\mathbf{F}(\mathbf{x}, \mathbf{b}) \simeq \mathbf{f}(\mathbf{x}, \mathbf{b}, \mathbf{b}') \tag{3.4}$$

where $\mathbf{b}$ has now been separated into $\mathbf{b}$ and $\mathbf{b}'$, where $\mathbf{b}'$ represents those forward function parameters which are ignored in the construction of the forward model.

To obtain a basic understanding of the transfer function, linearise it with respect to the various parameters involved, but first replace the forward function by the forward model, and include an error term to allow for this:

$$\hat{\mathbf{x}} = \mathbf{R}(\mathbf{F}(\mathbf{x}, \mathbf{b}) + \Delta\mathbf{f}(\mathbf{x}, \mathbf{b}, \mathbf{b}') + \boldsymbol{\epsilon}, \hat{\mathbf{b}}, \mathbf{x}_a, \mathbf{c}) \tag{3.5}$$

where $\Delta\mathbf{f}$ is the error in the forward model relative to the real physics:

$$\Delta\mathbf{f} = \mathbf{f}(\mathbf{x}, \mathbf{b}, \mathbf{b}') - \mathbf{F}(\mathbf{x}, \mathbf{b}) \tag{3.6}$$

Now linearise the forward model about $\mathbf{x} = \mathbf{x}_a$, $\mathbf{b} = \hat{\mathbf{b}}$:

$$\hat{\mathbf{x}} = \mathbf{R}(\mathbf{F}(\mathbf{x}_a, \hat{\mathbf{b}}) + \mathbf{K}_x(\mathbf{x} - \mathbf{x}_a) + \mathbf{K}_b(\mathbf{b} - \hat{\mathbf{b}}) + \Delta\mathbf{f}(\mathbf{x}, \mathbf{b}, \mathbf{b}') + \boldsymbol{\epsilon}, \hat{\mathbf{b}}, \mathbf{x}_a, \mathbf{c}) \tag{3.7}$$

where the matrix $\mathbf{K}_x$ is the sensitivity of the forward model to the state, $\partial\mathbf{F}/\partial\mathbf{x}$, the weighting function or Jacobian matrix, and $\mathbf{K}_b$ is the sensitivity of the forward model to the forward model parameters, $\partial \mathbf{F}/\partial \mathbf{b}$. Next, linearise the inverse method with respect to its first argument, $\mathbf{y}$:

$$\begin{aligned}\hat{\mathbf{x}} &= \mathbf{R}[\mathbf{F}(\mathbf{x}_a, \hat{\mathbf{b}}), \hat{\mathbf{b}}, \mathbf{x}_a, \mathbf{c}] \\ &+ \mathbf{G}_y[\mathbf{K}_x(\mathbf{x} - \mathbf{x}_a) + \mathbf{K}_b(\mathbf{b} - \hat{\mathbf{b}}) + \Delta \mathbf{f}(\mathbf{x}, \mathbf{b}, \mathbf{b}') + \epsilon]\end{aligned} \qquad (3.8)$$

where $\mathbf{G}_y = \partial \mathbf{R}/\partial \mathbf{y}$ is the sensitivity of the retrieval to the measurement, which is the same as its sensitivity to measurement error. Rearrange this to give:

$$\begin{aligned}\hat{\mathbf{x}} - \mathbf{x}_a &= \mathbf{R}[\mathbf{F}(\mathbf{x}_a, \hat{\mathbf{b}}), \hat{\mathbf{b}}, \mathbf{x}_a, \mathbf{c}] - \mathbf{x}_a & &\ldots bias \\ &+ \mathbf{A}(\mathbf{x} - \mathbf{x}_a) & &\ldots smoothing \\ &+ \mathbf{G}_y \epsilon_y & &\ldots retrieval\ error\end{aligned} \qquad (3.9)$$

where

$$\mathbf{A} = \mathbf{G}_y \mathbf{K}_x = \frac{\partial \hat{\mathbf{x}}}{\partial \mathbf{x}} \qquad (3.10)$$

is the sensitivity of the retrieval to the true state and

$$\epsilon_y = \mathbf{K}_b(\mathbf{b} - \hat{\mathbf{b}}) + \Delta \mathbf{f}(\mathbf{x}, \mathbf{b}, \mathbf{b}') + \epsilon \qquad (3.11)$$

is the total error in the measured signal relative to the forward model.

3.1.5 *Interpretation*

The first term on the right hand side of Eq. (3.9), the bias, is the error that would result from a simulated retrieval using a simulated error-free measurement of the *a priori* state computed with the forward model. To understand the first term, all we need to know about the meaning of *a priori* is that it represents knowledge of the state before the measurement is made. If the measurement were not made, *a priori* would be our only knowledge. Thus if the measurements are consistent with the state being equal to the *a priori* then any well behaved inverse method should return the *a priori*. Therefore the first term should be zero, and if it is nonzero for some particular inverse method, then this is probably a good reason for modifying the method!

The linearisation has been carried out about the *a priori* state mainly so that the above argument can be made to eliminate the bias term. Consequently it is not applicable to methods that do not use an explicit *a priori* in the above sense. In these cases an arbitrary profile could be used for the linearisation point, but the argument about the bias term would no longer apply. Alternatively if a state can be found which is unchanged by a simulated noiseless measurement and retrieval, then that can be taken as the linearisation point.

The second term on the right hand side of Eq. (3.9) represents the way in which the observing system smooths the profile. The difference between the retrieval and

Characterisation 47

the linearisation point is obtained by operating on the difference between the true state and the linearisation point with the matrix $\mathbf{A}$:

$$\hat{\mathbf{x}} = \mathbf{x}_a + \mathbf{A}(\mathbf{x} - \mathbf{x}_a) + \mathbf{G}_y \boldsymbol{\epsilon}_y = (\mathbf{I}_n - \mathbf{A})\mathbf{x}_a + \mathbf{A}\mathbf{x} + \mathbf{G}_y \boldsymbol{\epsilon}_y \qquad (3.12)$$

In the case where the state vector represents a profile, then rows $\mathbf{a}_i^T$ of $\mathbf{A}$ can be regarded as smoothing functions: the averaging kernels or model resolution functions. In the ideal inverse method, $\mathbf{A}$ would be a unit matrix. In reality, rows of $\mathbf{A}$ are generally peaked functions, peaking at the appropriate level, and with a half-width which is a measure of the spatial resolution of the observing system, thus providing a simple characterisation of the relationship between the retrieval and the true state. The averaging kernel also has an area, which is found to be approximately unity at levels where the retrieval is accurate, and in general can be thought of as a rough measure of the fraction of the retrieval that comes from the data, rather than the *a priori*. The area of $\mathbf{a}_i$ is the sum of its elements, $\mathbf{a}_i^T \mathbf{u}$, where $\mathbf{u}$ is a vector with unit elements, so a vector of areas is $\mathbf{A}\mathbf{u}$, which can also be regarded as the response of the retrieval to a unit perturbation in all elements of the state vector.

The columns of $\mathbf{A}$ give the response of the retrieval to a δ-function perturbation in the state vector, and may be described as the δ-function response or point spread function. This provides a convenient, if slow, method of computing the averaging kernel matrix in cases where the algebra is complicated. The δ-function response can be calculated numerically by finding the change in the retrieval which results when each element of the state vector in turn is perturbed by some suitably small amount. The perturbation should be small enough that the response is linear in the size of the perturbation, but large enough that rounding errors are unimportant. The response is placed in the appropriate column of the $\mathbf{A}$ matrix, whose rows will then provide the averaging kernels.

The third term on the right hand side of Eq. (3.9) is the error in the retrieval due to the total measurement error, which we will call *retrieval error*, meaning the error in the retrieval due to $\boldsymbol{\epsilon}_y$, rather than error of the retrieval process. This will be discussed in more detail in section 3.2.

⇒ ***Exercise*** 3.1: Using an eigenvector expansion of $\mathbf{A}$, give an interpretation of Eq. (3.12) in which the retrieval is expressed using scalar rather than matrix weights.

3.1.6 *Retrieval method parameters*

The linearisation has ignored the retrieval method parameters $\mathbf{x}_a$ and $\mathbf{c}$. The argument about *a priori* and good behaviour of the retrieval method led to the requirement that

$$\mathbf{R}[\mathbf{F}(\mathbf{x}_a, \hat{\mathbf{b}}), \hat{\mathbf{b}}, \mathbf{x}_a, \mathbf{c}] = \mathbf{x}_a \qquad (3.13)$$

This must be true when $\mathbf{x}_a$, $\hat{\mathbf{b}}$ or $\mathbf{c}$ are perturbed, and in particular, by taking the derivative with respect to $\mathbf{c}$, we must have $\partial \mathbf{R}/\partial \mathbf{c} = 0$ for any well behaved inverse method. Thus a well behaved inverse method can have no retrieval method parameters that matter apart from $\mathbf{x}_a$! It is clear that if bias is zero at $\mathbf{x} = \mathbf{x}_a$ then it cannot change, to first order, if $\mathbf{c}$ is changed. Thus if retrieval method parameters are identified which actually affect the retrieval, then there is a good reason to modify the method.

Similarly by taking the derivative of (3.13) with respect to $\hat{\mathbf{b}}$ we obtain

$$\frac{\partial \mathbf{R}}{\partial \mathbf{y}}\frac{\partial \mathbf{F}}{\partial \mathbf{b}} + \frac{\partial \mathbf{R}}{\partial \mathbf{b}} = \mathbf{G}_y \mathbf{K}_b + \mathbf{G}_b = 0 \quad (3.14)$$

which simply says that the effect of $\mathbf{b}$ in the inverse method should compensate exactly for variations in values of $\mathbf{b}$ in the forward model.

From the linearised expression for the retrieved state Eq. (3.12) we can obtain the following expression for its sensitivity to the *a priori*:

$$\frac{\partial \mathbf{R}}{\partial \mathbf{x}_a} = \frac{\partial \hat{\mathbf{x}}}{\partial \mathbf{x}_a} = \mathbf{I}_n - \mathbf{A} \quad (3.15)$$

3.2 Error Analysis

An expression for the error in $\hat{\mathbf{x}}$ can be obtained by some further rearrangement of Eqs. (3.9) and (3.11):

$$\begin{aligned}
\hat{\mathbf{x}} - \mathbf{x} =\ & (\mathbf{A} - \mathbf{I}_n)(\mathbf{x} - \mathbf{x}_a) & &\ldots \text{smoothing error} \\
& + \mathbf{G}_y \mathbf{K}_b (\mathbf{b} - \hat{\mathbf{b}}) & &\ldots \text{model parameter error} \\
& + \mathbf{G}_y \Delta \mathbf{f}(\mathbf{x}, \mathbf{b}, \mathbf{b}') & &\ldots \text{forward model error} \\
& + \mathbf{G}_y \boldsymbol{\epsilon} & &\ldots \text{retrieval noise}
\end{aligned} \quad (3.16)$$

3.2.1 Smoothing error

For the purpose of carrying out an error analysis, the retrieval can either be regarded as an estimate of a state smoothed by the averaging kernel rather than an estimate of the true state, or as an estimate of the true state, but with an error contribution due to smoothing. The error analysis will be different in the two cases, because in the second case there is an extra term, the smoothing error.

Because the true state is not normally known, we cannot estimate the *actual* smoothing error, $(\mathbf{A} - \mathbf{I}_n)(\mathbf{x} - \mathbf{x}_a)$. What is really required is a description of the statistics of the error, which must be calculated from the mean and covariance over some appropriate ensemble of states, which may or may not be that described by $\mathbf{x}_a$ and $\mathbf{S}_a$. The mean is $(\mathbf{A} - \mathbf{I}_n)(\bar{\mathbf{x}} - \mathbf{x}_a)$, which will be zero if an ensemble has been chosen for which $\bar{\mathbf{x}} = \mathbf{x}_a$.

The covariance of the smoothing error about $\bar{\mathbf{x}}$ is:

$$\begin{aligned}
\mathbf{S}_s &= \mathcal{E}\{(\mathbf{A}-\mathbf{I}_n)(\mathbf{x}-\bar{\mathbf{x}})\cdot(\mathbf{x}-\bar{\mathbf{x}})^T(\mathbf{A}-\mathbf{I}_n)^T\} \\
&= (\mathbf{A}-\mathbf{I}_n)\mathcal{E}\{(\mathbf{x}-\bar{\mathbf{x}})(\mathbf{x}-\bar{\mathbf{x}})^T\}(\mathbf{A}-\mathbf{I}_n)^T \\
&= (\mathbf{A}-\mathbf{I}_n)\mathbf{S}_e(\mathbf{A}-\mathbf{I}_n)^T
\end{aligned} \qquad (3.17)$$

where $\mathbf{S}_e$ is the covariance of the ensemble of states about the mean state. Thus

> to estimate the smoothing error covariance, the covariance matrix of a real ensemble of states must be known.

Many remote observing systems cannot see spatial fine structure, the loss of which contributes to the smoothing error. To estimate it correctly, the actual statistics of the fine structure must be known. It is not enough to simply use some *ad hoc* matrix that has been constructed as a reasonable *a priori* constraint in the retrieval. If the real covariance is not available, it may be better to abandon the estimation of the smoothing error, and consider the retrieval as an estimate of a smoothed version of the state, rather than an estimate of the complete state.

This component of the error budget was described incorrectly as 'null-space error' in Rodgers (1990). The term null space error should properly be used to describe the contribution to the error budget from those components of the state that lie in the null space of $\mathbf{K}$, and are consequently not seen by the retrieval. Smoothing error is only equal to null space error for some types of retrieval method. A general method may succeed in estimating some null-space components because of the correlations provided by the *a priori* covariance, so that the contribution from that source is reduced, and there may be poorly determined components in the 'near-null space' that appear in the retrieval at reduced amplitude, and therefore contribute to the smoothing error.

3.2.2 Forward model parameter error

The error in the retrieval due to errors in the forward model parameters is $\mathbf{G}_y\mathbf{K}_b(\mathbf{b}-\hat{\mathbf{b}})$, and is straightforward to evaluate, in principle. If the forward model parameters have been estimated properly and the model is linear as far as they are concerned, then their individual errors will be unbiassed, and the expected error will be zero. The evaluation of the sensitivities $\mathbf{G}_y$ and $\mathbf{K}_b$ is straightforward in principle, either by evaluating the derivatives algebraically or by perturbation from the inverse method and forward model respectively. Note that if the sensitivity for some particular element of $\mathbf{b}$ is unacceptably large given its accuracy, then there are three possibilities:

(1) make a better (e.g. laboratory) measurement of that element;
(2) consider whether it should be an element of the state vector, and be retrieved from the measurements;

(3) redesign the observing system so it is not sensitive to that element.

Note that possibility (2) does not help an individual retrieval if the error due to this parameter has been correctly accounted for in the measurement error covariance (see section 4.1.2), but an improved estimate may be obtainable from multiple measurements. The error covariance for this contribution is:

$$\mathbf{S}_f = \mathbf{G}_y \mathbf{K}_b \mathbf{S}_b \mathbf{K}_b^T \mathbf{G}_y^T \tag{3.18}$$

where $\mathbf{S}_b = \mathcal{E}\{(\mathbf{b} - \hat{\mathbf{b}})(\mathbf{b} - \hat{\mathbf{b}})^T\}$ is the error covariance matrix of $\mathbf{b}$. However remember that $\mathbf{b}$ may contain both random and systematic components on various time and/or space scales. It is sensible to evaluate them separately.

3.2.3 Forward model error

The modelling error is given by $\mathbf{G}_y \Delta \mathbf{f} = \mathbf{G}_y [\mathbf{f}(\mathbf{x}, \mathbf{b}, \mathbf{b}') - \mathbf{F}(\mathbf{x}, \mathbf{b})]$. Note that this is evaluated at the true state, and with the true value of $\mathbf{b}$, rather than at $\hat{\mathbf{x}}$ and $\hat{\mathbf{b}}$, but the sensitivity to these quantities should not be large provided the problem is not grossly nonlinear. Modelling error can be hard to evaluate, because it requires a model for $\mathbf{f}$ which includes the correct physics. If the correct physics is known and can be modelled accurately, and $\mathbf{F}$ is simply a numerical approximation for efficiency's sake, then evaluating modelling error is straightforward. But if $\mathbf{f}$ is not known in detail, or so horrendously complex that no proper model is feasible, then modelling error can be tricky to estimate. Modelling error is likely to be systematic.

3.2.4 Retrieval noise

Retrieval noise is given by $\mathbf{G}_y \boldsymbol{\epsilon}$ and is usually the easiest component to evaluate. Measurement noise $\boldsymbol{\epsilon}$ is usually random, and is normally unbiassed and often uncorrelated between channels, and has a known covariance matrix. The covariance of the retrieval noise is:

$$\mathbf{S}_m = \mathbf{G}_y \mathbf{S}_\epsilon \mathbf{G}_y^T \tag{3.19}$$

3.2.5 Random and systematic error

Errors are traditionally classified as systematic or random according to whether they are constant between consecutive measurements, or vary randomly. Related terms are widely used are 'precision' and 'accuracy', where precision measures the variability between repeated measurements of the same state, and accuracy measures the total difference between the measurement and the truth, and includes both random and systematic errors. In practice the distinction is somewhat vague, because error sources may have time variability on a range of scales, and a source which is random on one scale may be systematic on another. The retrieval noise is normally a truly random quantity, uncorrelated in time. Some model parameter errors (for example spectral data) may be truly systematic errors, unchanged in time, but others (for example calibration parameters) may vary from day to day, but be constant over minutes or hours. In any case the effect of a source of error on a retrieval, even if the source is constant, may vary with the state because of non-linearity. There is another category of systematic error which varies with state, of which smoothing error is the prime example, which we can describe as a *gain error*. The smoothing error is correlated with the state, and will have variations on the same time scale as changes of the state.

3.2.6 *Representing covariances*

Presenting spatially correlated errors to users of retrieved data can be a problem. Error bars on scalar quantities are straightforward, but for a vector quantity **x**, where errors are expressed by a covariance matrix **S**, the equivalent of an error bar is a closed surface (usually a hyper-ellipsoid) in state space. If errors are uncorrelated then the major axes of the ellipsoid are aligned with the coordinates of the state space, but if they are correlated the ellipsoid may be oriented at some angle to the coordinate system, and it becomes more difficult to visualise and to understand the implications. The diagonal elements of **S** are the familiar error variances of the elements of **x**, but if the off-diagonal elements show that the errors are correlated, then we can have more information about **x** than is the case if the off-diagonal elements are negligible, so it is important not to ignore them.

This is illustrated for two-element vectors in Fig. 3.1, where the error bounds are presented as contours of a *pdf*. The point $\mathbf{x}_1$ has uncorrelated errors, so its error bounds are described by circular contours. Point $\mathbf{x}_2$ has correlated errors, but with the same error variances as $\mathbf{x}_1$. Its error bounds are described by elliptical contours. It is clear that the uncertainty of $\mathbf{x}_2$ encompasses a smaller area of state space than that of $\mathbf{x}_1$, and therefore $\mathbf{x}_2$ provides more information. One way of conceptualising the error covariance matrix is to diagonalise it. In the case in the figure, this would be done by transforming the basis to the principal axes of the ellipse. In this basis, it is even clearer that $\mathbf{x}_2$ is a better measurement than $\mathbf{x}_1$. Another way is to consider the determinant of the covariance matrix. This is clearly always smaller for the case with off-diagonal elements, $|\mathbf{S}| = S_{11}S_{22} - S_{12}^2$, compared with $S_{11}S_{22}$, so that the Shannon information content must be greater.

> ⇒ **Exercise** 3.2: Show that the determinant of a covariance matrix of any order with given diagonal elements is greatest if the off-diagonal elements are all zero

In the general case we can find a basis in which the errors are independent by diagonalising the covariance matrix, i.e. finding the eigenvalues λ_i and eigenvectors

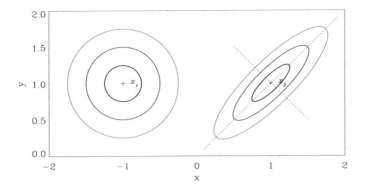

Fig. 3.1 Illustrating correlated and uncorrelated errors

$\mathbf{l}_i$ of the covariance matrix, $\mathbf{S}\mathbf{l}_i = \lambda_i \mathbf{l}_i$. A covariance matrix is symmetric, so that the matrix of eigenvectors $\mathbf{L}$ is orthogonal and can be normalized: $\mathbf{L}^{-1} = \mathbf{L}^T$. Consequently, $\mathbf{S}$ can be decomposed as

$$\mathbf{S} = \sum_i \lambda_i \mathbf{l}_i \mathbf{l}_i^T = \sum_i \mathbf{e}_i \mathbf{e}_i^T \qquad (3.20)$$

where the orthogonal vectors $\mathbf{e}_i$ are $\lambda_i^{\frac{1}{2}} \mathbf{l}_i$ and can be thought of as 'error patterns', in the sense that the error $\boldsymbol{\epsilon}_\mathbf{x}$ in $\mathbf{x}$ can be expressed a sum of these error patterns, each multiplied by a random factor a_i having unit variance

$$\boldsymbol{\epsilon}_\mathbf{x} = \sum_{i=1}^{n} a_i \mathbf{e}_i \qquad (3.21)$$

To illustrate this concept the most significant ten error patterns of the *a priori* covariance used in section 2.6 are shown in Fig. 3.2. They are all truncated sinusoids with the largest amplitude associated with the longest wavelength.

3.3 Resolution

'Resolution', like 'information', is a word with a multiplicity of meanings, and tends to be used differently in different contexts. To arrive at a satisfactory definition for our purposes, we need to consider the aim of the measurement. In astronomy or spectroscopy the resolution of a telescope or a spectrometer is a measure of its ability to distinguish two point sources in one or two dimensions. This is not directly useful in the case of profile retrieval, because we are not normally dealing with states subject to δ-function perturbations. More useful would be some measure of how well we can see the type of structures that might be present in the atmosphere. Possibilities include characteristics of the averaging kernel or state resolution matrix, such as the width of the averaging kernel or the point spread function, where

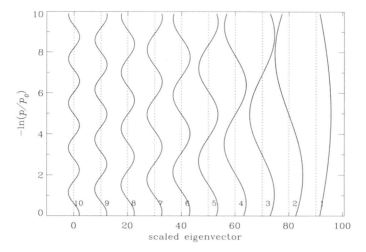

Fig. 3.2 The most significant ten error patterns of the *a priori* covariance matrix $S_{ij} = \sigma_a^2 \exp(-|i-j|\delta z/h)$, for $\sigma_a \simeq 3\,\mathrm{K}$ and $h=1$

'width' has many possible interpretations, the response of the retrieval to sine wave perturbations in the state, and the range of heights covered divided by number of independent quantities measured.

Resolution should be clearly distinguished from the grid spacing used to represent a profile. The choice of a practical grid spacing depends on various features of the problem, and the resolution of the measurement is one of them. This will be discussed in more detail in Chapter 10.

A measure of resolution is not useful unless there is an associated error. The ability to reproduce a sine wave perturbation is not useful if the response to a reasonable atmospheric amplitude is so small that it is less than the noise. Possible characterisations of resolution that include error are:

(i) For each element of $\hat{\mathbf{x}}$, the 'width' of the corresponding averaging kernel and the corresponding error on the element. It is also helpful to know the prior error, i.e. the expected variation of that quantity before the measurement is made. The error on the element may or may not be reduced by combination with the prior information, depending on the retrieval method.

(ii) The 'trade-off plot', i.e. the error (and prior error) of an estimate of an average of the profile over some given averaging function, together with the width of that function. This information can be in the form of a plot of noise against width, for example the error in mean temperature over a layer, as a function of layer width.

(iii) A sine wave imposed on the input profile in a simulation will be retrieved at some amplitude that may or may not be greater than noise, depending on its amplitude and wavelength. A trade-off curve would give the input amplitude of the smallest sine wave which is just detectable in the retrieval, as a function of wavelength. This is likely to vary with altitude, and the output perturbation is unlikely to be strictly sinusoidal.

(iv) The degrees of freedom for signal, d_s is the trace of the averaging kernel matrix. Consequently the diagonal of $\mathbf{A}$ may be thought of as a measure of the number of degrees of freedom per level, and its reciprocal as the number of levels per degree of freedom, and thus a measure of resolution. In a simplistic way we can see that if an averaging kernel is mostly positive and has an area of about unity, then the reciprocal of its peak is a measure of its width. Purser and Huang (1993) have developed this idea as a concept of 'data density'.

Even Lord Rayleigh's original definition is not really as simple as it sounds. He was considering the case of viewing objects through optical systems, for which it might appear that the resolution is related to the diffraction limit of the system. But if it is known that the field really does consist of for example delta functions (stars), the diffraction pattern is known, and the noise on the measurement is low, then a higher resolution representation of the field can be retrieved, or the positions of the stars could be retrieved using for example a least squares fit or an optimal estimator, even if the image appears to the eye as a single star. The diffraction pattern is in effect a weighting function for an indirect measurement, and retrieval methods can be applied.

Width of an averaging kernel

How do we define the width of a peaked function? There are several possibilities, but some definitions are more useful than others. The 'full width at half height' is satisfactory for some purposes, but is not very helpful if algebraic manipulation of the width is needed, and can be difficult to define for functions with significant lobes, positive or negative. For mathematical convenience it should be possible to use the definition algebraically for optimisation. It should also produce sensible results if the function has negative lobes.

The second moment about the mean of a function $A(z, z')$ of height z' whose nominal peak is at z is

$$w(z) = \left(\frac{\int A(z, z')(z' - \bar{z})^2 dz'}{\int A(z, z') dz'} \right)^{\frac{1}{2}} \tag{3.22}$$

where the mean $\bar{z}(z) = \int z' A(z, z') dz' / \int A(z, z') dz'$. This may be a reasonable definition for positive $A(z, z')$, but can give problems with negative lobes. It is also quite possible to obtain a negative or indeterminate second moment. Consider for example $A(z, z') = \text{sinc}(z' - z) = \sin(z' - z)/(z' - z)$, the amplitude of the diffraction pattern of a single slit. The integrand of (3.22) oscillates with ever growing amplitude as $|z' - z| \to \infty$! Even the intensity of the same diffraction pattern, $A(z, z') = \text{sinc}^2(z' - z)$, gives an indeterminate integral.

Backus and Gilbert (1970) defined a quantity they called the 'spread'

$$s(z) = 12 \int (z - z')^2 A^2(z, z') \, dz' / (\int A(z, z') \, dz')^2. \tag{3.23}$$

The original definition was for functions with unit area, but Eq. (3.23) has been slightly generalised to include the normalising area integral. The factor 12 is chosen so that a simple slit function has a spread equal to its full width. A related concept is the 'resolving length', which can be defined as the spread about the 'centre', defined as the mean of $A^2(z, z')$:

$$c(z) = \int z' A^2(z, z') \, dz' / \int A^2(z, z') \, dz' \tag{3.24}$$

These definitions were designed for optimisation, and not as a general purpose definition of resolution. A retrieval method which minimises spread (section 4.5) is very successful in suppressing sidelobes in the averaging kernel, but spread is not a satisfactory general-purpose definition of resolution for functions which decay more slowly than about $O\{(z - z')^{-3/2}\}$.

3.4 The Standard Example: Linear Gaussian Case

The concepts described above are more easily understood with a concrete example. For this purpose we will use the linear problem first introduced in Chapter 1, with the weighting functions shown in Fig. 1.1, together with the maximum *a posteriori* solution for Gaussian *pdf*'s which will be discussed in more detail in Chapter 4, but is no more than the peak of the Bayesian solution $P(\mathbf{x}|\mathbf{y})$ as given by Eq. (2.30).

$$\hat{\mathbf{x}} = \mathbf{x}_a + (\mathbf{K}^T \mathbf{S}_\epsilon^{-1} \mathbf{K} + \mathbf{S}_a^{-1})^{-1} \mathbf{K}^T \mathbf{S}_\epsilon^{-1} (\mathbf{y} - \mathbf{K}\mathbf{x}_a) \tag{3.25}$$

The covariance of $P(\mathbf{x}|\mathbf{y})$ given by Eq. (2.27) will not be used here. Instead the error analysis will be used to determine the accuracy of $\hat{\mathbf{x}}$.

For numerical illustration the measurement error covariance matrix is taken to be proportional to a unit matrix, of the form $\mathbf{S}_\epsilon = \sigma_\epsilon^2 \mathbf{I}_m$ with $\sigma_\epsilon = 0.5$, and the prior covariance by the matrix introduced in section 2.6:

$$S_{ij} = \sigma_a^2 \exp\left(-|i - j|\frac{\delta z}{h}\right) \tag{3.26}$$

with the same parameters.

Fig. 3.3 shows a simulated retrieval under these conditions. The *a priori* is the U.S. Standard Atmosphere, shown dotted, and the simulated true profile (solid) has been constructed by adding error patterns of the prior covariance to this, multiplied by Gaussian random coefficients, as described by Eq. (3.21). There is perhaps too much fine structure in comparison with real atmospheric profiles, but otherwise

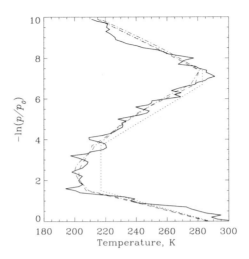

Fig. 3.3 Simulated retrieval. A priori: dotted; temperature profile: solid; retrieval without noise: dashed; retrieval with noise: Dash-dot.

this is fairly realistic. Radiances are simulated by multiplying the profile by the weighting function matrix, Fig. 1.1, and Gaussian noise with a standard deviation of 0.5 K is added. The retrieval without added noise is shown dashed, and the retrieval with noise is dash-dot. Some features to be expected are seen in the retrieval: the fine structure is not retrieved because of the width of the weighting functions; the discontinuities of gradient in the retrievals are present because they are present in the *a priori*, not because they have been retrieved; the fit of the retrieval to the profile is poorer at the top and bottom, where it approaches the *a priori*.

From Eq. (3.25) the retrieval gain matrix is seen to be:

$$\mathbf{G}_y = (\mathbf{K}^T \mathbf{S}_\epsilon^{-1} \mathbf{K} + \mathbf{S}_a^{-1})^{-1} \mathbf{K}^T \mathbf{S}_\epsilon^{-1} \tag{3.27}$$

The contribution functions, or columns of $\mathbf{G}_y$, for each channel are shown in Fig. 3.4. It is easy to see that the bias for this retrieval is zero. If the prior measurement computed from the forward model for the *a priori* state, $\mathbf{y}_a = \mathbf{K}\mathbf{x}_a$, is substituted for $\mathbf{y}$ in Eq. (3.25), then the result is clearly $\hat{\mathbf{x}} = \mathbf{x}_a$.

3.4.1 *Averaging kernels*

The averaging kernel matrix is, from (3.10) and (3.27):

$$\mathbf{A} = \mathbf{G}_y \mathbf{K} = (\mathbf{K}^T \mathbf{S}_\epsilon^{-1} \mathbf{K} + \mathbf{S}_a^{-1})^{-1} \mathbf{K}^T \mathbf{S}_\epsilon^{-1} \mathbf{K} \tag{3.28}$$

Fig. 3.5 shows the averaging kernels (rows of $\mathbf{A}$) in this case for a subset of levels, at intervals of unity in $-\ln(p/p_0)$. Also shown as the dotted line on the right is the area of the averaging kernels, scaled by a factor of 0.1. The area is approximately unity from $-\ln(p/p_0) \simeq 1.5$ to 8, indicating that the retrieval is sensitive to the

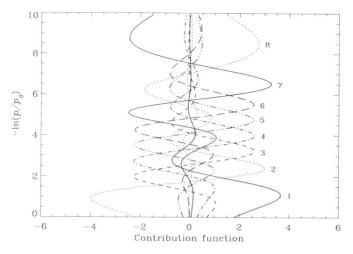

Fig. 3.4 Contribution functions for the standard simulated nadir temperature sounding case. Curves are labelled with channel number.

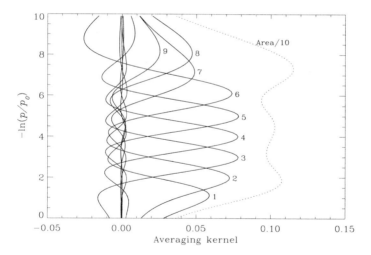

Fig. 3.5 Averaging kernels for the standard simulated nadir temperature sounding case (solid), and their area (dotted).

true profile over this height range. The peaks of the averaging kernels are located at approximately the right level from about $-\ln(p/p_0) \simeq 1$ to 8. The widths are more or less uniform, and comparable with or narrower than those of the original weighting functions from $-\ln(p/p_0) \simeq 2$ to 6. The general indication is that the retrieval is reasonable from the point of view of the transfer function over this latter range of heights.

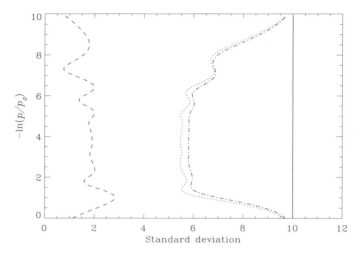

Fig. 3.6 The prior standard deviation (solid), the residual standard deviation (dash-dot) and the contributions from measurement error (dashed) and smoothing error (dotted)

3.4.2 Error components

The covariance of smoothing error for an ensemble of states corresponding to the *a priori* is, from Eq. (3.17), $(\mathbf{A} - \mathbf{I}_n)\mathbf{S}_a(\mathbf{A} - \mathbf{I}_n)^T$, and on substituting (3.28), we obtain:

$$\mathbf{S}_s = (\mathbf{K}^T\mathbf{S}_\epsilon^{-1}\mathbf{K} + \mathbf{S}_a^{-1})^{-1}\mathbf{S}_a^{-1}(\mathbf{K}^T\mathbf{S}_\epsilon^{-1}\mathbf{K} + \mathbf{S}_a^{-1})^{-1} \quad (3.29)$$

The covariance of retrieval noise is, from Eq. (3.19), $\mathbf{G}_y\mathbf{S}_\epsilon\mathbf{G}_y^T$, which gives for this case, on substituting (3.27):

$$\mathbf{S}_m = (\mathbf{K}^T\mathbf{S}_\epsilon^{-1}\mathbf{K} + \mathbf{S}_a^{-1})^{-1}\mathbf{K}^T\mathbf{S}_\epsilon^{-1}\mathbf{K}(\mathbf{K}^T\mathbf{S}_\epsilon^{-1}\mathbf{K} + \mathbf{S}_a^{-1})^{-1} \quad (3.30)$$

and the two add to give the total error from these two sources of:

$$\hat{\mathbf{S}} = (\mathbf{K}^T\mathbf{S}_\epsilon^{-1}\mathbf{K} + \mathbf{S}_a^{-1})^{-1} \quad (3.31)$$

which is identical to the covariance of $P(\mathbf{x}|\mathbf{y})$ as given by Eq. (2.27).

A covariance matrix as a description of the error of a measurement is not easy to visualise. The simplest component of the error is the variance, i.e. the diagonal elements of the covariance matrix, and this is often the only error description that is presented with a retrieval. Such a set is illustrated in Fig. 3.6. The prior standard deviation is the square root of the diagonal elements of the prior covariance matrix $\mathbf{S}_a$, taken to be constant 10 K in this example. The residual standard deviation is the same function of the total retrieval error covariance $\hat{\mathbf{S}}$, and the curves for smoothing error and retrieval noise are similarly obtained from $\mathbf{S}_s$ and $\mathbf{S}_m$. Note that the retrieval noise standard deviation is small outside the range of the weighting functions, as the retrieval tends to the *a priori* in this region, and the measurements

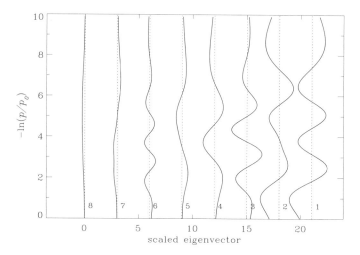

Fig. 3.7 The retrieval noise error patterns, spaced by 3. Curves are labelled in descending order of eigenvalue.

contribute little. For the same reason the smoothing error and the total retrieval error tend towards the *a priori* standard deviation.

Decomposition of the covariance matrix into error patterns, as described in section 3.2.6 does not clarify the nature of the errors as much as one might wish when applied to the total error covariance $\mathbf{S}_s + \mathbf{S}_m + \mathbf{S}_f$, but can give a useful insight into the error characteristics if applied to each component singly. Fig. 3.7 shows all eight error patterns for the retrieval noise covariance $\mathbf{S}_m$. Errors in the retrieved state due to measurement noise can be regarded as a linear combination of these vectors, with random coefficients having unit variance. Note also that the retrieval contains only these shapes plus the *a priori* profile. Patterns with broader scale structure (8, 7 and 5) generally correspond to smaller errors, as there is information in the measurements to determine these components well. Patterns with short vertical scale of variation, like 6, can also give small error, as the inverse method does not attempt to determine these, so that information at these scales comes from the *a priori* state. Larger errors are associated with intermediate scale patterns, with variation on a scale comparable with the weighting function or averaging kernel width.

The ten largest error patterns of the smoothing error covariance are shown in Fig. 3.8. Smoothing error may be of full rank, so there can be n smoothing error patterns. They are proportional to eigenvectors of $\mathbf{S}_s$, and show orthogonal structure that the observing system cannot see, or sees at a reduced amplitude. The vectors can have larger values at the top and bottom of the profile, where there is little information about the true state in the measurements, or can have a fairly small scale vertical structure, somewhat shorter than the width of the averaging kernels. There is less contribution from the very small scales, because the prior

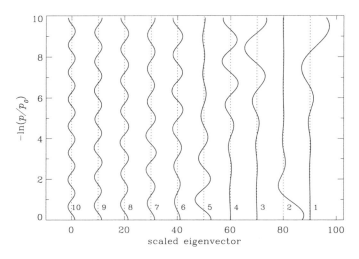

Fig. 3.8 The first eight smoothing error patterns, spaced by 10. Curves are labelled in descending order of eigenvalue.

covariance does not contain significant very small scale structure. The ones not shown are smaller in both amplitude and scale.

3.4.3 *Modelling error*

With as simple a simulation as we are using here, it is difficult to illustrate modelling error comprehensively. We will select a few typical kinds of modelling error for a simple illustration:

Calibration error. This can be simulated by using a forward model for channel i of the form $y_i = gF_i(\mathbf{x}) + y_z$, and assuming that there is a small error, y_z, in the zero level of the measurement and a small error in the gain, g, the errors being the same in all channels.

Transmittance modelling error. This can be simulated by assuming that the transmittance function for channel i is really of the form:

$$\tau_i(p) = \exp[-\alpha(p/p_i)^\beta] \qquad (3.32)$$

where α, which accounts for error in the absorption coefficient, and β, which accounts for its pressure dependence, are both approximately unity.

Fig. 3.9(a) shows the sensitivity of the retrieved state to each of these simulated forward model parameters, when this source of error has not been allowed for. The sensitivities are given for quite small perturbations in the parameters, namely 1% in gain, 1 K in measurement offset, 1% in absorption coefficient or absorber amount and 1% in the exponent of pressure dependence of absorption coefficient. The perturbation in the retrieval when they are ignored in the error covariance can

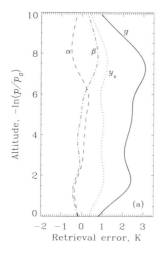 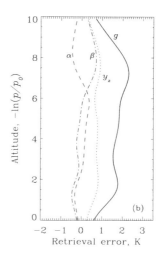

Fig. 3.9 Simulation of modelling error sensitivities. Error in retrieved temperature profile due to: 1% gain error (solid); 1 K measurement offset (dotted); 1% error in absorption coefficient or absorber amount (dashed); 1% error in the exponent of pressure dependence of absorption coefficient (dash-dot). (a) When modelling error is not allowed for in the measurement error covariance. (b) When modelling error is allowed for.

be significant. If it is, the proper procedure is either to include the parameters in the state vector, or to treat them as extra sources of error, and replace the error covariance by $\mathbf{S}_\epsilon + \mathbf{K}_b^T \mathbf{S}_b \mathbf{K}_b$. These two approaches are equivalent, as will be seen in section 4.1.2 and in this case reduces the sensitivity a little, Fig. 3.9(b).

3.4.4 Resolution

Fig. 3.10 shows the resolution of the retrieval according to four of the definitions introduced in section 3.3, in order to illustrate their properties. They are all based on the averaging kernels of Fig. 3.5.

The simplest definition, the full width at half height, is shown as dotted, plotted against the nominal height rather than against the peak or the mean. This has qualitatively reasonable behaviour, but the resolution at the highest and lowest altitudes is misleading, because the corresponding averaging kernels do not have their peaks at the right place, the width is about where the peak is, not where it should be.

The second moment about the mean is shown as dash-dot, also plotted against the nominal height rather than against the mean. It is unsatisfactory, as there are regions where the integrand is negative, so the square root cannot be taken. The alternative definition of the second moment about the nominal height is shown dash-dot-dot-dot. This is close to the curve for the first definition except at the extremes where it becomes large.

The reciprocal of Purser and Huang's data density, based on the diagonal of the

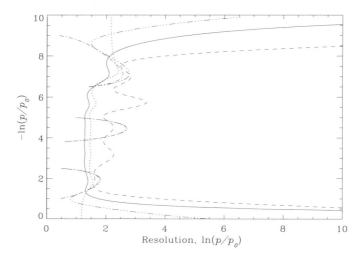

Fig. 3.10 Resolution according to several definitions: Full width at half height (dotted). Second moment about the mean: Dash-dot. Second moment about the nominal height: Dash-dot-dot-dot. Data density reciprocal: solid. Backus–Gilbert spread: dashed.

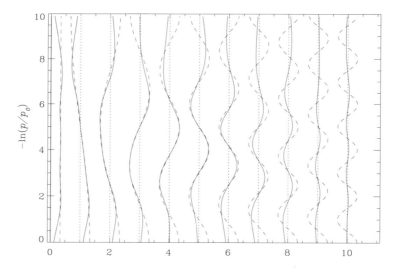

Fig. 3.11 Response of the retrieval to sinusoidal disturbances of various wavelengths.

averaging kernel matrix is shown solid. This has the features to be hoped for from a definition of resolution. It is reasonably smooth, and is large only at the extremes where there is little information.

The Backus and Gilbert spread is shown dashed. This is similar to the data density curve, but is everywhere larger, mainly because it emphasises the negative lobes of the averaging kernels.

It is less straightforward to quantify a definition of resolution based on the response to a wavelike disturbance, so the response itself is shown in Fig. 3.11. For each wavelength the original sinusoid is shown dashed, and the response is shown solid. The scale is arbitrary. It can be seen that the response is not necessarily sinusoidal, and the local amplitude of the response, and its shape, vary with height. At the extremes it is not even locally a sinusoid of the same wavelength. This makes it difficult to give a simple quantitative definition of resolution based on sine-wave response.

Chapter 4
Optimal Linear Inverse Methods

The Bayesian approach provides us with a framework within which we can understand the inverse problem. Given a measurement together with a description of its error statistics, a forward model describing the relation between the measurement and the unknown state, and any *a priori* information that might be available, it allows us to identify the class of possible states that are consistent with the available information, and to assign a probability density function to them.

However for most purposes we wish to select just one of the possible states as 'the solution' to the inverse problem, and to assign it some error estimate, rather than to provide the more general but less directly useful complete description of the ensemble of possible solutions given by the conditional probability density function $P(\mathbf{x}|\mathbf{y})$. Thus we need an objective way of selecting just one state, and a logical approach is to select one that optimises something. The approach to error analysis and characterisation described in Chapter 3 enables us to consider a range of possible quantities that might be optimised in order to produce an appropriate solution to the problem.

Some possible approaches to optimal inverse methods are

(i) Using the posterior *pdf* of the state vector, we can represent the solution as either the expected value of the state or the most likely state, together with the second moment matrix as a measure of the width of the distribution, or the uncertainty of the solution.

(ii) From the error analysis we have identified the smoothing error, the measurement error and the modelling error as quantities that might be minimised in a solution.

(iii) Some measure such as the width of the averaging kernel might be minimised to optimise the spatial resolution of a retrieval.

In this chapter we will discuss these possibilities, and show how they are related. We will deal only with linear problems here, in order to concentrate on the estimation theory involved. Non-linearity, which will be dealt with in Chapter 5, brings another set of questions, but the estimation theory aspects are not very different from those of the linear problem.

The forward model for the elementary linear problem is $\mathbf{y} = \mathbf{Kx} + \boldsymbol{\epsilon}$, with a constant $m \times n$ weighting function matrix $\mathbf{K}$. The slightly more general form $\mathbf{y} - \mathbf{y}_0 = \mathbf{K}(\mathbf{x} - \mathbf{x}_0) + \boldsymbol{\epsilon}$, where $(\mathbf{x}_0, \mathbf{y}_0)$ is some linearisation point, can be reduced to the elementary form by redefining $\mathbf{x}$ and $\mathbf{y}$ as e.g. $\mathbf{x} - \mathbf{x}_0$ and $\mathbf{y} - \mathbf{y}_0$. For the purposes of this chapter no distinction will be made between the noise $\boldsymbol{\epsilon}$ and the total measurement error $\boldsymbol{\epsilon}_y$ of Chapter 3. Anticipating that the linear problem has a linear solution, we expect that the elementary linear solution is of the general form

$$\hat{\mathbf{x}} = \mathbf{x}_o + \mathbf{Gy} \qquad (4.1)$$

where $\mathbf{G}$ is a constant $n \times m$ retrieval gain matrix and $\mathbf{x}_o$ is some constant offset. Our task here is in effect to identify appropriate forms for $\mathbf{G}$ and $\mathbf{x}_o$ for various types of optimality.

4.1 The Maximum a *Posteriori* Solution

The most straightforward way of selecting a state from an ensemble described by a probability density function is to choose either the most likely state, i.e. the one for which $P(\mathbf{x}|\mathbf{y})$ is maximum, or the expected value solution, i.e. the mean state averaged over the *pdf*:

$$\hat{\mathbf{x}} = \int \mathbf{x} P(\mathbf{x}|\mathbf{y}) \, d\mathbf{x}. \qquad (4.2)$$

In either case some measure of the width of the *pdf* provides the error estimate. For the linear problem with Gaussian density functions these two solutions are identical because of the symmetry of the *pdf*. A derivation of this solution has been given in Chapter 2 as an illustration of the Bayesian approach.

The state that maximises the posterior *pdf* $P(\mathbf{x}|\mathbf{y})$ is known as the *maximum a posteriori* solution (MAP). In the atmospheric remote sounding literature it is widely and incorrectly referred to as the *maximum likelihood* (ML) solution, by me as well as almost everyone else. As mentioned in section 2.5, the term likelihood was originally defined by Fisher [1921] to mean the conditional *pdf* $P(\mathbf{y}|\mathbf{x})$, considered as a function of $\mathbf{x}$ for a given value of $\mathbf{y}$, so strictly the ML solution would only be the same as the MAP solution in the absence of prior information.

For non-Gaussian statistics the MAP and expected value solutions will give different results in the general case when the *pdf* is skew or asymmetric. In these circumstances the covariance matrix is not an adequate description, and higher order moments of the *pdf* are needed. In the extreme case of a multiply-peaked *pdf* it is quite possible for the expected value solution to be very unlikely.

The maximum *a posteriori* solution given in Eq. (2.30) can be written in several different forms, for example:

The Maximum a Posteriori Solution

$$\begin{align}
\hat{\mathbf{x}} &= (\mathbf{K}^T\mathbf{S}_\epsilon^{-1}\mathbf{K} + \mathbf{S}_a^{-1})^{-1}(\mathbf{K}^T\mathbf{S}_\epsilon^{-1}\mathbf{y} + \mathbf{S}_a^{-1}\mathbf{x}_a) & (4.3)\\
&= (\mathbf{S}_a\mathbf{K}^T\mathbf{S}_\epsilon^{-1}\mathbf{K} + \mathbf{I}_n)^{-1}(\mathbf{S}_a\mathbf{K}^T\mathbf{S}_\epsilon^{-1}\mathbf{y} + \mathbf{x}_a) & (4.4)\\
&= \mathbf{x}_a + (\mathbf{K}^T\mathbf{S}_\epsilon^{-1}\mathbf{K} + \mathbf{S}_a^{-1})^{-1}\mathbf{K}^T\mathbf{S}_\epsilon^{-1}(\mathbf{y} - \mathbf{K}\mathbf{x}_a) & (4.5)\\
&= \mathbf{x}_a + \mathbf{S}_a\mathbf{K}^T(\mathbf{K}\mathbf{S}_a\mathbf{K}^T + \mathbf{S}_\epsilon)^{-1}(\mathbf{y} - \mathbf{K}\mathbf{x}_a). & (4.6)
\end{align}$$

Its covariance as given in Eq. (2.27) can be written as, for example:

$$\begin{align}
\hat{\mathbf{S}} &= (\mathbf{K}^T\mathbf{S}_\epsilon^{-1}\mathbf{K} + \mathbf{S}_a^{-1})^{-1} & (4.7)\\
&= \mathbf{S}_a - \mathbf{S}_a\mathbf{K}^T(\mathbf{K}\mathbf{S}_a\mathbf{K}^T + \mathbf{S}_\epsilon)^{-1}\mathbf{K}\mathbf{S}_a & (4.8)
\end{align}$$

and its averaging kernel matrix as:

$$\begin{align}
\mathbf{A} &= (\mathbf{K}^T\mathbf{S}_\epsilon^{-1}\mathbf{K} + \mathbf{S}_a^{-1})^{-1}\mathbf{K}^T\mathbf{S}_\epsilon^{-1}\mathbf{K} & (4.9)\\
&= \mathbf{S}_a\mathbf{K}^T(\mathbf{K}\mathbf{S}_a\mathbf{K}^T + \mathbf{S}_\epsilon)^{-1}\mathbf{K}. & (4.10)
\end{align}$$

There are two basically different forms here, distinguished by the matrix to be inverted. In the m-form it is $m \times m$ and in the n-form it is $n \times n$, so that the choice of formulation is determined in part by the relative sizes of the state vector and the measurement vector. The equivalence of (4.5) and (4.6) can be shown as follows. Consider

$$\mathbf{K}^T\mathbf{S}_\epsilon^{-1}(\mathbf{S}_\epsilon + \mathbf{K}\mathbf{S}_a\mathbf{K}^T) = (\mathbf{S}_a^{-1} + \mathbf{K}^T\mathbf{S}_\epsilon^{-1}\mathbf{K})\mathbf{S}_a\mathbf{K}^T \qquad (4.11)$$

which can be seen to be valid by multiplying out. Premultiplying both sides by $(\mathbf{S}_a^{-1} + \mathbf{K}^T\mathbf{S}_\epsilon^{-1}\mathbf{K})^{-1}$ and postmultiplying by $(\mathbf{S}_\epsilon + \mathbf{K}\mathbf{S}_a\mathbf{K}^T)^{-1}$ gives

$$(\mathbf{S}_a^{-1} + \mathbf{K}^T\mathbf{S}_\epsilon^{-1}\mathbf{K})^{-1}\mathbf{K}^T\mathbf{S}_\epsilon^{-1} = \mathbf{S}_a\mathbf{K}^T(\mathbf{S}_\epsilon + \mathbf{K}\mathbf{S}_a\mathbf{K}^T)^{-1}, \qquad (4.12)$$

which shows that the two forms are indeed equivalent.

We can interpret the form of the solution given in Eq. (4.3) in terms of concepts that should be familiar from the scalar least squares problem. In the underconstrained case there must exist at least one 'exact' solution $\mathbf{x}_e = \mathbf{G}\mathbf{y}$ in the sense that $\mathbf{K}\mathbf{x}_e$ is exactly equal to the measurement $\mathbf{y}$, i.e. that $\mathbf{K}\mathbf{G} = \mathbf{I}$. For example we could choose $\mathbf{G} = \mathbf{K}^T(\mathbf{K}\mathbf{K}^T)^{-1}$. Replacing $\mathbf{y}$ by $\mathbf{K}\mathbf{x}_e$ in Eq. (4.3) gives

$$\hat{\mathbf{x}} = (\mathbf{K}^T\mathbf{S}_\epsilon^{-1}\mathbf{K} + \mathbf{S}_a^{-1})^{-1}(\mathbf{K}^T\mathbf{S}_\epsilon^{-1}\mathbf{K}\mathbf{x}_e + \mathbf{S}_a^{-1}\mathbf{x}_a) \qquad (4.13)$$

In the overconstrained case there is a weighted least squares solution $\mathbf{x}_l$ which satisfies $\mathbf{K}^T\mathbf{S}_\epsilon^{-1}\mathbf{K}\mathbf{x}_l = \mathbf{K}^T\mathbf{S}_\epsilon^{-1}\mathbf{y}$. Substituting this in Eq. (4.3) gives

$$\hat{\mathbf{x}} = (\mathbf{K}^T\mathbf{S}_\epsilon^{-1}\mathbf{K} + \mathbf{S}_a^{-1})^{-1}(\mathbf{K}^T\mathbf{S}_\epsilon^{-1}\mathbf{K}\mathbf{x}_l + \mathbf{S}_a^{-1}\mathbf{x}_a) \qquad (4.14)$$

Both these equations are of the same form, and can be interepreted as the weighted mean of an exact or least squares solution with the *a priori*, with matrix weights $\mathbf{K}^T\mathbf{S}_\epsilon^{-1}\mathbf{K}$ and $\mathbf{S}_a^{-1}$ respectively. This is exactly like the familiar combination of scalar measurements x_1 and x_2 of an unknown x, with variances σ_1^2 and σ_2^2 respectively, i.e.,

$$\hat{x} = (1/\sigma_1^2 + 1/\sigma_2^2)^{-1}(x_1/\sigma_1^2 + x_2/\sigma_2^2). \tag{4.15}$$

⇒ **Exercise** 4.1: Let **x** be represented as a linear combination of singular vectors of $\tilde{\mathbf{K}}$ (section 2.4.1). Show that in this representation the MAP method combines the coefficients of an exact solution and the *a priori* independently as scalars.

4.1.1 Several independent measurements

We can generalise the above discussion to the case when we have several independent measurements $\mathbf{y}_i$ of a particular state, for $i = 1\ldots l$:

$$\mathbf{y}_i = \mathbf{K}_i \mathbf{x} + \boldsymbol{\epsilon}_i \tag{4.16}$$

each with an independent error covariance $\mathbf{S}_{\epsilon_i}$. The $\mathbf{y}_i$ may be different channels of one instrument, different instruments, or perhaps the *a priori* considered as a 'virtual measurement' with $\mathbf{K} = \mathbf{I}_n$ and $\mathbf{S}_\epsilon = \mathbf{S}_a$. Bayes theorem gives

$$P(\mathbf{x}|\mathbf{y}_1, \mathbf{y}_2, \ldots, \mathbf{y}_l) = \frac{P(\mathbf{y}_1, \mathbf{y}_2, \ldots, \mathbf{y}_l|\mathbf{x}) P(\mathbf{x})}{P(\mathbf{y}_1, \mathbf{y}_2, \ldots, \mathbf{y}_l)}. \tag{4.17}$$

If the measurement errors are independent we can expand this as

$$P(\mathbf{x}|\mathbf{y}_1, \mathbf{y}_2, \ldots, \mathbf{y}_l) = \frac{P(\mathbf{y}_1|\mathbf{x}) P(\mathbf{y}_2|\mathbf{x}) \ldots P(\mathbf{y}_l|\mathbf{x}) P(\mathbf{x})}{P(\mathbf{y}_1, \mathbf{y}_2, \ldots, \mathbf{y}_l)} \tag{4.18}$$

$$= \frac{P(\mathbf{x})}{P(\mathbf{y}_1, \mathbf{y}_2, \ldots, \mathbf{y}_l)} \prod_i P(\mathbf{y}_i|\mathbf{x}). \tag{4.19}$$

To find the maximum probability solution (i.e. MAP if there is *a priori*, or ML if there is not) we maximise with respect to **x**, so the term $P(\mathbf{y}_1, \mathbf{y}_2, \ldots, \mathbf{y}_l)$ can be omitted as it is independent of **x**. Also, if we know nothing about **x** other than the values of **y**, e.g. if we have taken *a priori* to be a virtual measurement, one of the $\mathbf{y}_i$, then $P(\mathbf{x})$ is an infinitesimal constant. In the linear Gaussian case, this corresponds to minimising minus the logarithm:

$$\frac{\partial}{\partial \mathbf{x}} \sum_i (\mathbf{y}_i - \mathbf{K}_i \mathbf{x})^T \mathbf{S}_{\epsilon_i}^{-1} (\mathbf{y}_i - \mathbf{K}_i \mathbf{x}) = 0 \tag{4.20}$$

giving

$$\hat{\mathbf{x}} = \left(\sum_i \mathbf{K}_i^T \mathbf{S}_{\epsilon_i}^{-1} \mathbf{K}_i \right)^{-1} \left(\sum_i \mathbf{K}_i^T \mathbf{S}_{\epsilon_i}^{-1} \mathbf{y}_i \right). \tag{4.21}$$

The inverse covariance of the solution, the Fisher information matrix,

$$\hat{\mathbf{S}}^{-1} = \sum_i \mathbf{K}_i^T \mathbf{S}_{\epsilon_i}^{-1} \mathbf{K}_i \tag{4.22}$$

is the sum of the information matrices for the independent sources $\mathbf{y}_i$. The solution only exists (the problem is not underconstrained) if the information matrix is non-singular. We can always ensure that this is the case by including a prior estimate $\mathbf{x}_a$ with a positive-definite covariance matrix $\mathbf{S}_a$.

4.1.2 Independent components of the state vector

Another generalisation arises when the state vector is composed of several independent components. We may include as components of a *full state vector* all of the uncertain quantities that contribute to the measurement, including the desired atmospheric quantities, the measurement noise, and instrumental and physics parameters. We will consider the common case when the priors for the various components are not correlated and will investigate retrieving some or all of these components.

Let the full state vector $\mathbf{x}$ be composed of k independent components, $\mathbf{x}_1, \ldots, \mathbf{x}_k$, with $\mathbf{x}_i$ having prior *pdf* $P(\mathbf{x}_i)$. Then we can write

$$P(\mathbf{x}|\mathbf{y}) = \frac{P(\mathbf{y}|\mathbf{x})P(\mathbf{x})}{P(\mathbf{y})} = \frac{P(\mathbf{y}|\mathbf{x}_1, \mathbf{x}_2, \ldots, \mathbf{x}_k)}{P(\mathbf{y})} \prod_i P(\mathbf{x}_i). \tag{4.23}$$

If we include all the random variables as part of $\mathbf{x}$, then the forward model is of the form

$$\mathbf{y} = \sum_i \mathbf{K}_i \mathbf{x}_i, \tag{4.24}$$

no longer including an explicit random noise term because the measurement noise is now an element of the state vector, with a unit weighting function. Consequently $P(\mathbf{y}|\mathbf{x})$ is only non-zero when this equation is exactly satisfied, and maximising $P(\mathbf{x}|\mathbf{y})$ with respect to $\mathbf{x}$, or any part of $\mathbf{x}$, becomes a matter of maximising $P(\mathbf{x})$ with Eq. (4.24) as a constraint. We need only consider those parts $\mathbf{x}_j$ of the state vector that we are interested in, for example if the full state vector includes the temperature and ozone profiles plus the measurement noise, we might only be interested in ozone. For a Gaussian *pdf* we solve

$$\frac{\partial}{\partial \mathbf{x}_j} \left[\left(\sum_i \mathbf{x}_i^T \mathbf{S}_{ai}^{-1} \mathbf{x}_i \right) + \boldsymbol{\lambda}^T \left(\mathbf{y} - \sum_i \mathbf{K}_i \mathbf{x}_i \right) \right] = 0 \tag{4.25}$$

where $\mathbf{S}_{ai}$ is the prior covariance of $\mathbf{x}_i$ and $\boldsymbol{\lambda}$ is a vector of Lagrangian multipliers. Taking the derivative we obtain

$$2\mathbf{x}_j^T \mathbf{S}_{aj}^{-1} - \boldsymbol{\lambda}^T \mathbf{K}_j = 0 \tag{4.26}$$

i.e.

$$x_j = \tfrac{1}{2} S_{aj} K_j^T \lambda. \qquad (4.27)$$

On substituting into the constraint Eq. (4.24), we find

$$\lambda = 2\left(\sum_j K_j S_{aj} K_j^T\right)^{-1} y \qquad (4.28)$$

so the solution is

$$\hat{x}_j = S_{aj} K_j^T \left(\sum_i K_i S_{ai} K_i^T\right)^{-1} y \qquad (4.29)$$

with covariance

$$\hat{S}_{x_j} = S_{aj} - S_{aj} K_j^T \left(\sum_i K_i S_{ai} K_i^T\right)^{-1} K_j S_{aj}. \qquad (4.30)$$

This is clearly a generalisation of the m-form Eq. (4.6). Note that the quantity in brackets is the prior covariance of the measurement:

$$S_y = \sum_i K_i S_{ai} K_i^T. \qquad (4.31)$$

When retrieving a particular part x_j of the state vector we may interpret this as being the sum of $K_j S_{aj} K_j^T$ and the covariance of the effective total error in the measurement,

$$S_\epsilon = \sum_{i \neq j} K_i S_{ai} K_i^T. \qquad (4.32)$$

For example if we are interested in ozone only, then we can interpret the measurement error covariance as being the sum of the noise covariance plus a contribution due to temperature uncertainty.

If we write this generalisation in terms of our previous notation of the original state vector x, the forward model parameters, b and the noise ϵ, we find

$$\hat{x} = S_a K^T (K S_a K^T + K_b S_b K_b^T + S_\epsilon)^{-1} y \qquad (4.33)$$

and we see that we can interpret this as either a retrieval of x with an effective noise covariance $K_b S_b K_b^T + S_\epsilon$, or as part of an extended retrieval of x and b with noise covariance S_ϵ:

$$\begin{pmatrix} \hat{x} \\ \hat{b} \end{pmatrix} = \begin{pmatrix} S_a & O \\ O & S_b \end{pmatrix} \begin{pmatrix} K^T \\ K_b^T \end{pmatrix} \left[(K \; K_b) \begin{pmatrix} S_a & O \\ O & S_b \end{pmatrix} \begin{pmatrix} K^T \\ K_b^T \end{pmatrix} + S_\epsilon \right]^{-1} y. \qquad (4.34)$$

The full solution for the full state vector will give the same result for each of the partial state vectors as their individual solutions, so it does not help for example in a linear ozone retrieval to retrieve temperature at the same time as ozone. However to calculate the correct contribution to the error covariance it is necessary to calculate the temperature weighting functions, and if the Jacobian calculation is the main computational cost then the temperature retrieval is available as a by-product. Similarly, if forward model parameters are allowed for in the error covariance, then they too can be retrieved with little extra cost. In the nonlinear case, however, it may be necessary to retrieve everything in order to obtain the correct weighting functions.

It might appear, then, that there is little point in attempting to reduce systematic errors by retrieving the forward model errors responsible, as illustrated in section 3.4.3. However this is only the case for a single retrieval, if systematic errors are unchanging, as for example spectroscopic parameters, and several independent measurements are made, then their retrieval error can be improved.

4.2 Minimum Variance Solutions

As an alternative to MAP, we could look for a solution which minimises retrieval error in some sense. There is obviously no way of minimising $\mathbf{x} - \hat{\mathbf{x}}$ for a particular case because $\mathbf{x}$ is unknown, but we can look for the solution *method* as described by $\mathbf{x}_o$ and $\mathbf{G}$ in Eq. (4.1) which minimises the expected value of the retrieval error variance. In other words, we carry out a least squares fit or multiple regression of $\mathbf{x}$ on $\mathbf{y}$ for some ensemble of cases, and obtain $\mathbf{x}_o$ and $\mathbf{G}$ as regression coefficients. Let us choose an ensemble represented by a mean $\mathbf{x}_a$ with covariance $\mathbf{S}_a$, and express the retrieval as

$$\hat{\mathbf{x}} = \mathbf{x}_o + \mathbf{G}\mathbf{y}. \tag{4.35}$$

In this case we can minimise the error in each element $\hat{x}_i$ of $\hat{\mathbf{x}}$ separately, by finding the values of x_{oi} and the row $\mathbf{g}_i^T$ of the matrix $\mathbf{G}$ which minimise $\mathcal{E}\{(\hat{x}_i - x_i)^2\}$:

$$\mathcal{E}\{(\hat{x}_i - x_i)^2\} = \mathcal{E}\{(x_{oi} - x_i + \mathbf{g}_i^T \mathbf{y})^2\}. \tag{4.36}$$

Setting the derivative with respect to x_{oi} zero gives

$$\mathcal{E}\{x_{oi} - x_i + \mathbf{g}_i^T \mathbf{y}\} = 0 \tag{4.37}$$

so that

$$x_{oi} = \mathcal{E}\{x_i - \mathbf{g}_i^T \mathbf{y}\} = x_{ai} - \mathbf{g}_i^T \mathbf{y}_a \tag{4.38}$$

where $\mathbf{y}_a = \mathbf{K}\mathbf{x}_a$ is the ensemble mean of $\mathbf{y}$. Setting the derivative of Eq. (4.36) with respect to $\mathbf{g}_i$ zero gives

$$0 = \mathcal{E}\{(x_{oi} - x_i + \mathbf{g}_i^T \mathbf{y})\mathbf{y}^T\} = \mathcal{E}\{[x_{ai} - x_i + \mathbf{g}_i^T(\mathbf{y} - \mathbf{y}_a)](\mathbf{y} - \mathbf{y}_a)^T\} \tag{4.39}$$

so that

$$\mathbf{g}_i^T = \mathcal{E}\{(x_i - x_{ai})(\mathbf{y} - \mathbf{y}_a)^T\} \left[\mathcal{E}\{(\mathbf{y} - \mathbf{y}_a)(\mathbf{y} - \mathbf{y}_a)^T\}\right]^{-1}. \tag{4.40}$$

Assembling the $\mathbf{g}_i$'s into the matrix $\mathbf{G}$ and substituting $\mathcal{E}\{(\mathbf{y}-\mathbf{y}_a)(\mathbf{y}-\mathbf{y}_a)^T\} = \mathbf{KS}_a\mathbf{K}^T + \mathbf{S}_\epsilon$ and $\mathcal{E}\{(\mathbf{x}-\mathbf{x}_a)(\mathbf{y}-\mathbf{y}_a)^T\} = \mathbf{S}_a\mathbf{K}^T$, we obtain

$$\mathbf{G} = \mathbf{S}_a\mathbf{K}^T(\mathbf{KS}_a\mathbf{K}^T + \mathbf{S}_\epsilon)^{-1} \tag{4.41}$$

so that

$$\hat{\mathbf{x}} = \mathbf{x}_a + \mathbf{S}_a\mathbf{K}^T(\mathbf{KS}_a\mathbf{K}^T + \mathbf{S}_\epsilon)^{-1}(\mathbf{y} - \mathbf{Kx}_a). \tag{4.42}$$

This is exactly the same as the MAP solution for Gaussian statistics, Eq. (4.6), but here the form of the *pdf* has not been used, only knowledge of $\mathbf{x}_a$ and $\mathbf{S}_a$ is required.

The covariance of the retrieval error is $\hat{\mathbf{S}} = \mathcal{E}\{(\mathbf{x}-\hat{\mathbf{x}})(\mathbf{x}-\hat{\mathbf{x}})^T\}$. It is straightforward to obtain

$$\mathbf{x} - \hat{\mathbf{x}} = \mathbf{x} - \mathbf{x}_a - \mathbf{GK}(\mathbf{x} - \mathbf{x}_a) - \mathbf{G}\boldsymbol{\epsilon} \tag{4.43}$$

from which, assuming that the measurement error is uncorrelated with the state, we obtain

$$\hat{\mathbf{S}} = (\mathbf{I}_n - \mathbf{GK})\mathbf{S}_a(\mathbf{I}_n - \mathbf{GK})^T + \mathbf{GS}_\epsilon\mathbf{G}^T. \tag{4.44}$$

Substituting for $\mathbf{G}$ from (4.41) this reduces to

$$\hat{\mathbf{S}} = \mathbf{S}_a - \mathbf{S}_a\mathbf{K}^T(\mathbf{KS}_a\mathbf{K}^T + \mathbf{S}_\epsilon)^{-1}\mathbf{KS}_a \tag{4.45}$$

which is also the same as for the Gaussian MAP solution.

> ⇒ ***Exercise*** 4.2: The minimum variance solution has been found by minimising only the diagonal elements of $\hat{\mathbf{S}}$. Examine whether it also minimises the off-diagonal elements.

An alternative approach to the minimum variance solution uses the Bayesian solution *pdf*. Given $P(\mathbf{x}|\mathbf{y})$, find the state $\hat{\mathbf{x}}$ such that the variance about $\hat{\mathbf{x}}$ is minimised:

$$\frac{\partial}{\partial \hat{\mathbf{x}}} \int (\mathbf{x} - \hat{\mathbf{x}})^T(\mathbf{x} - \hat{\mathbf{x}}) P(\mathbf{x}|\mathbf{y}) \, \mathrm{d}\mathbf{x} = 0 \tag{4.46}$$

which gives

$$\hat{\mathbf{x}} = \int \mathbf{x} P(\mathbf{x}|\mathbf{y}) \, \mathrm{d}\mathbf{x}, \tag{4.47}$$

i.e. the minimum variance solution is the conditional expected value. This is the case for an arbitrary *pdf*. Thus the minimum variance and the maximum likelihood solutions are equivalent when the expected value and maximum likelihood are the same, if (but not only if) the *pdf* is symmetric about the expected value.

The minimum variance concept can also be applied by direct multiple regression between measurements $\mathbf{y}$ and independently measured states $\mathbf{x}$, avoiding the need to know the forward model. This will be discussed further in Chapter 6.

4.3 Best Estimate of a Function of the State Vector

Often we do not wish to make a measurement of any part of a full state vector, but rather some function of it, such as the total column amount of a constituent or the atmospheric thickness between two pressure levels. The desired quantity is not directly useful in the forward model. The question then arises as to whether the best estimate of the desired quantity is to be found by retrieving the state and calculating the appropriate function, or whether some other approach gives a more accurate result.

In general we will be interested in some linear function $h(\mathbf{x}) = h_0 + \mathbf{h}^T(\mathbf{x} - \mathbf{x}_0)$ of the underlying state vector, the prime example being the total column of a constituent. Is the best estimate $\hat{h} = h_0 + \mathbf{h}^T(\hat{\mathbf{x}} - \mathbf{x}_0)$ where $\hat{\mathbf{x}}$ is the optimal retrieved state? The following simple argument, based on the Bayesian approach to retrieval, shows that it is for the case of 'best' meaning the expected value. The expected value of the function $h(\mathbf{x})$ is

$$\hat{h} = \int P(\mathbf{x}|\mathbf{y})[h_0 + \mathbf{h}^T(\mathbf{x} - \mathbf{x}_0)]\,\mathrm{d}\mathbf{x}. \tag{4.48}$$

Provided $\mathbf{h}$ does not depend on $\mathbf{x}$, this can be rewritten as

$$\begin{aligned}\hat{h} &= h_0 + \mathbf{h}^T \int P(\mathbf{x}|\mathbf{y})(\mathbf{x} - \mathbf{x}_0)\,\mathrm{d}\mathbf{x} \\ &= h_0 + \mathbf{h}^T \left(\int P(\mathbf{x}|\mathbf{y})\mathbf{x}\,\mathrm{d}\mathbf{x} - \mathbf{x}_0\right) \\ &= h_0 + \mathbf{h}^T(\hat{\mathbf{x}} - \mathbf{x}_0).\end{aligned} \tag{4.49}$$

It is easy to show that the error covariance of $\hat{h}$ is $\mathbf{h}^T\hat{\mathbf{S}}\mathbf{h}$, its averaging kernel $\mathbf{a}^T = \mathbf{h}^T\mathbf{A}$, and its measurement error is $\mathbf{h}^T\mathbf{S}_m\mathbf{h}$.

A common approach to retrieving a column amount is to choose a fixed profile shape with a known column amount, and to find the scaling factor which gives a best match to the measured signal. This does not give the best estimate of the column, nor does it give the correct retrieval error.

4.4 Separately Minimising Error Components

The error analysis provides us with separate estimates of smoothing error, the measurement error and the modelling error, so we might consider finding solutions which minimise each of these separately or in combination. However as we have seen above, modelling error and measurement error are conceptually very similar quantities, and there is little point in treating them separately. In fact we could quite reasonably treat measurement error as an additive model parameter with zero mean, and include it as an element of $\mathbf{b}$. Thus we will follow Chapter 3 and combine all these errors as a total measurement error ϵ_y.

First consider minimising retrieval noise. The error in the retrieval due to total measurement noise ϵ is $\mathbf{G}\epsilon$. The retrieval method which minimises the variance of this error is found by choosing $\mathbf{G}$ to minimise the covariance $\mathbf{GS}_\epsilon\mathbf{G}^T$. Unfortunately this is easily solved, giving $\mathbf{G} = \mathbf{O}$! The solution which is least sensitive to measurement error with no other constraint is, not surprisingly, one which completely ignores the measurements, and is therefore quite useless.

Next consider smoothing error, which is $(\mathbf{I}_n - \mathbf{A})(\mathbf{x} - \mathbf{x}_a)$ for an individual retrieval, and has covariance $(\mathbf{I}_n - \mathbf{A})\mathbf{S}_a(\mathbf{I}_n - \mathbf{A})^T$ for an ensemble with prior covariance $\mathbf{S}_a$. Put $\mathbf{A} = \mathbf{GK}$ and minimise with respect to $\mathbf{G}$ to find the retrieval method that minimises smoothing error:

$$\mathbf{S}_s = (\mathbf{I}_n - \mathbf{GK})\mathbf{S}_a(\mathbf{I}_n - \mathbf{GK})^T. \tag{4.50}$$

Using the same approach as the solution to exercise 4.2 we find that this is minimised at $(\mathbf{I}_n - \mathbf{GK})\mathbf{S}_a\mathbf{K}^T = \mathbf{O}$. In the case of the overconstrained problem, this can be satisfied by any $\mathbf{G}$ for which $\mathbf{I} = \mathbf{GK}$, such as $(\mathbf{K}^T\mathbf{K})^{-1}\mathbf{K}^T$, leading to zero smoothing error, and in the underconstrained case we find

$$\mathbf{G} = \mathbf{S}_a\mathbf{K}^T(\mathbf{K}\mathbf{S}_a\mathbf{K}^T)^{-1}. \tag{4.51}$$

This is the same as the MAP solution Eq. (4.6) with $\mathbf{S}_\epsilon$ omitted or replaced by $\mathbf{O}$, i.e. with no measurement error.

Minimising either component of error separately is not satisfactory, but we can minimise a weighted sum of the two terms, to trade off noise sensitivity against smoothing error:

$$\frac{\partial}{\partial \mathbf{G}}[(\mathbf{I}_n - \mathbf{GK})\mathbf{S}_a(\mathbf{I}_n - \mathbf{GK})^T + \gamma\mathbf{GS}_\epsilon\mathbf{G}^T] = 0. \tag{4.52}$$

This is identical to the MAP formulation if $\gamma = 1$, and the solution is

errata

$$\mathbf{G} = (\mathbf{KS}_a\mathbf{K}^T + \gamma\mathbf{S}_\epsilon)^{-1}\mathbf{S}_a\mathbf{K}^T. \tag{4.53}$$

$G = S_a K^T (K S_a K^T + \gamma S_\epsilon)^{-1}$

The trade-off between error and resolution may be useful, but it is at the expense of departure from optimality in the MAP or minimum variance sense. Another approach to this trade-off is described in section 4.5

4.5 Optimising Resolution

Backus and Gilbert (1970) studied the inverse problem as applied to sounding the solid earth using for example seismic waves to retrieve density as a function of depth all the way to the core. This is a case where there is very little available in the way of prior information, so concepts like information content and degrees of freedom for signal are not useful. Therefore rather than attempt to make the best estimate of the actual density profile, they posed a somewhat different question: 'What useful functions of the density profile can be derived from the measurements?' The ideal function would be the identity, but with weighting functions of a finite width, and significant experimental error, this is clearly impossible. So they accepted that they could only retrieve a smoothed version of the profile, and examined ways of optimising the smoothing function, rather than the departure of the retrieved profile from the true profile. In this case it is clear that the profile retrieved will not be an unbiassed estimate of the true profile. Conrath (1972) applied the method to the atmospheric problem of nadir sounding, and the approach used here follows his development.

The spread as defined in Eq. (3.23) was invented for this purpose. As spread is most conveniently defined for continuous functions, we will carry out the derivations with continuous function rather than vectors. The measurements y_i, $i = 1 \ldots m$, are linear functions of the state $x(z)$, with weighting functions $K_i(z)$:

$$y_i = \int K_i(z') x(z') \, dz' + \epsilon_i. \tag{4.54}$$

The retrieved profile $\hat{x}(z)$ is a linear combination of the measurements y_i expressed in terms of contribution functions $G_i(z)$:

$$\hat{x}(z) = \sum_{i=1}^{m} G_i(z) y_i = \sum_{i=1}^{m} G_i(z) \left[\int K_i(z') x(z') \, dz' + \epsilon_i \right] \tag{4.55}$$

$$= \int A(z, z') x(z') \, dz' + \sum_{i=1}^{m} G_i(z) \epsilon_i \tag{4.56}$$

where the averaging kernel function $A(z, z')$ corresponding to the retrieved value at z is $\sum_{i=1}^{m} G_i(z) K_i(z')$. The requirements for the Backus–Gilbert solution are that the averaging kernels should have unit area,

$$\int A(z, z') \, dz' = 1 \tag{4.57}$$

and that their spread should be minimised about z, so the resolution of the retrieval is as high as possible. Substituting the expression for the averaging kernel in Eq. (3.23) gives the spread in terms of the contribution functions:

$$s(z) = 12 \int \left[\sum_{i=1}^{m} G_i(z) K_i(z') \right]^2 (z - z')^2 \, dz'. \tag{4.58}$$

This can be rearranged in the form

$$s(z) = \sum_{ij} G_i(z) Q_{ij}(z) G_j(z) \tag{4.59}$$

where the matrix of functions Q_{ij} depends only on the weighting functions:

$$Q_{ij}(z) = 12 \int (z - z')^2 K_i(z') K_j(z') \, dz'. \tag{4.60}$$

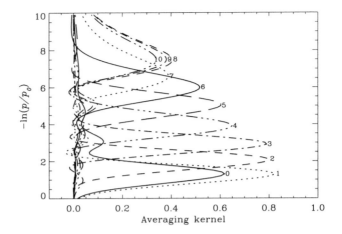

Fig. 4.1 Backus–Gilbert averaging kernels with minimum spread for the standard case, for $z = 0.0, 1.0, \ldots, 10.0$, and labelled with z.

From the unit area constraint on A we obtain:

$$1 = \int A(z, z') \, dz' = \int \sum_i G_i(z) K_i(z') \, dz' = \sum_i k_i G_i(z) \qquad (4.61)$$

where $k_i = \int K_i(z') \, dz'$. We now have all we need to minimise the spread with respect to the contribution functions subject to the unit area constraint, using a Lagrangian multiplier $\lambda(z)$:

$$\frac{\partial}{\partial G_k(z)} \left[\sum_{ij} G_i(z) Q_{ij}(z) G_j(z) + \lambda(z) \sum_i G_i(z) k_i \right] = 0. \qquad (4.62)$$

We can solve this more simply using matrix notation from this point:

$$\frac{\partial}{\partial \mathbf{g}(z)} \left[\mathbf{g}^T(z) \mathbf{Q}(z) \mathbf{g}(z) + \lambda(z) \mathbf{g}^T(z) \mathbf{k} \right] = 0 \qquad (4.63)$$

where the vector-valued function $\mathbf{g}(z)$ has elements $G_j(z)$ and the matrix-valued function $\mathbf{Q}(z)$ has elements $Q_{ij}(z)$. The solution is

$$\mathbf{g}(z) = \frac{\mathbf{Q}^{-1}(z) \mathbf{k}}{\mathbf{k}^T \mathbf{Q}^{-1}(z) \mathbf{k}}. \qquad (4.64)$$

⇒ **Exercise** 4.3: Derive this expression

The spread, from Eq. (3.23), is $s(z) = \mathbf{g}^T(z) \mathbf{Q}(z) \mathbf{g}(z)$. Substituting the solution in this gives:

$$s(z) = \frac{\mathbf{k}^T \mathbf{Q}^{-1}(z)}{\mathbf{k}^T \mathbf{Q}^{-1}(z) \mathbf{k}} \mathbf{Q}(z) \frac{\mathbf{Q}^{-1}(z) \mathbf{k}}{\mathbf{k}^T \mathbf{Q}^{-1}(z) \mathbf{k}} = \frac{1}{\mathbf{k}^T \mathbf{Q}^{-1}(z) \mathbf{k}}. \qquad (4.65)$$

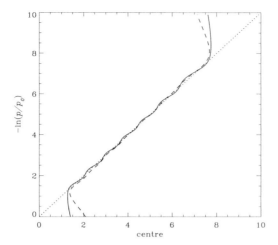

Fig. 4.2 Averaging kernel centres (solid line) and mean heights (dashed) plotted against nominal height.

The standard example

Averaging kernels computed this way for a range of heights using the weighting functions for standard example are shown in Fig. 4.1. It can be seen that they are considerably narrower than the original weighting functions (Fig. 1.1), but cover the same range of altitudes. The kernels for $z = 0$, and for $z = 7 \ldots 10$ are not located at their nominal levels, and the spread clearly varies with z.

Minimisation of the spread alone does not place any constraint on the location of the peak or of the center of gravity of the averaging kernel, which may in principle be located anywhere. The kernel with the smallest spread about a given level may not be centred on that level, and the *resolving length*, i.e. the spread about the centre:

$$r(z) = 12 \int [z' - c(z)]^2 A^2(z, z') \, dz' / (\int A(z, z') \, dz')^2, \tag{4.66}$$

where the centre, $c(z)$ is give by Eq. (3.24), may be smaller than the spread itself. The variation of both the centre and the centre of gravity of the averaging kernel are illustrated in Fig. 4.2 for the standard case, as a function of the altitude for which the averaging kernel is computed. The two quantities are very similar, and are located at the nominal position over the range of altitudes covered by the peaks of the weighting functions, but outside that range the spread is minimised by a function which has its centre or mean height in the wrong place.

The noise of the Backus–Gilbert solution due to measurement error is

$$\sigma^2(z) = \mathbf{g}^T(z) \mathbf{S}_\epsilon \mathbf{g}(z) = \frac{\mathbf{k}^T \mathbf{Q}^{-1}(z) \mathbf{S}_\epsilon \mathbf{Q}^{-1}(z) \mathbf{k}}{(\mathbf{k}^T \mathbf{Q}^{-1}(z) \mathbf{k})^2}. \tag{4.67}$$

As one might expect, because resolution has been minimised without any constraint

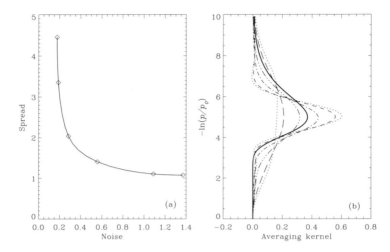

Fig. 4.3 (a) Trade-off between measurement error and resolution. (b) Averaging kernels for a range of trade-off parameters, corresponding to the symbols in panel (a). The narrowest is for $\mu = 0$, and the broadest for $\mu = \infty$. The solid line is the weighting function which peaks at $z = 5$.

on the noise in the solution, the noise is likely to be large, especially if the weighting functions have significant overlap. Fortunately it is quite easy to minimise a weighted sum of spread and noise variance, thus enabling us to trade off resolution for noise performance. We minimise

$$\frac{\partial}{\partial \mathbf{g}} \left[\mathbf{g}^T \mathbf{Q} \mathbf{g} + \lambda \mathbf{g}^T \mathbf{k} + \mu \mathbf{g}^T \mathbf{S}_\epsilon \mathbf{g} \right] = 0 \qquad (4.68)$$

where μ is a 'trade-off' parameter. The algebra is the same as in Eq. (4.62), except that $\mathbf{Q}$ is replaced by $\mathbf{Q} + \mu \mathbf{S}_\epsilon$, so the solution is obviously

$$\mathbf{g} = \frac{(\mathbf{Q} + \mu \mathbf{S}_\epsilon)^{-1} \mathbf{k}}{\mathbf{k}^T (\mathbf{Q} + \mu \mathbf{S}_\epsilon)^{-1} \mathbf{k}}. \qquad (4.69)$$

The spread is given by $s(z) = \mathbf{g}^T(z) \mathbf{Q}(z) \mathbf{g}(z)$, as before, but this does not simplify in the same way as Eq. (4.65). A trade-off plot for the standard case for the level $z = 5.0$, showing noise plotted against spread, is given in Fig. 4.3(a). It can be seen that high noise corresponds to small spread, and vice-versa. Averaging kernels corresponding to the locations of the symbols on the trade-off curve are shown in Fig. 4.3(b). The original weighting function is also shown as the solid line for comparison. The narrow kernels corresponds to high noise levels, and the broad ones to low noise levels.

The spread and noise as a function of altitude for the case $\mu = 0$ are shown in Fig. 4.4(a). The spread is around unity for the range covered by the weighting functions, and increases outside that. Within this range, the noise is in the range 2–5 K. Above that range the noise decreases as the best possible spread is not very different from that of the highest weighting function alone. We can try to choose a

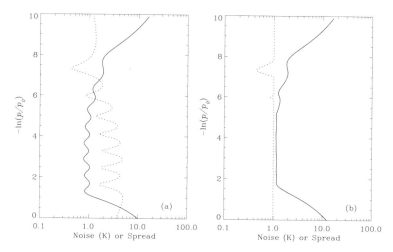

Fig. 4.4 (a) Spread (solid) and noise (dotted) as a function of altitude for the best spread case, $\mu = 0$. (b) Spread (solid) and noise (dotted) where $\mu(z)$ has been chosen to give a noise level of 1 K at all levels (except where it is already smaller).

trade-off parameter at each level to give a fixed measurement error, independent of height, but with the penalty of a poorer spread. This has been done in Fig. 4.4(b), which shows the spread and noise as a function of height when the measurement error has been fixed at 1 K. Note that there are a few altitudes where the worst noise was already less than 1 K.

Chapter 5
Optimal Methods for Non-linear Inverse Problems

A non-linear problem may be thought of simply as a problem in which the forward problem is non-linear, but in practice that is not all that matters. As we have seen, an inverse problem is usually solved by maximising a cost function for which the forward model provides only one term. There may be prior information of various kinds which must also be considered. In general a linear problem is one in which the cost function is quadratic in the state vector, so that the equations to be solved are then linear. A non-quadratic term due to a prior constraint would lead to a non-linear problem even if the forward model were linear. Any non-Gaussian *pdf* as prior information will lead to a non-linear problem.

We can make a qualitative classification of the linearity of inverse problems as follows:

Linear: when the forward model can be put in the form $\mathbf{y} = \mathbf{Kx}$, and any a priori is Gaussian; very few practical problems are truly linear.

Nearly linear: problems which are non-linear, but for which a linearisation about some prior state is adequate to find a solution. Problems which are linear to the accuracy of the measurements, or to the required accuracy of the solution, within the normal range of variation of the state.

Moderately non-linear: problems where linearisation is adequate for the error analysis, but not for finding a solution. Many problems are of this kind.

Grossly non-linear: problems which are non-linear even within the range of the errors. Problems with inequality constraints are of this kind if the solution is on a boundary.

Much of what has been described so far for linear problems applies directly to moderately non-linear problems when they are appropriately linearised. The main difference is that there is no general explicit expression for locating optimal solutions in the moderately non-linear case, as there is for linear and nearly linear problems, so that they must be found numerically and iteratively. Therefore the main topic of this chapter will be numerical methods for finding solutions to moderately non-linear problems. Nonlinear optimisation is a major industry in the numerical methods business, and I will not attempt an exhaustive the subject. I will describe the basics of the most straightforward and commonly used methods so that the reader will have some idea about which methods in computer subroutine libraries are most appropriate, and what their advantages and limitations might be. For a review of subroutine libraries on optimisation see Moré and Wright (1993)

Once a solution has been found, the error analysis and information content can be described in terms of a linearisation of the problem about the solution. The forward model is now a non-linear mapping from state space into measurement space. The measurement maps the state into measurement space, with an uncertainty described by the measurement error, which is often Gaussian. The mapping may be non-unique, as there may still be a null space. However the null space is not described by the same basis throughout measurement space, as it is in the linear case, because the weighting function matrix $\mathbf{K}$ now depends on the state, so its null space also depends on the state. Nevertheless once the retrieval has been found the null space can be identified in the same way as for the linear case. It is still spanned by the singular vectors of $\mathbf{K}$ that have zero singular value.

The inverse mapping from measurement space into state space will map the *pdf* of the measurement error into a *pdf* in state space. If the problem is no worse than moderately nonlinear, and the measurement error is Gaussian, then the retrieval error will be Gaussian, and the linear error analysis will apply.

Qualitatively, the well determined components of the row space of $\mathbf{K}$, defined by the eigenvectors of $\mathbf{S}_\epsilon^{-\frac{1}{2}}\mathbf{K}$ with larger eigenvalues, will map state space into relatively large regions of measurement space, and are likely to be more linear. The components corresponding to small eigenvalues, the near-null space, will map large ranges of state space into small ranges of measurement space, and so are likely to be more affected by nonlinearity.

5.1 Determination of the Degree of Nonlinearity

The degree of non-linearity for any particular problem can be examined by comparing the forward model with the linearised forward model within either the *a priori* variability or the solution error covariance. If we want to use a linear retrieval then the error in a retrieval due to nonlinearity will be about

$$\delta\hat{\mathbf{x}} = \mathbf{G}[\mathbf{F}(\hat{\mathbf{x}}) - \mathbf{F}(\mathbf{x}_a) - \mathbf{K}(\hat{\mathbf{x}} - \mathbf{x}_a)] \tag{5.1}$$

because we have assumed that $\hat{\mathbf{y}} - \mathbf{y}_a = \mathbf{K}(\hat{\mathbf{x}} - \mathbf{x}_a)$ whereas it is actually $\mathbf{F}(\hat{\mathbf{x}}) - \mathbf{F}(\mathbf{x}_a)$. This quantity can be estimated easily once a solution is found, or the degree of nonlinearity can be explored by evaluating $\delta\mathbf{y} = [\mathbf{F}(\mathbf{x}) - \mathbf{F}(\mathbf{x}_a) - \mathbf{K}(\mathbf{x} - \mathbf{x}_a)]$ for values of $\mathbf{x}$ at the equivalent of one standard deviation away from $\mathbf{x}_a$. This can be done systematically by using $\mathbf{x} = \mathbf{x}_a \pm \mathbf{e}_i$ where $\mathbf{e}_i$ is an error pattern of $\mathbf{S}_a$, Eq. (3.21). The size of $\delta\mathbf{y}$ can be evaluated by comparison with $\mathbf{S}_\epsilon$ using

$c^2 = \delta\mathbf{y}^T\mathbf{S}_\epsilon^{-1}\delta\mathbf{y}$, a quantity like a χ^2, or determining whether the corresponding $\delta\mathbf{x}$ is acceptable.

If we are using a non-linear retrieval but want to check whether the problem is moderately linear, so that a linear error analysis can be used, then a similar criterion can be used. The error in the retrieval due to nonlinearity is about

$$\delta\hat{\mathbf{x}} = \mathbf{G}[\mathbf{F}(\hat{\mathbf{x}}) - \mathbf{F}(\mathbf{x}) - \mathbf{K}(\hat{\mathbf{x}} - \mathbf{x})] \tag{5.2}$$

where $\mathbf{x}$ is the unknown true state. This cannot be estimated given only $\hat{\mathbf{x}}$, but the degree of nonlinearity can be explored by evaluating $\delta\mathbf{y} = [\mathbf{F}(\hat{\mathbf{x}}) - \mathbf{F}(\mathbf{x}) - \mathbf{K}(\hat{\mathbf{x}} - \mathbf{x})]$ for values of $\mathbf{x}$ at the equivalent of one standard deviation away from $\hat{\mathbf{x}}$ using error patterns of $\hat{\mathbf{S}}$. Again, the size of $\delta\mathbf{y}$ can be evaluated by comparison with $\mathbf{S}_\epsilon$ using $\delta\mathbf{y}^T\mathbf{S}_\epsilon^{-1}\delta\mathbf{y}$, or by determining whether the corresponding $\delta\mathbf{x}$ is acceptable.

The standard example has so far used temperature in place of the Planck function in order to make it linear in temperature units. It can be formulated as a nonlinear problem by returning to Eq. (1.2) and using the Planck function, a nonlinear function of temperature. To examine the nonlinearity of the original problem in a typical situation, I have used the Planck function at $667\,\mathrm{cm}^{-1}$, the centre of the main CO_2 band commonly used for temperature sounding. I have evaluated $c^2 = \delta\mathbf{y}^T\mathbf{S}_\epsilon^{-1}\delta\mathbf{y}$ for the ten largest error patterns of the non-diagonal *a priori* covariance matrix, as shown in Fig. 3.2, The state vector is temperature and the measurement vector is radiance. Measurement noise is $0.5\,\mathrm{mW\,m^{-2}\,ster^{-1}/cm^{-1}}$, corresponding to approximately the same number in Kelvin, as used for the linear examples. The values of the nonlinearity parameter are all less than unity, as shown in table 5.1, indicating that within the range of the *a priori* the effect of nonlinearity is less than that of measurement error, so the problem is nearly linear. It could of course be made even more linear by using brightness temperature for the measurement vector, or Planck function for the state vector.

Table 5.1 The nonlinearity parameter for the ten largest error patterns in the standard example.

Pattern	Nonlinearity	Pattern	Nonlinearity
1	0.4435	6	0.0186
2	0.1555	7	0.0115
3	0.0838	8	0.0073
4	0.0519	9	0.0048
5	0.0308	10	0.0033

5.2 Formulation of the Inverse Problem

The primary task of a linear retrieval method is to select a state satisfying some criterion of optimality from an ensemble of states which agree with the measurement within experimental error. Locating the ensemble in state space is straightforward in the linear case, but in the non-linear case it may no longer be possible to write down an explicit solution. The expected value could be found in principle by the straightforward but very expensive procedure of integrating Eq. (4.47) over all of state space, but this is not usually practicable.

The optimality criterion may also need reconsideration in the nonlinear case. If the Bayesian solution $P(\mathbf{x}|\mathbf{y})$ is described by a Gaussian *pdf* (or at least a symmetric *pdf*) then the maximum probability and expected value criteria still lead to the same solution. If it is not symmetric then they will lead to different solutions in general, but the interpretation is unchanged, the expected value is still the minimum variance solution, because Eq. (4.47) applies to any problem, however nonlinear. The Backus–Gilbert maximum resolution approach relies on the linear concept of averaging kernels, for which the transfer function should be linear about some predetermined reference state. Thus it is only applicable to nearly linear problems.

For nonlinear problems we will consider primarily the maximum *a posteriori* approach. The Bayesian solution for the linear problem, Eq. (2.24), can be modified straightforwardly for an inverse problem in which the forward model is a general function of the state, the measurement error is Gaussian, and there is a prior estimate with a Gaussian error:

$$-2\ln P(\mathbf{x}|\mathbf{y}) = [\mathbf{y} - \mathbf{F}(\mathbf{x})]^T \mathbf{S}_\epsilon^{-1} [\mathbf{y} - \mathbf{F}(\mathbf{x})] + [\mathbf{x} - \mathbf{x}_a]^T \mathbf{S}_a^{-1} [\mathbf{x} - \mathbf{x}_a] + c_3. \qquad (5.3)$$

The task in the nonlinear case is, as for the linear case, to find a best estimate $\hat{\mathbf{x}}$ and an error characteristic that describes this *pdf* well enough for practical purposes.

We should comment in passing on the more difficult class of non-linear problems not described by Eq. (5.3), namely problems in which the prior *pdf* is not well described by a Gaussian. An extreme example might be any problem involving clouds, such as the important question of sounding the troposphere in the nadir view with infrared instruments. Both prior and posterior *pdf*'s for cloud amount are likely to be non-Gaussian functions for cloud amounts between zero and one, with δ-function-like components at zero and one. In this case it is completely unclear whether a maximum probability or expected value solution would be more appropriate. Discussion of such problems will be postponed!

To find the maximum probability state $\hat{\mathbf{x}}$ we equate the derivative of Eq. (5.3) to zero:

$$\nabla_\mathbf{x}\{-2\ln(P(\mathbf{x}|\mathbf{y}))\} = -[\nabla_\mathbf{x}\mathbf{F}(\mathbf{x})]^T \mathbf{S}_\epsilon^{-1}[\mathbf{y} - \mathbf{F}(\mathbf{x})] + \mathbf{S}_a^{-1}[\mathbf{x} - \mathbf{x}_a] = 0. \qquad (5.4)$$

Note that the gradient $\nabla_\mathbf{x}$ of a vector-valued function is a matrix-valued function. Putting $\mathbf{K}(\mathbf{x}) = \nabla_\mathbf{x}\mathbf{F}(\mathbf{x})$ this gives the following implicit equation for $\hat{\mathbf{x}}$:

$$-\hat{\mathbf{K}}^T(\hat{\mathbf{x}})\mathbf{S}_\epsilon^{-1}[\mathbf{y} - \mathbf{F}(\hat{\mathbf{x}})] + \mathbf{S}_a^{-1}(\hat{\mathbf{x}} - \mathbf{x}_a) = 0. \qquad (5.5)$$

5.3 Newton and Gauss–Newton Methods

This equation must be solved numerically, and the difficulty involved depends on the non-linearity of the forward model.

If the problem is not too non-linear, Newtonian iteration is a straightforward numerical method for finding the zero of the gradient of the cost function J, i.e. the first two terms on the r.h.s. of (5.3). For the general vector equation $\mathbf{g}(\mathbf{x}) = 0$, the iteration is analogous to Newton's method for the scalar case and can be written:

$$\mathbf{x}_{i+1} = \mathbf{x}_i - [\nabla_{\mathbf{x}}\mathbf{g}(\mathbf{x}_i)]^{-1}\mathbf{g}(\mathbf{x}_i) \tag{5.6}$$

where the inverse is a matrix inverse. Using the l.h.s. of Eq. (5.5) for $\mathbf{g}$, the matrix is:

$$\nabla_{\mathbf{x}}\mathbf{g} = \mathbf{S}_a^{-1} + \mathbf{K}^T\mathbf{S}_\epsilon^{-1}\mathbf{K} - [\nabla_{\mathbf{x}}\mathbf{K}^T]\mathbf{S}_\epsilon^{-1}[\mathbf{y} - \mathbf{F}(\mathbf{x})]. \tag{5.7}$$

The function $\mathbf{g}$ is the derivative of the cost function (5.3) and $\nabla_{\mathbf{x}}\mathbf{g}$ is the second derivative, known as the *Hessian*. Consequently this is also known as the inverse Hessian method. The Hessian involves both the Jacobian $\mathbf{K}$, the first derivative of the forward model, and $\nabla_{\mathbf{x}}\mathbf{K}^T$, the second derivative of the forward model. The latter is a complicated object, because it is a vector whose elements are matrices, which are to be post-multiplied by the vector $\mathbf{S}_\epsilon^{-1}[\mathbf{y} - \mathbf{F}(\mathbf{x})]$ before the resulting vector of vectors is assembled into a matrix. It is also expensive to evaluate in many cases. Fortunately the resulting product is small in the moderately linear case, as we shall see, and the term involving it becomes smaller as the solution proceeds and $\mathbf{y} - \mathbf{F}(\mathbf{x})$ becomes similar to the noise. Note, therefore, that both nonlinearity and noise contribute to the size of this term. Problems for which this term can be ignored are called *small residual* in the numerical methods literature. Ignoring the term gives the Gauss–Newton method, obtained by substituting Eqs. (5.5) and (5.7) into the Newtonian iteration (5.6) and omitting this term,

$$\mathbf{x}_{i+1} = \mathbf{x}_i + (\mathbf{S}_a^{-1} + \mathbf{K}_i^T\mathbf{S}_\epsilon^{-1}\mathbf{K}_i)^{-1}[\mathbf{K}_i^T\mathbf{S}_\epsilon^{-1}(\mathbf{y} - \mathbf{F}(\mathbf{x}_i)) - \mathbf{S}_a^{-1}(\mathbf{x}_i - \mathbf{x}_a)] \tag{5.8}$$

where $\mathbf{K}_i = \mathbf{K}(\mathbf{x}_i)$. By expressing $\mathbf{x}_{i+1}$ as a departure from $\mathbf{x}_a$ rather than $\mathbf{x}_i$, we can rearrange (5.8) in both the n-form and the m-form, giving two other useful iterative solutions:

$$\begin{aligned}\mathbf{x}_{i+1} &= \mathbf{x}_a + (\mathbf{S}_a^{-1} + \mathbf{K}_i^T\mathbf{S}_\epsilon^{-1}\mathbf{K}_i)^{-1}\mathbf{K}_i^T\mathbf{S}_\epsilon^{-1}[\mathbf{y} - \mathbf{F}(\mathbf{x}_i) + \mathbf{K}_i(\mathbf{x}_i - \mathbf{x}_a)] \tag{5.9}\\ &= \mathbf{x}_a + \mathbf{S}_a\mathbf{K}_i^T(\mathbf{K}_i\mathbf{S}_a\mathbf{K}_i^T + \mathbf{S}_\epsilon)^{-1}[\mathbf{y} - \mathbf{F}(\mathbf{x}_i) + \mathbf{K}_i(\mathbf{x}_i - \mathbf{x}_a)]. \tag{5.10}\end{aligned}$$

It is convenient, but not essential, to start the iteration with $\mathbf{x}_0 = \mathbf{x}_a$.

5.4 An Alternative Linearisation

The definition of moderately linear implies that the *pdf* of the solution is Gaussian, because the forward model is linear within the region where the solution *pdf* is not negligible. Thus an alternative approach to deriving a solution is to replace the forward model in Eq. (5.5) by a linearisation:

$$\mathbf{F}(\mathbf{x}) = \mathbf{F}(\mathbf{x}_l) + \nabla_\mathbf{x}\mathbf{F}(\mathbf{x}_l)(\mathbf{x} - \mathbf{x}_l) = \mathbf{y}_l + \mathbf{K}_l(\mathbf{x} - \mathbf{x}_l) \qquad (5.11)$$

thus defining $\mathbf{y}_l$ and $\mathbf{K}_l$, where $\mathbf{x}_l$ is an arbitrary linearisation point. On substituting in Eq. (5.5) and solving for $\mathbf{x}$ we get

$$\hat{\mathbf{x}} = \mathbf{x}_a + (\mathbf{K}_l^T \mathbf{S}_\epsilon^{-1} \mathbf{K}_l + \mathbf{S}_a^{-1})^{-1} \mathbf{K}_l^T \mathbf{S}_\epsilon^{-1}[\mathbf{y} - \mathbf{y}_l + \mathbf{K}_l(\mathbf{x}_l - \mathbf{x}_a)]. \qquad (5.12)$$

⇒ ***Exercise*** 5.1: Derive this expression

This equation looks as if it might form the basis for an iterative solution. If we change the interpretation of the subscript l from 'linearisation' to 'iteration counter', we obtain the same iteration as Eq. (5.9). Therefore omitting the $\nabla_\mathbf{x}\mathbf{K}^T$ term in Eq. (5.7) is equivalent to a Gaussian posterior *pdf*, or to the assumption of linearity within the error bars of the retrieval, which is implicit in the assumption that the problem is not grossly non-linear.

The covariance of the solution can be found in exactly the same way as for the linear problem, Eq. (4.7), giving the same alternative forms:

$$\begin{aligned} \hat{\mathbf{S}} &= (\hat{\mathbf{K}}^T \mathbf{S}_\epsilon^{-1} \hat{\mathbf{K}} + \mathbf{S}_a^{-1})^{-1} & (5.13) \\ &= \mathbf{S}_a - \mathbf{S}_a \hat{\mathbf{K}}^T (\mathbf{S}_\epsilon + \hat{\mathbf{K}} \mathbf{S}_a \hat{\mathbf{K}}^T)^{-1} \hat{\mathbf{K}} \mathbf{S}_a. & (5.14) \end{aligned}$$

Eqs. (5.13) and (5.14) can be rewritten in a variety of forms, the most appropriate depending on the circumstances. Note that at the solution the Hessian is the same as the inverse covariance of the solution or the Fisher information matrix.

5.5 Error Analysis and Characterisation

The error analysis and characterisation of Chapter 3 are derived under the assumption that the forward model and retrieval method are linear over the range between the *a priori* the true state and the retrieved state, i.e. the problem is not worse than nearly linear. It might seem that the concept of the averaging kernel would therefore not apply to the moderately linear case. Fortunately this is not the case for the optimal estimator, and the concept is of a wider validity. The reason is that the prior state appears only linearly in the retrieval method, and we would obtain the same solution from a linear problem corresponding to the linearisation at the retrieval.

We can show this by considering the limit of Eq. (5.9) as $i \to \infty$ and $\mathbf{x}_i \to \hat{\mathbf{x}}$. Express Eq. (5.9) as

$$\mathbf{x}_{i+1} = \mathbf{x}_a + \mathbf{G}_i[\mathbf{y} - \mathbf{F}(\mathbf{x}_i) + \mathbf{K}_i(\mathbf{x}_i - \mathbf{x}_a)] \tag{5.15}$$

where $\mathbf{G}_i = (\mathbf{S}_a^{-1} + \mathbf{K}_i^T \mathbf{S}_\epsilon^{-1} \mathbf{K}_i)^{-1} \mathbf{K}_i^T \mathbf{S}_\epsilon^{-1} = \mathbf{S}_a \mathbf{K}_i^T (\mathbf{K}_i \mathbf{S}_a \mathbf{K}_i^T + \mathbf{S}_\epsilon)^{-1}$. Now let $i \to \infty$:

$$\hat{\mathbf{x}} = \mathbf{x}_a + \hat{\mathbf{G}}[\mathbf{y} - \mathbf{F}(\hat{\mathbf{x}}) + \hat{\mathbf{K}}(\hat{\mathbf{x}} - \mathbf{x}_a)]. \tag{5.16}$$

Substituting $\mathbf{y} = \mathbf{F}(\mathbf{x}) + \boldsymbol{\epsilon}$ and using Eq. (5.11) to expand $\mathbf{F}(\mathbf{x})$ about $\hat{\mathbf{x}}$, valid within $\boldsymbol{\epsilon}$ in the moderately non-linear case, gives

$$\hat{\mathbf{x}} = \mathbf{x}_a + \hat{\mathbf{G}}[\hat{\mathbf{K}}(\mathbf{x} - \mathbf{x}_a) + \boldsymbol{\epsilon}], \tag{5.17}$$

so we can write

$$\hat{\mathbf{x}} - \mathbf{x}_a = \hat{\mathbf{A}}(\mathbf{x} - \mathbf{x}_a) + \hat{\mathbf{G}}\boldsymbol{\epsilon} \tag{5.18}$$

$$\hat{\mathbf{x}} - \mathbf{x} = (\hat{\mathbf{A}} - \mathbf{I}_n)(\mathbf{x} - \mathbf{x}_a) + \hat{\mathbf{G}}\boldsymbol{\epsilon} \tag{5.19}$$

where $\hat{\mathbf{A}} = \hat{\mathbf{G}}\hat{\mathbf{K}}$, corresponding to Eq. (3.9) and (3.16), but now derived for the moderately non-linear case. A full treatment which includes forward modelling errors will give the same result as Eq. (3.16) for the error analysis.

5.6 Convergence

5.6.1 *Expected convergence rate*

The convergence of the Newton method is second order, which can be seen qualitatively as follows. Ignoring for the moment the variation of $\mathbf{G}_i$ and $\mathbf{K}_i$ with i, an equation for the convergence of the difference between $\mathbf{x}_i$ and the final solution can be found by subtracting (5.16) from (5.15),

$$\mathbf{x}_{i+1} - \hat{\mathbf{x}} = -\mathbf{G}\{[\mathbf{F}(\mathbf{x}_i) - \mathbf{F}(\hat{\mathbf{x}})] + \mathbf{K}[\mathbf{x}_i - \hat{\mathbf{x}}]\}, \tag{5.20}$$

and expanding $\mathbf{F}(\mathbf{x})$ about $\hat{\mathbf{x}}$ to obtain:

$$\mathbf{x}_{i+1} - \hat{\mathbf{x}} = \mathbf{G}\{O(\mathbf{x}_i - \hat{\mathbf{x}})^2\}. \tag{5.21}$$

An iteration of the form of Eq. (5.15) converges quadratically, regardless of the value of $\mathbf{G}$! However it only converges to the optimal solution if $\mathbf{G}$ has the appropriate value. If the convergence analysis is treated in detail, allowing for the variation of $\mathbf{G}$ and $\mathbf{K}$ with i, it is found that Newton's method is second order, and Gauss–Newton is first order theoretically. However if the problem is less than grossly nonlinear, i.e. the term involving the second derivative of the forward model, $\nabla_\mathbf{x} \mathbf{K}^T$, is small, Gauss–Newton will converge quickly, and as the minimum is approached the ignored term becomes smaller and convergence approaches second order.

5.6.2 A popular mistake

A very popular mistake in carrying out this kind of iteration is to confuse the prior state $\mathbf{x}_a$ with the current iteration. It may be thought that in carrying out an iteration cycle, the current estimate can with advantage be used as the *a priori* for the next iteration. This is clearly inappropriate in the context of an optimal estimator, where the prior represents our knowledge of the state before the measurements are made, so should not be replaced by something that is the result of using the measurements. To do that would be equivalent to using the measurements more than once in the analysis.

The following three concepts are often confused. It is worth making the distinctions clear:

> *Climatology* is the mean and covariance of an ensemble of states.
> *A Priori* is the best estimate of the state before the measurement is made. Climatology often provides a convenient *a priori*, but is not the only source.
> *First Guess* is the starting point of an iteration. The *a priori* is often used as a first guess, but is not the only source.

It may also be thought that using the current iteration for the *a priori* state will speed up the convergence. In fact, as can easily be shown, the effect is to converge more slowly, towards a non-optimal 'exact' solution. If we replace $\mathbf{x}_a$ by $\mathbf{x}_i$ in Eq. (5.15), the iteration becomes

$$\begin{aligned}\mathbf{x}_{i+1} &= \mathbf{x}_i + \mathbf{G}_i[\mathbf{y} - \mathbf{F}(\mathbf{x}_i) + \mathbf{K}_i(\mathbf{x}_i - \mathbf{x}_i)] \\ &= \mathbf{x}_i + \mathbf{G}_i[\mathbf{y} - \mathbf{F}(\mathbf{x}_i)].\end{aligned} \quad (5.22)$$

This is a linear relaxation of the kind that will be discussed in section 6.7, except that $\mathbf{G}_i$ may vary with iteration number. If it converges the solution will satisfy

$$\hat{\mathbf{x}} = \hat{\mathbf{x}} + \hat{\mathbf{G}}[\mathbf{y} - \mathbf{F}(\hat{\mathbf{x}})], \quad (5.23)$$

clearly one for which $\mathbf{y} = \mathbf{F}(\hat{\mathbf{x}})$ if $\mathbf{G}$ is of rank $m < n$. Close to the solution, as $\mathbf{G}_n \to \hat{\mathbf{G}}$, the convergence will follow

$$\begin{aligned}\mathbf{x}_{i+1} - \hat{\mathbf{x}} &= \mathbf{x}_i - \hat{\mathbf{x}} + \hat{\mathbf{G}}[\mathbf{F}(\hat{\mathbf{x}}) - \mathbf{F}(\mathbf{x}_i)] \\ &= \mathbf{x}_i - \hat{\mathbf{x}} + \hat{\mathbf{G}}\mathbf{K}(\hat{\mathbf{x}} - \mathbf{x}_i) \\ &= (\mathbf{I}_n - \hat{\mathbf{G}}\mathbf{K})(\mathbf{x}_i - \hat{\mathbf{x}}),\end{aligned} \quad (5.24)$$

which will be first order, unless $\mathbf{GK} = \mathbf{I}_n$. For the maximum *a posteriori* method $\mathbf{G}$ is given by Eq. (3.27), and $\mathbf{GK} \neq \mathbf{I}_n$, so convergence is first order in this case.

5.6.3 Testing for convergence

The convergence analysis is required so that we can determine the correct criterion for stopping the iteration. It is clearly unnecessary to continue until there is no change in the solution at machine precision, but we do want the solution to differ from the true maximum probability state by a quantity which is ignorably small, say an order of magnitude smaller than the error in the solution.

A commonly used but inadequate test is to compare the difference between the fit and the measurement with the expected experimental error, e.g:

$$\chi^2[\mathbf{y} - \mathbf{F}(\mathbf{x}_i)] = [\mathbf{y} - \mathbf{F}(\mathbf{x}_i)]^T \mathbf{S}_\epsilon^{-1}[\mathbf{y} - \mathbf{F}(\mathbf{x}_i)] \simeq m. \quad (5.25)$$

Something like this should be more or less true on the average for an ensemble of retrievals, but for any particular case this χ^2 may be greater or smaller than m. It might be expected to follow a χ^2 distribution with m degrees of freedom—if it were not the case that $\mathbf{S}_\epsilon$ is the wrong covariance to use here. The expected value of this χ^2 at the solution is of course the degrees of freedom for noise. The right one would be the covariance of the difference between the fit and the measurement, i.e. the expected value of $[\mathbf{y} - \mathbf{F}(\hat{\mathbf{x}})][\mathbf{y} - \mathbf{F}(\hat{\mathbf{x}})]^T$:

$$\begin{aligned}\mathbf{S}_{\delta\hat{y}} &= \mathcal{E}\{[\mathbf{y} - \mathbf{F}(\hat{\mathbf{x}})][\mathbf{y} - \mathbf{F}(\hat{\mathbf{x}})]^T\} \\ &= \mathcal{E}\{[\mathbf{F}(\mathbf{x}) + \boldsymbol{\epsilon} - \mathbf{F}(\hat{\mathbf{x}})][\mathbf{F}(\mathbf{x}) + \boldsymbol{\epsilon} - \mathbf{F}(\hat{\mathbf{x}})]^T\}.\end{aligned} \quad (5.26)$$

⇒ **Exercise** 5.2: Show that the covariance of $\delta\hat{\mathbf{y}} = \mathbf{y} - \mathbf{F}(\hat{\mathbf{x}})$ is

$$\mathbf{S}_{\delta\hat{y}} = \mathbf{S}_\epsilon(\hat{\mathbf{K}}\mathbf{S}_a\hat{\mathbf{K}}^T + \mathbf{S}_\epsilon)^{-1}\mathbf{S}_\epsilon \quad (5.27)$$

for an optimal estimator and, by considering an eigenvector or singular vector expansion, show how it differs from $\mathbf{S}_\epsilon$.

The correct convergence test must be based on the convergence analysis. There are three different kinds of test that can be used. At each iteration we can check for the smallness of (a) the reduction of the cost function (b) the gradient of the cost function or (c) the size of the step, in state space or in measurement space. The absolute size of the cost function or of $\delta\hat{\mathbf{y}}$ does not say anything about whether the iteration has converged. However it is useful in determining whether the solution obtained is sensible.

Probably the most straightforward test is on the cost function being minimised. Around its minimum it must be close to quadratic in $\mathbf{x} - \hat{\mathbf{x}}$, and if convergence is second order, both $\mathbf{x}_i - \hat{\mathbf{x}}$ and the cost function error $J_i - \hat{J}$ are roughly squared on each iteration. Its expected value at the minimum is m (section 2.4.2), so a change between iterations of $\ll m$, or even $\ll 1$ would be appropriate. In the n-form iteration Eq. (5.8) J_i can be computed from available quantities in only $n + m$ operations.

To test the size of the step in $\mathbf{x}$ or $\mathbf{y}$, we note that as the solution error is squared at each stage, then close to the solution the difference between the estimates at stages n and $n+1$ is a reasonable measure of the error at stage n, generally much greater than the error at stage $n+1$. A conservative criterion is to stop when the difference is an order of magnitude smaller than the estimated error. To do this so we must scale the change in the solution by its estimated error. This test can also be carried out in terms of either the retrieved state or the computed measurement, depending on the form of the equations being used, and which is the more efficiently computed. In the case where $n \lesssim m$, we would expect to be using the n-form. In terms of the state

$$\chi^2(\hat{\mathbf{x}} - \mathbf{x}) = (\hat{\mathbf{x}} - \mathbf{x})^T \hat{\mathbf{S}}^{-1}(\hat{\mathbf{x}} - \mathbf{x}) \simeq n \qquad (5.28)$$

so that an appropriate test would be

$$d_i^2 = (\mathbf{x}_i - \mathbf{x}_{i+1})^T \hat{\mathbf{S}}^{-1}(\mathbf{x}_i - \mathbf{x}_{i+1}) \ll n. \qquad (5.29)$$

As the iteration converges, we note that

$$(\mathbf{S}_a^{-1} + \mathbf{K}_i^T \mathbf{S}_\epsilon^{-1} \mathbf{K}_i)^{-1} \to \hat{\mathbf{S}} \qquad (5.30)$$

and putting this in Eq. (5.8) we obtain

$$\hat{\mathbf{S}}^{-1}(\mathbf{x}_{i+1} - \mathbf{x}_i) \simeq \mathbf{K}_i^T \mathbf{S}_\epsilon^{-1}[\mathbf{y} - \mathbf{F}(\mathbf{x}_i)] - \mathbf{S}_a^{-1}(\mathbf{x}_i - \mathbf{x}_a). \qquad (5.31)$$

This vector of dimension n is evaluated in any case as part of the iteration, and calculating d_i^2 only requires it to be multiplied by $(\mathbf{x}_{i+1} - \mathbf{x}_i)$, taking n operations.

In the case where $m \lesssim n$, the m-form is likely to be used, and this trick is not available. Noting that

$$\chi^2[\mathbf{y} - \mathbf{F}(\hat{\mathbf{x}})] = [\mathbf{y} - \mathbf{F}(\hat{\mathbf{x}})]^T \mathbf{S}_{\delta\hat{y}}^{-1}[\mathbf{y} - \mathbf{F}(\hat{\mathbf{x}})] \simeq m \qquad (5.32)$$

an appropriate test would be

$$d_i^2 = [\mathbf{F}(\mathbf{x}_{i+1}) - \mathbf{F}(\mathbf{x}_i)]^T \mathbf{S}_{\delta\hat{y}}^{-1}[\mathbf{F}(\mathbf{x}_{i+1}) - \mathbf{F}(\mathbf{x}_i)] \ll m. \qquad (5.33)$$

$\mathbf{S}_{\delta\hat{y}}$ is of a smaller order then $\hat{\mathbf{S}}$ and can be easily evaluated from items computed as part of the iteration. $\mathbf{S}_{\delta\hat{y}}^{-1}$ can be computed with m^2 operations if $\mathbf{S}_\epsilon$ is diagonal, see Eq. (5.26), and the evaluation of d_i^2 takes m^2 more.

5.6.4 Testing for correct convergence

Once the iteration has converged, we must test whether it has converged to the correct answer. With a nonlinear problem it is quite possible to find a spurious minimum of the cost function being minimised. There may be multiple minima, only one of which is the required solution.

This is the right place to use the comparison of the retrieval with the measurement, Eq. (5.32), and to carry out a standard χ^2 test to determine whether the difference is statistically significant at some appropriate level. A very significant result may indicate a spurious convergence. It should be remembered that if retrievals are being carried out for large ensembles, for example for global satellite data, then $N\%$ of the good retrievals should be significant at the $N\%$ level, so N should not be set too high. A 0.1% significance test would discard one retrieval in 10^3 even if nothing were amiss, and these might well be cases of particular interest. It is helpful to examine the distribution of χ^2 actually found.

It is also possible to carry out a χ^2 test against any *a priori* data that is being used in the retrieval. Considering the *a priori* as a 'virtual measurement', the test could be carried out against the real and virtual measurements jointly, but as these two sources of information are uncorrelated, treating them separately allows one to distinguish possible sources of trouble. This will be discussed further in Chapter 11 where validation methods are dealt with in more detail.

5.6.5 *Recognising and dealing with slow convergence*

The cost function for a moderately linear problem may be seriously non-quadratic away from the solution, and a simple Gauss–Newton method may not be suitable. It is quite possible for the cost function to increase as a result of a Gauss–Newton (or Newton) step. In this case some other numerical method, such as that described in the section 5.7, is required.

It may also happen that the process converges, but not as quickly as you might like. The convergence can be followed by examining the size of steps taken and the rate of decrease (or otherwise!) of the cost function or some other χ^2. The quantity d_i^2 defined by Eq. (5.29) is a measure of a step taken, conveniently scaled by the expected solution error covariance. For second order convergence, the step size (or the change in the cost function) should roughly be squared each iteration, or its (negative) logarithm should roughly double. If the decrease is much slower than this, then there is a problem that needs treatment—the iteration has started a long way from the solution, or the problem is seriously nonlinear, and other numerical methods may be needed. Sometimes relatively simple *ad hoc* treatments can help a great deal. For example:

Transform the problem to a more linear form. E.g. for nadir thermal emission sounding of the temperature profile, let the state vector be the Planck function rather than the temperature, or the measurement be the brightness temperature rather than radiance.

Start with a better first guess. E.g. use an *ad hoc* non-optimal retrieval method to find a close first guess, or, if retrieving a sequence of geographically close profiles, as along a satellite measurement track, use one retrieval as the starting point for the next.

5.7 Levenberg–Marquardt Method

Both Newton's method and Gauss–Newton will find the minimum in one step for a cost function which is exactly quadratic in **x**, and will get close if the function is nearly quadratic. However if the true solution is sufficiently far from the current iteration point it is quite possible that a quadratic represents the surface so poorly that the step taken is completely meaningless, and may even increase rather than decrease the residual. For the nonlinear least squares problem, Levenberg (1944) proposed the iteration

$$\mathbf{x}_{i+1} = \mathbf{x}_i + (\mathbf{K}^T\mathbf{K} + \gamma_i\mathbf{I})^{-1}\mathbf{K}^T[\mathbf{y} - \mathbf{F}(\mathbf{x}_i)] \qquad (5.34)$$

where γ_i is chosen at each step to minimise the cost function. It can be seen that for $\gamma_i \to 0$ the step tends to Gauss–Newton, and for $\gamma_i \to \infty$ the step direction tends to steepest descent, but the step size tends to zero. The cost function will clearly initially decrease as γ_i is decreased from infinity, therefore there will be an optimum value (possibly zero) which maximally reduces the cost function. Unfortunately the computation needed for choosing γ_i is significant, as $F(\mathbf{x})$ must be evaluated for each γ_i tried. Marquardt (1963) simplified the choice of γ_i, by not searching for the best γ_i for each iteration, but by starting a new iteration step as soon as a value is found for which the cost function is reduced. An initial arbitrary value of γ is updated at each iteration. A simplified version of Marquardt's strategy is given by Press *et al.* (1995):

- If χ^2 increases as a result of a step, increase γ, don't update $\mathbf{x}_i$ and try again.
- If χ^2 decreases as a result of a step, update $\mathbf{x}_i$ and decrease γ for the next step.

The factor by which γ is increased or decreased is a matter for experiment in particular cases. Marquardt suggested ten. He also pointed out that as the elements of the state vector may have different magnitudes and dimensions, they should be scaled. A convenient way to do this is to replace $\gamma\mathbf{I}$ by $\gamma\mathbf{D}$ where $\mathbf{D}$ is a diagonal scaling matrix.

Fletcher (1971) found that Marquardt's strategy had inadequacies, and proposed a strategy of updating γ based on the ratio R of the change in the cost function computed properly to that computed with the linear approximation to the forward model. This ratio will be unity if the linear approximation is satisfactory, and negative if χ^2 has increased rather than decreased. The aim is to find a value of γ which restricts the new value of $\mathbf{x}$ to lie within linear range of the previous estimate, in the so called *trust region*. The strategy is basically:

- If the ratio is greater than 0.75, reduce γ.
- If the ratio is less than 0.25 increase γ.
- Otherwise make no change.

- If γ is less than some critical value, use zero.

The numbers 0.75 and 0.25 were found by experiment, and are not crucial. He suggested a factor of two for reducing γ, and a factor between 2 and 10 for increasing it.

Applying this modification to the Gauss–Newton method in the n-form, the iteration equation (5.8) becomes

$$\mathbf{x}_{i+1} = \mathbf{x}_i + (\mathbf{S}_a^{-1} + \mathbf{K}_i^T \mathbf{S}_\epsilon^{-1} \mathbf{K}_i + \gamma \mathbf{D}_n)^{-1} \{\mathbf{K}_i^T \mathbf{S}_\epsilon^{-1}[\mathbf{y} - \mathbf{F}(\mathbf{x}_i)] - \mathbf{S}_a^{-1}[\mathbf{x}_i - \mathbf{x}_a]\} \quad (5.35)$$

The m-form cannot be easily used in the Levenberg–Marquardt method because it does not explicitly use the inverse of the Hessian. We are at liberty to choose the scaling matrix $\mathbf{D}$, which does not have to be diagonal, although that can make the numerical work more efficient, but must be positive definite. In our case the simplest choice of a matrix whose elements have the right dimensions to scale the problem is $\mathbf{D} = \mathbf{S}_a^{-1}$, when Eq. (5.35) becomes a little simpler:

$$\mathbf{x}_{i+1} = \mathbf{x}_i + [(1+\gamma)\mathbf{S}_a^{-1} + \mathbf{K}_i^T \mathbf{S}_\epsilon^{-1} \mathbf{K}_i]^{-1} \{\mathbf{K}_i^T \mathbf{S}_\epsilon^{-1}[\mathbf{y} - \mathbf{F}(\mathbf{x}_i)] - \mathbf{S}_a^{-1}[\mathbf{x}_i - \mathbf{x}_a]\} \quad (5.36)$$

The computation required in a single step of this iteration is little different from the Gauss–Newton method, but more steps are likely to be needed, if only because this method is used for more difficult problems.

Many more sophisticated developments of the Levenberg-Marquardt approach are available, the most popular being those based on using the radius of the trust region rather than γ as the parameter which is adjusted at each iteration. These are beyond the scope of this book, and the reader is referred to the numerical methods literature, e.g Moré (1978), Scales (1985), and Fletcher (1987). Standard routines are available in computer subroutine libraries.

5.8 Numerical Efficiency

A significant part of the skill of designing retrieval methods is in making them computationally efficient, particularly when large quantities of data are involved. This involves several aspects, including the selection of the state vector, the efficiency of the forward model computation and the evaluation of its Jacobian, which will be discussed in chapter 9, and the strategy for carrying out the inverse calculations, and the numerical methods used.

5.8.1 *Which formulation for the linear algebra?*

There are two basically different formulations of the linear algebra for the Gauss–Newton method, one of which involves the solution of an $n \times n$ matrix equation, and the other an $m \times m$ matrix. The decision on which to use will depend in part on the relative sizes of n and m, although in some cases the cost of computing the linear algebra will be far outweighed by the cost of evaluating the forward model and the nm elements of its derivative matrix.

5.8.1.1 The n-form

The n-form is fundamentally a non-linear weighted least squares calculation, about which there is a vast quantity of literature. I will not attempt to reproduce all the details here. Least squares routines of many kinds are available in computer subroutine libraries, (Moré and Wright, 1993) so you should not need to write your own other than as an exercise, but you should have enough understanding of the fundamentals to make an informed choice. Useful texts on linear and nonlinear least squares fitting include Scales (1985), Fletcher (1987) and Gill, Murray and Wright (1981, 1990).

Each iteration of the Gauss–Newton method requires the following operations:

$$\mathbf{x}_{i+1} = \mathbf{x}_a + (\mathbf{K}_i^T \mathbf{S}_\epsilon^{-1} \mathbf{K}_i + \mathbf{S}_a^{-1})^{-1} \mathbf{K}_i^T \mathbf{S}_\epsilon^{-1} [\mathbf{y} - \mathbf{F}(\mathbf{x}_i) + \mathbf{K}_i(\mathbf{x}_i - \mathbf{x}_a)] \quad (5.37)$$

and after the final iteration:

$$\hat{\mathbf{S}} = (\mathbf{S}_a^{-1} + \mathbf{K}_i^T \mathbf{S}_\epsilon^{-1} \mathbf{K}_i)^{-1} \quad (5.38)$$

The Levenberg–Marquardt method is virtually identical as far as the numerical solution of a single iteration is concerned. The operation count in this case depends on some details of the problem. We will consider first the 'normal' circumstances in which $\mathbf{S}_\epsilon$ is diagonal, $\hat{\mathbf{S}}$ is required and $\mathbf{S}_a^{-1}$ has been be precomputed. The beginner's method is usually something like:

$$\begin{aligned}
\mathbf{W}_1 &= \mathbf{S}_\epsilon^{-1} \mathbf{K} & mn \\
\mathbf{W}_2 &= \mathbf{S}_a^{-1} + \mathbf{K}^T \mathbf{W}_1 & n^2 m \\
\mathbf{w}_1 &= \mathbf{K}(\mathbf{x}_i - \mathbf{x}_a) & nm \\
\mathbf{w}_2 &= \mathbf{W}_1^T [\mathbf{y} - \mathbf{F}(\mathbf{x}_i) - \mathbf{w}_1] & nm \\
\mathbf{W}_3 &= \mathbf{W}_2^{-1} & n^3 \\
\hat{\mathbf{x}} &= \mathbf{x}_a + \mathbf{W}_3 \mathbf{w}_2 & n^2
\end{aligned} \quad (5.39)$$

The numbers on the right are the number of numerical operations required at each stage, each operation comprising a floating point multiply and add. The numerical cost of evaluating the forward model and its Jacobian are ignored here, to be considered in Chapter 9. If we take both m and n to be much greater than unity, only two steps are significant, the matrix product $\mathbf{K}^T \mathbf{W}_1$ and the inverse $\mathbf{W}_2^{-1}$, totalling $n^2 m + n^3$ operations. If $\mathbf{S}_\epsilon$ is not diagonal then the first step will also be significant, taking a further $m^2 n$ operations if $\mathbf{S}_\epsilon^{-1}$ is precomputed, and will be the largest component when $m > n$. There are two fundamental problems with this approach, firstly that it can be done more quickly, and secondly that it can suffer from numerical problems of rounding error.

The main numerical techniques you should be aware of are Gaussian triangulation and back substitution, Cholesky decomposition, QR decomposition and singular value decomposition. These are all methods which can be used for solving linear equations of the form $\mathbf{Bx} = \mathbf{y}$, and replace the last two stages in the above list. Matrix inversion should never be used for this purpose as it is normally performed by solving simultaneous equations with a unit matrix on the right hand side, and will therefore take longer than other methods. There are many texts which discuss the details of these techniques, for example Wilkinson (1965), Noble (1969), Atkinson (1989) and Golub and Van Loan (1996) are some of the more well known.

Gaussian triangulation involves operating on both sides of the equation in such a way that $\mathbf{B}$ is transformed to an upper triangular form $\mathbf{T}$, giving an equation of the form $\mathbf{Tx} = \mathbf{z}$. The last row of the transformed equation only involves x_n, and is trivially solved. It is then substituted into row $n-1$, giving x_{n-1}, and so on back through the matrix. The triangulation takes $n^3/3$ operations, and the back substitution takes $n^2/2$ operations per column on the right hand side. Thus the speed of the beginner's method can be improved significantly by replacing the last two stages by a Gaussian elimination and back substitution. If $\mathbf{S}_\epsilon$ is not diagonal then the first stage can be treated similarly, taking $m^2n/2$ operations if the triangulation of $\mathbf{S}_\epsilon$ ($m^3/3$ operations) is precomputed. Also notice that as $\mathbf{W}_2$ is symmetric, only half of the elements need to be computed, taking $n^2m/2$ operations.

Each step of the Gaussian triangulation involves a process of the following type:

$$\begin{pmatrix} 1 & 0 & 0 & 0 \\ -B_{21}/B_{11} & 1 & 0 & 0 \\ -B_{31}/B_{11} & 0 & 1 & 0 \\ -B_{41}/B_{11} & 0 & 0 & 1 \end{pmatrix} \begin{pmatrix} B_{11} & B_{12} & B_{13} & B_{14} \\ B_{21} & B_{22} & B_{23} & B_{24} \\ B_{31} & B_{32} & B_{33} & B_{34} \\ B_{41} & B_{42} & B_{43} & B_{44} \end{pmatrix} = \begin{pmatrix} B_{11} & B_{12} & B_{13} & B_{14} \\ 0 & C_{22} & C_{23} & C_{24} \\ 0 & C_{32} & C_{33} & C_{34} \\ 0 & C_{42} & C_{43} & C_{44} \end{pmatrix}$$
(5.40)

That is, multiples of the first row are subtracted from the others to eliminate all except the first element of the first column. The right hand side of $\mathbf{Bx} = \mathbf{y}$ is operated upon in the same way. The elimination is then carried out on the smaller submatrix $\mathbf{C}$, and so on. Care needs to be exercised so that the number to be divided by the *pivot* (B_{11} in this case) is not small. The elimination does not have to be carried out in the original row order, so the pivot can be chosen to be the largest number in the column at each stage (called *pivotting*).

Cholesky decomposition is even faster. It applies to symmetric matrices, the kind we have, and decomposes the matrix into the product of a lower and an upper triangular matrix, $\mathbf{B} = \mathbf{T}^T\mathbf{T}$ in $n^3/6$ operations. The solution of $\mathbf{T}^T\mathbf{Tx} = \mathbf{y}$ is then carried out by two successive back substitutions. Back substitution has the same effect as multiplying by $\mathbf{T}^{-1}$, and is faster than matrix multiplication by $\mathbf{T}^{-1}$ would be, even if were precomputed.

⇒ **Exercise** 5.3: By considering how the elements of $\mathbf{B}$ are related to those of $\mathbf{T}$ one at a time in a suitable order, construct a straightforward algorithm for Cholesky decomposition. Ignore pivotting.

Cholesky decomposition is the fastest way of solving symmetric equations, but it does not deal with the possible loss of precision involved in evaluating $\mathbf{K}_i^T \mathbf{S}_\epsilon^{-1} \mathbf{K}_i$, which squares the condition number of the problem, essentially because the eigenvalues of $\mathbf{K}^T \mathbf{K}$ are the squares of the singular values of $\mathbf{K}$. Adding $\mathbf{S}_a^{-1}$ does help the conditioning, but we also need to be able to deal with unconstrained least squares when $\mathbf{S}_a$ is not used. Fortunately QR decomposition allows us to solve the equations without even evaluating $\mathbf{W}_2$, saving even more time. Consider first the simplest form of the normal equations for least squares, $(\mathbf{K}^T \mathbf{K})\mathbf{x} = \mathbf{K}^T \mathbf{y}$, where $m > n$. The QR decomposition allows $\mathbf{K}$ to be put into the form $\mathbf{K} = \mathbf{QT}$ where $\mathbf{Q}$ is an $m \times n$ orthogonal matrix, $\mathbf{Q}^T \mathbf{Q} = \mathbf{I}_n$, and $\mathbf{T}$ is an upper triangular matrix*. Thus the normal equations become

$$\mathbf{T}^T \mathbf{T} \mathbf{x} = \mathbf{T}^T \mathbf{Q}^T \mathbf{y} \tag{5.41}$$

We can see that $\mathbf{T}$ is the Cholesky decomposition of $\mathbf{K}^T \mathbf{K}$, but computed without explicitly evaluating $\mathbf{K}^T \mathbf{K}$. If $\mathbf{K}$ is of full rank n, $\mathbf{T}$ will be nonsingular so we can write this as $\mathbf{T}\mathbf{x} = \mathbf{Q}^T \mathbf{y}$ and solve it by back substitution. Like the Gaussian triangulation, the matrix $\mathbf{Q}$ is implemented as a sequence of n transformations (Householder transformations) which are applied to both $\mathbf{K}$ and $\mathbf{y}$, and does not have to be stored explicitly as a matrix. The QR decomposition of $\mathbf{K}$ takes $mn^2 - n^3/3$ operations, and the solution of the normal equations takes $2mn^2 - 2n^3/3$, while Cholesky takes $mn^2 + n^3/3$, including the $mn^2/2$ required to evaluate $\mathbf{K}^T \mathbf{K}$. QR is faster for $m < n$, but the speeds are comparable, and QR is more robust for the least squares problem.

Our problem needs a little transformation to put it into the normal form, because of the *a priori* and the covariance matrices. We can write Eq. (5.37) in the form

$$(\mathbf{K}_i^T \mathbf{S}_\epsilon^{-1} \mathbf{K}_i + \mathbf{S}_a^{-1})(\mathbf{x}_{i+1} - \mathbf{x}_a) = \mathbf{K}_i^T \mathbf{S}_\epsilon^{-1} \delta \mathbf{y}_i \tag{5.42}$$

where $\delta \mathbf{y}_i = \mathbf{y} - \mathbf{F}(\mathbf{x}_i) + \mathbf{K}_i(\mathbf{x}_i - \mathbf{x}_a)$, or as the $(m+n) \times n$ problem

$$\begin{pmatrix} \mathbf{S}_\epsilon^{-\frac{1}{2}} \mathbf{K}_i \\ \mathbf{S}_a^{-\frac{1}{2}} \end{pmatrix} (\mathbf{x}_{i+1} - \mathbf{x}_a) = \begin{pmatrix} \mathbf{S}_\epsilon^{-\frac{1}{2}} \delta \mathbf{y}_i \\ 0 \end{pmatrix} \tag{5.43}$$

to be solved by least squares. Note that for a nondiagonal $\mathbf{S}$ it is more efficient to use the Cholesky decomposition for the square roots, $\mathbf{S}^{-\frac{1}{2}} = \mathbf{T}^{-T}$ (using an obvious notation for the inverse of a transpose) rather than the eigenvector expression. The least squares problem is then solved by QR with pivoting.

After the iteration has converged we will have available a Cholesky decomposition of $\hat{\mathbf{S}}^{-1}$ which can be used to produce $\hat{\mathbf{S}}$ in n^3 operations.

---

*The notation $\mathbf{T}$ is used, rather than the traditional $\mathbf{R}$, for consistency and because $\mathbf{R}$ is used here for right eigenvectors.

It may be that the problem is still poorly conditioned, in which case the last resort of singular vector decomposition may be required, which significantly more expensive than any of the above approaches. However this is a sign of a fundamentally ill-posed problem for which there is no satisfactory solution, and it is preferable to consider whether the problem is better regularised with the help of *a priori* constraints.

5.8.1.2 *The m-form*

In the m-form each iteration requires the evaluation of:

$$\mathbf{x}_{i+1} = \mathbf{x}_a + \mathbf{S}_a \mathbf{K}_i^T (\mathbf{S}_\epsilon + \mathbf{K}_i \mathbf{S}_a \mathbf{K}_i^T)^{-1} [\mathbf{y} - \mathbf{F}(\mathbf{x}_i) + \mathbf{K}_i (\mathbf{x}_i - \mathbf{x}_a)] \qquad (5.44)$$

and the final stage also requires the computation of the covariance:

$$\hat{\mathbf{S}} = \mathbf{S}_a - \mathbf{S}_a \mathbf{K}_i^T (\mathbf{S}_\epsilon + \mathbf{K}_i \mathbf{S}_a \mathbf{K}_i^T)^{-1} \mathbf{K}_i \mathbf{S}_a \qquad (5.45)$$

It is clear that the fastest method will involve a Cholesky solution of the equation

$$(\mathbf{S}_\epsilon + \mathbf{K}_i \mathbf{S}_a \mathbf{K}_i^T) \mathbf{z} = [\mathbf{y} - \mathbf{F}(\mathbf{x}_i) + \mathbf{K}_i (\mathbf{x}_i - \mathbf{x}_a)] \qquad (5.46)$$

taking $m^3/6$ operations. The calculation of $\mathbf{K}_i \mathbf{S}_a \mathbf{K}_i^T$, which must be done first, is more expensive, taking $n^2 m + nm^2/2$. After the decomposition, $\mathbf{x}_{i+1} = \mathbf{x}_a + \mathbf{S}_a \mathbf{K}_i^T \mathbf{z}$ will take only a further nm operations, because $\mathbf{S}_a \mathbf{K}_i^T$ will already be available. After convergence of the iteration, the evaluation of $\hat{\mathbf{S}}$ is straightforward if the decomposition of $(\mathbf{S}_\epsilon + \mathbf{K}_i \mathbf{S}_a \mathbf{K}_i^T)$ from the last stage has been preserved.

5.8.1.3 *Sequential updating*

If the measurement error covariance is diagonal, as it often is, then there is a further economy available in the case of the m-form. Within each iteration cycle, the state estimate can be updated sequentially, one measurement at a time, thus replacing the matrix inverse in Eq. (5.44) by a scalar reciprocal. The process is as follows, where the column vector $\mathbf{k}_j$ is the weighting function for the j-th channel, i.e. the j-th row of $\mathbf{K}$, and superscript i refers to the iteration cycle.

$$\begin{aligned}
&\mathbf{x}_0^{i+1} := \mathbf{x}_a \\
&\mathbf{S}_0 := \mathbf{S}_a \\
&\text{for } j := 1 \text{ to } m \text{ do}: \\
&\quad \begin{cases} \mathbf{x}_j^{i+1} := \mathbf{x}_{j-1}^{i+1} + \mathbf{S}_{j-1} \mathbf{k}_j [y_j - F_j(\mathbf{x}_l) - \mathbf{k}_i^T (\mathbf{x}_l - \mathbf{x}_a)] / (\mathbf{k}_j^T \mathbf{S}_{j-1} \mathbf{k}_j + \sigma_j^2) \\ \mathbf{S}_j := \mathbf{S}_{j-1} - \mathbf{S}_{j-1} \mathbf{k}_j \mathbf{k}_j^T \mathbf{S}_{j-1} / (\mathbf{k}_j^T \mathbf{S}_{j-1} \mathbf{k}_j + \sigma_j^2) \end{cases} \\
&\mathbf{x}^{i+1} := \mathbf{x}_m^{i+1}
\end{aligned} \qquad (5.47)$$

For each iteration cycle, the estimate starts with the *a priori*, and updates it channel by channel. The choice of linearisation point $\mathbf{x}_l$ for each of these updates determines the convergence rate. The best linearisation point would of course be the final solution, but this is not known initially. For the first iteration cycle, the best we can do for channel j is the current estimate $\mathbf{x}^1_{j-1}$. For subsequent cycles, the end point of the previous cycle, $\mathbf{x}^i$ is likely to be better.

This has two potential advantages over updating with a vector of measurements. Some measurements are likely to be more linearly related to the state vector than others; if they are assimilated first in the initial iteration cycle, then the intermediate state will be closer to the final solution when the more non-linear measurements are used, so the linearisation will be more accurate, and fewer iterations may be needed. The linear algebra computations may be faster, depending on the number of iterations required, because we no longer need to solve linear equations (or invert a matrix). As described, the derivatives of the forward model are computed channel by channel, at least in iteration cycle 1, which may be less efficient than computing them all together outside the m loop, as time can often be saved when computing several derivatives in parallel. If this is significant they can be computed in parallel outside the m loop, less accurately, but no worse than the parallel methods described in the previous two sections.

⇒ *Exercise* 5.4: Find the number of operations required, and determine when a sequential update is faster than updating with a vector of measurements

Straightforward sequential updating does not help in the n-form of the equations, mainly because it does not eliminate the explicit solution of linear equations. However for the m-form, the sequential updating operation count does not contain any m^2 or m^3 terms, so is in any case comparable with the non-sequential n-form method for the case $m > n$.

5.8.2 Computation of derivatives

In most cases the computation of the forward model and its derivatives will take far longer than the linear algebra, so it is important to pay careful attention to this aspect. Strategies for computing Jacobians or derivatives will be discussed in Chapter 9, where it will be shown that in most circumstances it is preferable to evaluate the algebraic derivative of the forward model code rather than to perturb the forward model for each element of the state vector and recompute the forward model m times.

If the derivative computation is particularly expensive, it may be possible to approximate it or to minimise the number of times it is evaluated. For example:

- Start at $\mathbf{x}_0 = \mathbf{x}_a$, and use $\mathbf{K}_0$ for $\mathbf{K}_i$.
- Compute $\mathbf{G}_0$ once, according to the m- or n-form as appropriate, and use it for every iteration.
- $\mathbf{F}(\mathbf{x}_i)$ will converge to approximately $\mathbf{y}$, but $\mathbf{x}_i$ will not converge to the correct optimal solution. For further discussion of the penalty for this approximation, see section 6.5.3.
- Convergence will be first order.

However, this non-optimal solution may be close enough for practical purposes, or it may be close enough that one more iteration, evaluating the derivative this time, will give the optimal solution.

5.8.3 *Optimising representations*

For numerical efficiency the state should be represented in terms of as small a number of parameters as possible, consistent with the rank of the weighting function matrix, and the number of degrees of freedom for signal. However the representation should be chosen so that the result does not have non-physical or misleading features, particularly in the case of profiles. The kind of representations which are usually unsuitable are polynomials, which are poorly constrained at the extremes, or low order linear splines (linear interpolation between a small number of levels) which have discontinuous changes of gradient at arbitrary altitudes.

One way of constructing a meaningful efficient representation is to start with a high resolution representation for which an *a priori* state and covariance can be obtained, and to 'represent the representation'. For example, consider a profile represented initially by a vector $\mathbf{x}_N$ of values at a large number N of levels together with an even larger $N \times N$ prior covariance matrix $\mathbf{S}_N$, where N is chosen so that any departure of the true profile from an interpolation between the values of the representation is physically insignificant. Consider transforming $\mathbf{x}_N$ to a representation $\mathbf{z}$ in terms of the prior state and eigenvectors of $\mathbf{S}_N$:

$$\mathbf{z} = \mathbf{L}^T(\mathbf{x}_N - \mathbf{x}_a) \quad \text{and} \quad \mathbf{x}_N = \mathbf{x}_a + \mathbf{L}\mathbf{z} \qquad (5.48)$$

where $\mathbf{S}_N \mathbf{L} = \mathbf{L}\mathbf{\Lambda}$. In this representation the prior covariance of $\mathbf{z}$ is

$$\mathbf{S}_z = \mathbf{L}^T \mathbf{S}_N \mathbf{L} = \mathbf{\Lambda} \qquad (5.49)$$

a diagonal matrix. It is often the case that many of the eigenvalues λ_i are small, which may be interpreted to mean that the components of $\mathbf{x}_N$ due to the corresponding eigenvectors have little variability in the *a priori*, and are already well defined, so that they need not be retrieved. These will often refer to fine scale structure in the profile. A representation in terms of the elements of $\mathbf{z}$ which corresponding to the larger eigenvalues may be adequate. This is known as an *empirical orthogonal function* (EOF) representation, with the eigenvectors as the orthogonal functions, empirical because the covariance is normally based on prior measurements of the state, and not on its theoretical behaviour. The total variance of the *a priori* is $\text{tr}(\mathbf{S}_N) = \text{tr}(\mathbf{\Lambda}) = \sum_1^N \lambda_i$, and the variance explained by the first n orthogonal functions is $\sum_1^n \lambda_i$, which may be as close as desired to the total variance by appropriate choice of n.

The empirical orthogonal function representation provides the most accurate representation of the true profile with a given number of parameters, provided the prior statistics are known. However it is possible to go further because significant components of the true profile may reside in the null space of the measurement, so they need not be included in the representation, as they cannot be measured. An even more efficient representation is in terms of the eigenvectors of the covariance of the ensemble of *retrievals* that corresponds to the *a priori* ensemble, using the characterisation of the solution given in Eq. (5.18). In the nearly-linear case, they can be evaluated from the weighting functions and the prior and noise covariances, but in the moderately non-linear case the ensemble covariance must be evaluated numerically because the averaging kernel depends on the state.

⇒ **Exercise** 5.5: Show that the covariance of an ensemble of retrievals corresponding to the *a priori* is, in the nearly-linear case:

$$\hat{\mathbf{S}}_a = \mathbf{S}_a \mathbf{K}^T (\mathbf{K} \mathbf{S}_a \mathbf{K}^T + \mathbf{S}_\epsilon)^{-1} \mathbf{K} \mathbf{S}_a \tag{5.50}$$

and relate it to the singular vectors of $\tilde{\mathbf{K}}$.

There are often good practical reasons for preferring a more direct representation of a profile in terms of values at some set of levels with a specified interpolation rule. The question then arises as to the minimum number of levels that is reasonable, consistent with being able to represent all of the information that is present in the original measurement. This is best done by trying out various options and comparing their error analyses and characterisations, but guidance can be found by examining the averaging kernel matrix. The degrees of freedom for signal is the trace of the averaging kernel matrix, and if calculated for a representation with a large number of levels, will give an indication of the number of layers which it is reasonable to use. The diagonal elements are a measure of resolution of the observing system, and can be used to indicate an approximate level spacing. This will be discussed further in Chapter 10.

Chapter 6
Approximations, Short Cuts and Ad-hoc Methods

In most circumstances the maximum amount of information will be extracted from a set of measurements when we use a full nonlinear retrieval which minimises a cost function based on all of the data and appropriate *a priori*, but these methods can be time consuming when done properly. There are occasions when an approximate and numerically fast method can be useful, for example:

(i) when the measurement contains more information than is needed to retrieve the state to the required accuracy and resolution;
(ii) for quick-look retrievals as a check in the field that an instrument is working correctly, when detailed retrievals will be carried out later;
(iii) to generate a good first guess for a detailed retrieval in cases where the cost function has a complicated topology, and a full retrieval would take a long time, and/or be likely to find a false minimum;
(iv) when the problem is so large that the computer time required is prohibitive.

A wide variety of suboptimal methods are to be found in the literature, some of which are useful, and some are not. Some are good approximations to optimal methods, and may have advantages of numerical efficiency. This chapter discusses several possibilities, to describe the nature of the solutions produced, and to point out the advantages and disadvantages of the various methods.

6.1 The Constrained Exact Solution

The exact solution was introduced in chapter 1 as a warning that simple solutions may cause problems, but there I did not discuss why, or in what circumstances. I describe this kind of solution as 'constrained' because the state is constrained to be represented by a p parameter linear representation, i.e. it is constrained to lie in an p-dimensional sub-space of state space, where p is the rank of $\mathbf{K}$ when expressed in a high resolution representation.

Writing the notation of chapter 1 in terms of vectors and matrices, we start with a high resolution representation $\mathbf{x}$, and constrain it with the aid of an $n \times m$ representation $\mathbf{W}$. Eq. (1.3) becomes:

$$\mathbf{x} = \mathbf{W}\mathbf{w} \qquad (6.1)$$

so that the forward model, Eq. (1.4), becomes

$$\mathbf{y} = \mathbf{K}\mathbf{x} + \boldsymbol{\epsilon} = \mathbf{K}\mathbf{W}\mathbf{w} + \boldsymbol{\epsilon} = \mathbf{C}\mathbf{w} + \boldsymbol{\epsilon} \qquad (6.2)$$

The solution in terms of $\mathbf{w}$ is

$$\hat{\mathbf{w}} = \mathbf{C}^{-1}\mathbf{y} = \mathbf{w} + \mathbf{C}^{-1}\boldsymbol{\epsilon} \qquad (6.3)$$

and in terms of $\mathbf{x}$, the equivalent of Eq. (1.5) is:

$$\hat{\mathbf{x}} = \mathbf{W}(\mathbf{K}\mathbf{W})^{-1}\mathbf{y} = \mathbf{W}(\mathbf{K}\mathbf{W})^{-1}\mathbf{K}\mathbf{x} + \mathbf{W}(\mathbf{K}\mathbf{W})^{-1}\boldsymbol{\epsilon} \qquad (6.4)$$

The problem lies in the solution of the matrix $\mathbf{C} = \mathbf{K}\mathbf{W}$. If this matrix is ill-conditioned, then the term $\mathbf{C}^{-1}\boldsymbol{\epsilon}$ may be large.

⇒ **Exercise** 6.1: Explain why the averaging kernel $\mathbf{A} = \mathbf{W}(\mathbf{K}\mathbf{W})^{-1}\mathbf{K}$ is likely to be well behaved in principle, apart from possible problems due to numerical rounding errors.

To examine the circumstances in which ill conditioning happens, express $\mathbf{C}$ in terms of its eigenvector decomposition, $\mathbf{C} = \mathbf{R}\boldsymbol{\Lambda}\mathbf{L}^T$, (a similar analysis can be carried out in terms of a singular vector decomposition):

$$\hat{\mathbf{x}} = \mathbf{W}(\mathbf{K}\mathbf{W})^{-1}\mathbf{K}\mathbf{x} + \mathbf{W}\mathbf{R}\boldsymbol{\Lambda}^{-1}\mathbf{L}^T\boldsymbol{\epsilon} \qquad (6.5)$$

If $\mathbf{C}$ has small eigenvalues, there will be potentially large contributions to the solution from the corresponding components of $\boldsymbol{\Lambda}^{-1}\mathbf{L}^T\boldsymbol{\epsilon}$. Strictly, because $\boldsymbol{\Lambda}$ has dimensions, 'small' and 'large' need qualifying. The important quantity for describing the stability of a matrix to inversion is the *condition number* which is the ratio of the absolute values of the largest and smallest eigenvalues. For the case discussed in chapter 1, an exact retrieval in terms of polynomials for the standard example, the smallest eigenvalue of $\mathbf{C}$ is 1.21×10^{-7} and its condition number is 1.29×10^7, so it is not surprising that the solution was unstable.

Exact solutions are only viable in cases where the matrix to be inverted is full rank and well conditioned. This will depend in detail on the weighting functions and the representation, and can only be determined by evaluating the condition number, but as a general guide there is likely to be a problem if there is significant overlap of the weighting functions so that there are linear combinations of weighting functions which are approximately zero.

The above analysis applies to the case of an arbitrary $n \times m$ representation. It is pertinent to consider the extent to which the representation is important, and whether there is a representation which minimises the instability of the solution. The polynomial representation of chapter 1 is clearly unsatisfactory because z^n

tends to infinity as $z \to \pm\infty$, and as a result the extremes of the retrieved profile can flop around wildly. We can choose a more satisfactory representation by minimising some measure of retrieval noise with respect to the representation function.

In practice, minimising with respect to $\mathbf{W}$ is tricky, because any non-singular linear combination $\mathbf{L}$ of the representation functions would give the same result, i.e. $\mathbf{G} = \mathbf{W}(\mathbf{KW})^{-1} = \mathbf{WL}(\mathbf{KWL})^{-1}$, so there is no unique solution. To avoid this mathematical difficulty, find instead the gain matrix $\mathbf{G}$ that minimises retrieval noise variance, subject to the condition that an exact solution is produced, i.e. $\mathbf{KG} = \mathbf{I}_m$. The retrieval noise variance at level i is $\sigma_i^2 = [\mathbf{GS}_\epsilon \mathbf{G}^T]_{ii}$, which simplifies to $\sigma_i^2 = \sigma^2 \sum_k G_{ik}^2$ on the assumption, for the purpose of illustration, that $\mathbf{S}_\epsilon = \sigma^2 \mathbf{I}_m$. The quantity minimised must be a scalar, so we use the total variance, and solve:

$$\frac{\partial}{\partial G_{ij}} \left[\sum_{ik} G_{ik}^2 + \sum_{lm} \gamma_{lm} \sum_n K_{ln} G_{nm} \right] = 0. \tag{6.6}$$

This clearly minimises each σ_i^2 separately, as each corresponds to a different row of $\mathbf{G}$. It has been written in components to clarify that a matrix of Lagrangian multipliers is needed, one for each element of the constraint. Carrying out the derivative we get

$$2G_{ij} + \sum_l \gamma_{lj} K_{li} = 0, \tag{6.7}$$

or in matrix notation

$$\mathbf{G} = \tfrac{1}{2}\mathbf{K}^T \mathbf{\Gamma}. \tag{6.8}$$

Substituting into the constraint $\mathbf{KG} = \mathbf{I}_m$, we obtain

$$\tfrac{1}{2}\mathbf{KK}^T \mathbf{\Gamma} = \mathbf{I}_m \tag{6.9}$$

so that the retrieval that minimises the solution error variance is

$$\hat{\mathbf{x}} = \mathbf{K}^T (\mathbf{KK}^T)^{-1} \mathbf{y}. \tag{6.10}$$

Note that this is the solution that would be obtained if the weighting functions, or any non-singular combination of them, were used as a representation, e.g. $\mathbf{W} = \mathbf{K}^T$. Surprisingly we obtain the same result in the case of a general measurement noise covariance matrix.

⇒ **Exercise 6.2**: Show this. *Hint*: transform the forward model so that the noise covariance is a unit matrix.

This form of solution might have been expected on the grounds that $\mathbf{K}$ defines the non-null space, and anything outside this does not contribute to the measurement.

Fig. 6.1(a) shows a set of contribution functions using this approach for our standard case, for comparison with the result for polynomials shown in Fig. 1.3(a).

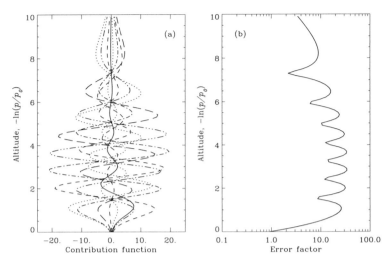

Fig. 6.1 Diagnostics for an exact retrieval in the standard case when the profile is represented as a linear combination of the weighting functions. (a) Contribution functions. (b) Error amplification factor.

Notice that the large values at the top and bottom are no longer present, but the values in the mid-range are of a similar size, and perhaps a little larger. Fig. 6.1(b) shows the corresponding error amplification factor, $\sum_k G_{ik}^2$. This is smaller than that in Fig. 1.3(b) at the extremes, at the expense of being somewhat larger in the mid-range. The integrated error amplification, however, is smaller. The matrix to be inverted, $\mathbf{KK}^T$, is better conditioned than in the polynomial case, the condition number being only(!) 2500. Fig. 6.2 shows the result of a retrieval simulation to compare with Fig. 1.2, for both the noise-free case and when a simulated noise of 0.5 K was added to the simulated measurements. It is clear that the noise sensitivity is considerably improved, at the top and bottom, and not in the middle, but the representation is not as successful as a polynomial at approximating the U.S. Standard Atmosphere even though it reproduces the measured signals exactly!

This noise sensitivity is a common feature of exact solutions to retrieval problems, especially when there is significant overlap between the weighting functions. In the extreme case of complete overlap, where two channels have identical weighting functions, noise will cause them to give slightly different signals, and no solution can possibly satisfy both pieces of data. When there is as much overlap of weighting functions as shown in Fig. 1.1, we would expect that the weighting function for signal i would be very similar to the average of the weighting functions for signals $i-1$ and $i+1$, so that the signals themselves ought to be similar. In this case a less extreme phenomenon will occur, the symptom usually being large noise sensitivity, rather than the non-existence of a solution.

There is clearly no sensible reason to try to find an exact solution. All that can be reasonably expected is that the solution will agree with the measurements within

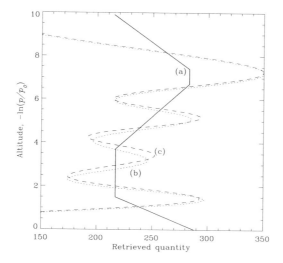

Fig. 6.2 Simulated retrievals using the weighting functions of Fig. 1.1 and representation as a linear combination of weighting functions. (a) The original profile; (b) An exact retrieval with no simulated experimental error; (c) An exact retrieval with a 0.5 K simulated experimental error.

experimental error. This gives more freedom in choosing a solution, for now there will be an infinite number of possibilities, so we should find some way of deciding which solution is best.

6.2 Least Squares Solutions

6.2.1 *The overconstrained case*

A common approach to an ill-conditioned problem is to reduce the number of variables that we are trying to fit, and to use a least squares approach. Thus the profile representation becomes, for our standard example:

$$B[\bar{\nu}, \theta(z)] = \sum_{j=1}^{n} x_j W_j(z) \quad n < m \tag{6.11}$$

and the problem to be solved is now overconstrained, with more equations (m) than unknowns (n):

$$y_i = L(\nu_i) = \sum_{j=1}^{n} x_j \int_0^\infty W_j(z) K_i(z) dz = \sum_{j=1}^{n} K_{ij} x_j \quad i = \ldots m, \tag{6.12}$$

thus defining the relation between the weighting function matrix **K** and the original weighting functions $K_i(z)$.

In the fundamentally underconstrained case, such as our standard example, the result is an improvement over the exact solution for a given representation because its noise sensitivity is better, at the expense of a reduced capacity to represent a profile. However, there are no obvious criteria for determining such things as how many terms to use, and what form of representation is suitable. The least squares method is really only appropriate for problems which are fundamentally overconstrained, where the measurement vector has considerably more elements than the state vector, and perhaps where the algebraic form of the solution is known from sound physical reasoning.

In the case where there are more measurements than unknowns, $m > n$, an exact solution is not possible in general. Therefore we look for a solution that minimises the sum of the squares of the differences between the actual measurements and those calculated from the forward model using the solution. That is, we minimise:

$$[\mathbf{y} - \mathbf{F}(\mathbf{x})]^T[\mathbf{y} - \mathbf{F}(\mathbf{x})] \quad \text{or} \quad (\mathbf{y} - \mathbf{Kx})^T(\mathbf{y} - \mathbf{Kx}). \tag{6.13}$$

In the linear case a derivative with respect to $\mathbf{x}$ leads immediately to the familiar *normal equations*:

$$\hat{\mathbf{x}} = (\mathbf{K}^T\mathbf{K})^{-1}\mathbf{K}^T\mathbf{y}. \tag{6.14}$$

⇒ ***Exercise*** 6.3: Derive this solution.

In the nonlinear case there are various ways of proceeding numerically. A simple one is to linearise about some estimate of the solution $\mathbf{x}_i$, and use the resulting normal equations to improve the solution, iterating as far as required:

$$\mathbf{x}_{i+1} = \mathbf{x}_i + (\mathbf{K}_i^T\mathbf{K}_i)^{-1}\mathbf{K}_i^T[\mathbf{y} - \mathbf{F}(\mathbf{x}_i)]. \tag{6.15}$$

Note the similarity to the optimal solutions. The normal equations can be obtained by putting $\mathbf{S}_a^{-1} = \mathbf{O}$ and $\mathbf{S}_\epsilon = \mathbf{I}_m$. If a weighted least squares is used, then the quantity to be minimised is

$$[\mathbf{y} - \mathbf{F}(\mathbf{x})]^T\mathbf{S}_\epsilon^{-1}[\mathbf{y} - \mathbf{F}(\mathbf{x})] \quad \text{or} \quad (\mathbf{y} - \mathbf{Kx})^T\mathbf{S}_\epsilon^{-1}(\mathbf{y} - \mathbf{Kx}) \tag{6.16}$$

and the similarity is even closer. It is the same as the optimal solution in the case when there is no *a priori* information, i.e. it is the maximum likelihood solution.

6.2.2 The underconstrained case

In the underconstrained case we can develop a different kind of least squares solution. If the rank of the weighting functions is smaller than the number of unknowns, $p < n$, there are an infinite number of exact solutions. In this case we can look for the exact solution whose length is shortest according to a least squares criterion. We can think of this as being the smoothest solution in the sense that

$$\mathbf{x}^T\mathbf{x} \quad \text{or} \quad (\mathbf{x} - \mathbf{x}_a)^T(\mathbf{x} - \mathbf{x}_a) \tag{6.17}$$

is minimum, subject to the constraint that

$$\mathbf{y} = \mathbf{K}\mathbf{x} \quad \text{or} \quad \mathbf{y} = \mathbf{F}(\mathbf{x}). \tag{6.18}$$

Therefore we evaluate in the linear case:

$$\frac{\partial}{\partial \mathbf{x}}\left[\mathbf{x}^T \mathbf{x} + \boldsymbol{\gamma}^T (\mathbf{y} - \mathbf{K}\mathbf{x})\right] = 0, \tag{6.19}$$

where $\boldsymbol{\gamma}$ is a vector of undetermined multipliers, giving

$$\mathbf{x} = \tfrac{1}{2}\mathbf{K}^T \boldsymbol{\gamma}. \tag{6.20}$$

Substituting back in the constraint gives

$$\hat{\mathbf{x}} = \mathbf{K}^T (\mathbf{K}\mathbf{K}^T)^{-1} \mathbf{y}, \tag{6.21}$$

and in the nonlinear case we can iterate

$$\mathbf{x}_{i+1} = \mathbf{x}_i + \mathbf{K}_i^T (\mathbf{K}_i \mathbf{K}_i^T)^{-1} [\mathbf{y} - \mathbf{F}(\mathbf{x}_i)]. \tag{6.22}$$

Again note the similarity to the optimal solutions. This form can be obtained by putting $\mathbf{S}_a = \mathbf{I}_n$ and $\mathbf{S}_\epsilon = \mathbf{O}$ in the optimal solution. Not also that this solution is identical to the exact solution with minimum solution error variance, Eq. (6.10). We have found a solution which lies entirely in the $\mathbf{K}$-space, its length is minimum because there is no contribution to the length from components in the null space.

We can generalise the concept of smoothness with a weighted least squares, minimising for example, $(\mathbf{x} - \mathbf{x}_a)^T \mathbf{S}_a^{-1} (\mathbf{x} - \mathbf{x}_a)$, in which case the similarity is even closer.

However there is still no good reason to try to fit noise exactly, even if we do it with the smoothest possible solution. We should be looking for the smoothest solution within measurement error of the measurement, which of course leads us back to the optimal solution of previous chapters.

6.3 Truncated Singular Vector Decomposition

The error analysis in section 6.1 leads to the suggestion of using the eigenvector expansion, approximating $\mathbf{C}^{-1}$ by a truncated eigenvector expansion:

$$\mathbf{C}^{-1} \simeq \mathbf{R}_t \mathbf{\Lambda}_t^{-1} \mathbf{L}_t^T \tag{6.23}$$

where the subscript t indicates that only the t largest eigenvalues and their corresponding eigenvectors have been retained, and the rest dropped. Thus $\mathbf{R}_t$ and $\mathbf{L}_t$ are not square. Eigenvalues to be dropped would be those which contribute more to the error term than to the profile.

It is more convenient to use the alternative formulation with singular vectors, mainly because singular vectors are orthogonal. Put

$$\mathbf{C}^{-1} \simeq \mathbf{U}'_t \mathbf{\Lambda}_t^{-2} \mathbf{V}'^T_t, \tag{6.24}$$

where $\mathbf{U}'_t$ and $\mathbf{V}'_t$ are matrices of left and right singular vectors truncated to t vectors, and Λ_t^2 is the truncated matrix of eigenvalues of $\mathbf{C}$. This gives

$$\hat{\mathbf{x}} = \mathbf{W}\mathbf{U}'_t \mathbf{\Lambda}_t^{-2} \mathbf{V}'^T_t \mathbf{y} \tag{6.25}$$
$$= \mathbf{W}\mathbf{U}'_t \mathbf{\Lambda}_t^{-2} \mathbf{V}'^T_t \mathbf{K}\mathbf{x} + \mathbf{W}\mathbf{U}'_t \mathbf{\Lambda}_t^{-1} \mathbf{V}'^T_t \boldsymbol{\epsilon}. \tag{6.26}$$

If $\boldsymbol{\epsilon}$ has covariance $\sigma_\epsilon^2 \mathbf{I}$ the error term will have covariance $\sigma_\epsilon^2 \mathbf{W}\mathbf{U}'_t \mathbf{\Lambda}_t^{-2} \mathbf{U}'_t \mathbf{W}^T$, each singular vector contributing an independent error pattern equal to a column of $\mathbf{W}\mathbf{U}'_t$ multiplied by a random normal variable with standard deviation $(\sigma_\epsilon / \lambda_i)$.

Rather than retaining a general representation function let us simplify matters by only considering $\mathbf{W} = \mathbf{K}^T$. In this case $\mathbf{C} = \mathbf{K}\mathbf{K}^T$ is symmetric and its singular vectors and eigenvectors are identical. Thus $\mathbf{U}'$ and $\mathbf{V}'$ are equal, and equal to the left singular vectors of $\mathbf{K}$, $\mathbf{K} = \mathbf{U}\mathbf{\Lambda}\mathbf{V}^T$. Diagonal elements of $\mathbf{\Lambda}^2$ are the squares of the singular values of $\mathbf{K}$. In this case

$$\hat{\mathbf{x}} = \mathbf{K}^T \mathbf{U}_t \mathbf{\Lambda}_t^{-2} \mathbf{U}_t^T \mathbf{y} \tag{6.27}$$
$$= \mathbf{V}_t \mathbf{\Lambda}_t^{-1} \mathbf{U}_t^T \mathbf{y} \tag{6.28}$$
$$= \sum_{i=1}^{t} \lambda_i \mathbf{v}_i \mathbf{u}_i^T \mathbf{y} \tag{6.29}$$

because $\mathbf{K}^T \mathbf{U}_t^T = \mathbf{V}_t \mathbf{\Lambda}_t$. Note that Eq. (6.28) is equivalent to using a truncation of the pseudo inverse $\mathbf{K}^* = \mathbf{V}\mathbf{\Lambda}^{-1}\mathbf{U}^T$ based on $\mathbf{K} = \mathbf{U}\mathbf{\Lambda}\mathbf{V}^T$. The relation of the retrieval to the state vector is therefore

$$\hat{\mathbf{x}} = \mathbf{V}_t \mathbf{\Lambda}_t^{-1} \mathbf{U}_t^T \mathbf{K}\mathbf{x} + \mathbf{V}_t \mathbf{\Lambda}_t^{-1} \mathbf{U}_t^T \boldsymbol{\epsilon} \tag{6.30}$$
$$= \mathbf{V}_t \mathbf{V}_t^T \mathbf{x} + \mathbf{V}_t \mathbf{\Lambda}_t^{-1} \mathbf{U}_t^T \boldsymbol{\epsilon} \tag{6.31}$$

showing that the averaging kernel matrix is $\mathbf{A} = \mathbf{V}_t \mathbf{V}_t^T$.

6.4 Twomey–Tikhonov

The first methods applied to the retrieval problem in which due consideration was given to error sensitivity and constraints were published at about the same time by Twomey (1963) and by Tikhonov (1963). Both methods were essentially the same, involving the minimisation of a cost function involving departures of the solution from the measurements and of the solution from some *a priori*:

$$(\mathbf{x} - \mathbf{x}_a)^T \mathbf{H}(\mathbf{x} - \mathbf{x}_a) + \gamma (\mathbf{y} - \mathbf{K}\mathbf{x})^T (\mathbf{y} - \mathbf{K}\mathbf{x}) \tag{6.32}$$

where the first term represents some weighted departure from an *a priori* $\mathbf{x}_a$, and the second term constrains the solution to approximately fit the measurements. The factor γ is chosen to give appropriate relative weighting to the two constraints.

In Twomey's method, the matrix $\mathbf{H}$ can be chosen to minimise for example the mean squared difference between $\mathbf{x}$ and $\mathbf{x}_a$ ($\mathbf{H} = \mathbf{I}_n$) or the mean squared second difference for a smooth solution. It could clearly also be chosen to minimise the logarithm of an *a priori* probability density function, making it similar to a statistically optimal method. It would only need γ to be replaced by an inverse error covariance matrix. For example the $\mathbf{H}$-matrix for the squared second difference is, for order 10, of the form

$$\begin{pmatrix} 1 & -2 & 1 & 0 & 0 & 0 & 0 & 0 & 0 & 0 \\ -2 & 5 & -4 & 1 & 0 & 0 & 0 & 0 & 0 & 0 \\ 1 & -4 & 6 & -4 & 1 & 0 & 0 & 0 & 0 & 0 \\ 0 & 1 & -4 & 6 & -4 & 1 & 0 & 0 & 0 & 0 \\ 0 & 0 & 1 & -4 & 6 & -4 & 1 & 0 & 0 & 0 \\ 0 & 0 & 0 & 1 & -4 & 6 & -4 & 1 & 0 & 0 \\ 0 & 0 & 0 & 0 & 1 & -4 & 6 & -4 & 1 & 0 \\ 0 & 0 & 0 & 0 & 0 & 1 & -4 & 6 & -4 & 1 \\ 0 & 0 & 0 & 0 & 0 & 0 & 1 & -4 & 5 & -2 \\ 0 & 0 & 0 & 0 & 0 & 0 & 0 & 1 & -2 & 1 \end{pmatrix} \qquad (6.33)$$

This matrix is singular (its columns sum to zero), so it cannot be thought of as the inverse of some covariance matrix, but it can be taken to be an information matrix. If the state vector is expressed in a polynomial representation, this constraint will be found not to constrain the constant and linear coefficient. The solution is

$$\hat{\mathbf{x}} = \mathbf{x}_a + (\gamma^{-1}\mathbf{H} + \mathbf{K}^T\mathbf{K})^{-1}\mathbf{K}^T(\mathbf{y} - \mathbf{K}\mathbf{x}_a) \qquad (6.34)$$

which bears a stronger resemblance to the MAP approach than either of the least squares methods, the only difference being the interpretation of the constraint matrices.

Some insight into how this method works can be obtained by using a singular vector decomposition (Mateer, 1965). Consider just the case $\mathbf{H} = \mathbf{I}_n$, and substitute the singular value decomposition $\mathbf{K} = \mathbf{U}\mathbf{\Lambda}\mathbf{V}^T$, noting that $\mathbf{K}$ need not be square:

$$\hat{\mathbf{x}} = \mathbf{x}_a + (\gamma^{-1}\mathbf{I}_n + \mathbf{V}\mathbf{\Lambda}^2\mathbf{V}^T)^{-1}\mathbf{V}\mathbf{\Lambda}\mathbf{U}^T(\mathbf{y} - \mathbf{U}\mathbf{\Lambda}\mathbf{V}^T\mathbf{x}_a). \qquad (6.35)$$

Now transform state space to the basis defined by the columns of $\mathbf{V}$, and measurement space to the basis defined by the columns of $\mathbf{U}$, i.e. $\mathbf{x}' = \mathbf{V}^T\mathbf{x}$ and $\mathbf{y}' = \mathbf{U}^T\mathbf{y}$. In these bases, the forward model becomes $\mathbf{y}' = \mathbf{\Lambda}\mathbf{x}' + \boldsymbol{\epsilon}'$, where $\boldsymbol{\epsilon}' = \mathbf{U}^T\boldsymbol{\epsilon}$, and the solution is

$$\hat{\mathbf{x}}' = \mathbf{x}'_a + (\gamma^{-1}\mathbf{I}_n + \mathbf{\Lambda}^2)^{-1}\mathbf{\Lambda}(\mathbf{y}' - \mathbf{\Lambda}\mathbf{x}'_a). \qquad (6.36)$$

This falls apart into separate equations, one for each element:

$$\hat{x}'_i = x'_{ai} + \frac{\lambda_i}{\gamma^{-1} + \lambda_i^2}(y'_i - \lambda_i x'_{ai}) \qquad (6.37)$$

and we can see that elements of **y** for which $\lambda_i^2 \ll \gamma^{-1}$ contribute at reduced weight, elements for which $\lambda_i^2 \gg \gamma^{-1}$ contribute at full weight.

This bears some resemblance to the truncated singular vector decomposition method, except that in the SVD method the weights are either one or zero, with the choice determined by a criterion not unlike a comparison of λ_i^2 with γ^{-1}. The Twomey–Tikhonov method is simpler, in that it requires only matrix solution and not singular vector calculations, and also provides a rationale for the choice of cut-off, especially when a statistical interpretation of the matrices is used.

6.5 Approximations for Optimal Methods

Although some form of optimal method is usually the method of choice in solving an inverse problem, there are circumstances in which approximations can be made with little loss in the quality of the result, and a considerable gain in the simplicity of the method or in the computer resources needed.

All practical implementations of optimal methods contain approximations that are hard to avoid, such as finite dimensional representations, Gaussian probability density functions, forward model numerical methods, etc. These will not be discussed here, but we will consider the effect of approximating the *a priori* state, its covariance, the measurement error covariance and the weighting functions.

In general we will find that it is only possible to assess the effects of these approximations by simulation.

6.5.1 *Approximate* a priori *and its covariance*

Any retrieval problem for which there is a null space will need some kind of *a priori* information in order to estimate the null-space components of the retrieval. If an optimal estimator is being used, then the prior data is notionally of the form of a probability density function for the state, typically an expected value and a covariance for a Gaussian distribution. Very often such data is hard to obtain, and must be constructed from a variety of unsatisfactory sources, including the experimenters prejudice about the prior state. The construction and use of *a priori* data and its covariance will be discussed in detail in Chapter 11. Here, we will briefly touch on the effect of incorrect or inappropriate values *a priori*.

The primary reason for including *a priori* information in a retrieval is to provide an estimate of null-space components, and constrain near-null-space components of the retrieval in a reasonable way. If the *a priori* is inappropriate, these components and their errors will be incorrect. The sensitivity of the retrieval to an incorrect prior state is easy to compute. From Eq. (3.12) we obtain

$$\frac{\partial \hat{\mathbf{x}}}{\partial \mathbf{x}_a} = \mathbf{I}_n - \mathbf{A}. \qquad (6.38)$$

The sensitivity to an incorrect prior covariance is a much more complicated matter. If a component of the covariance implies too tight a constraint, the corresponding component of the retrieval will be biassed towards that of the *a priori* state, and if too loose, the retrieval error for the component and its estimate may be too large if the measurement does not provide the information. The effect is best explored by simulation in any particular case. For a further discussion, see Strand (1974).

6.5.2 *Approximate measurement error covariance*

It should always be possible in principle to estimate the error covariance of the measured signal, as this follows from the design of the instrument. It is a basic principle of experimental physics that a measurement without a good error estimate has little value. However, error covariance is sometimes difficult to estimate well in practice, and it can be numerically more efficient to use a diagonal covariance matrix rather than the full matrix including its off-diagonal elements (Exercise 3.2). Simply ignoring the off-diagonal elements is equivalent to abandoning some of the information content of the measurement because the entropy of the diagonal matrix will always be larger than that of the full matrix.

A common approximation is simply to ignore components of measurement error due to errors in forward model parameters, i.e. $\mathbf{K}_b \mathbf{S}_b \mathbf{K}_b^T$ of Eq. (4.33) when $\mathbf{b}$ is not retrieved. My only advice here is that you should avoid doing this, but if you must, evaluate the effects on the retrieval and its estimated error numerically.

⇒ **Exercise** 6.4: Derive expressions for the error resulting from making this approximation.

Both $\mathbf{S}_a$ and $\mathbf{S}_\epsilon$ are commonly used as 'tuning parameters' and adjusted by arbitrary factors to obtain an aesthetically pleasing retrieval. It goes without saying that any such use should be justifiable, and please remember that e.g. doubling both $\mathbf{S}_a$ and $\mathbf{S}_\epsilon$ will not change the retrieval, but only the error estimates.

6.5.3 *Approximate weighting functions*

If the weighting functions used in a linear retrieval are incorrect, then the retrieval will be incorrect, approximately by an amount $\mathbf{G}\delta\mathbf{K}\mathbf{x}$ for small errors $\delta\mathbf{K}$. For a better estimate, simply simulate the error.

The situation for a nonlinear iterated retrieval is more subtle, because a forward model is also involved, which we take to be correct. The usual reason for using an approximate weighting function is to save processing time, as suggested in section 5.8.2; the weighting function is computed for the first guess, and is not updated each time around the iteration cycle. The Gauss–Newton method Eq. (5.9) is of the form

$$\mathbf{x}_{i+1} = \mathbf{x}_a + \mathbf{G}_i[\mathbf{y} - \mathbf{F}(\mathbf{x}_i) + \mathbf{K}_i(\mathbf{x}_i - \mathbf{x}_a)]. \tag{6.39}$$

If $\mathbf{K}_i$ and $\mathbf{G}_i$ are kept constant at $\mathbf{K}_0$ and $\mathbf{G}_0$, the iteration becomes

$$\mathbf{x}_{i+1} = \mathbf{x}_a + \mathbf{G}_0[\mathbf{y} - \mathbf{F}(\mathbf{x}_i) + \mathbf{K}_0(\mathbf{x}_i - \mathbf{x}_a)] \tag{6.40}$$

and the solution converges to $\mathbf{x}_\infty$, satisfying the equation

$$\mathbf{x}_\infty = \mathbf{x}_a + \mathbf{G}_0[\mathbf{y} - \mathbf{F}(\mathbf{x}_\infty) + \mathbf{K}_0(\mathbf{x}_\infty - \mathbf{x}_a)]. \tag{6.41}$$

Take the difference between these two equations, and expand $\mathbf{F}(\mathbf{x}_i)$ about $\mathbf{x}_\infty$ to get the convergence equation

$$\mathbf{x}_{i+1} - \mathbf{x}_\infty = \mathbf{G}_0(\mathbf{K}_0 - \mathbf{K}_\infty)(\mathbf{x}_i - \mathbf{x}_\infty) + O(\mathbf{x}_i - \mathbf{x}_\infty)^2. \tag{6.42}$$

Therefore convergence is of first order, but at a rate which depends on $\mathbf{K}_0 - \mathbf{K}_\infty$. It will converge to a slightly wrong answer. To estimate how wrong, substitute $\mathbf{y} = \mathbf{F}(\mathbf{x}) + \epsilon$ and $\mathbf{G}_0 = (\mathbf{S}_a^{-1} + \mathbf{K}_0^T \mathbf{S}_\epsilon^{-1} \mathbf{K}_0)^{-1} \mathbf{K}_0^T \mathbf{S}_\epsilon^{-1}$ in Eq. (6.41), expand the forward model about $\mathbf{x}_\infty$, and rearrange to obtain the characterisation of the retrieval:

$$\mathbf{x}_\infty - \mathbf{x}_a = (\mathbf{S}_a^{-1} + \mathbf{K}_0^T \mathbf{S}_\epsilon^{-1} \mathbf{K}_\infty)^{-1} \mathbf{K}_0^T \mathbf{S}_\epsilon^{-1} [\mathbf{K}_\infty(\mathbf{x} - \mathbf{x}_a) + \epsilon] \tag{6.43}$$

showing that for the iteration as a whole the gain and averaging kernel matrices are

$$\begin{aligned}\mathbf{G}_\infty &= (\mathbf{S}_a^{-1} + \mathbf{K}_0^T \mathbf{S}_\epsilon^{-1} \mathbf{K}_\infty)^{-1} \mathbf{K}_0^T \mathbf{S}_\epsilon^{-1} \\ \mathbf{A}_\infty = \mathbf{G}\mathbf{K}_\infty &= (\mathbf{S}_a^{-1} + \mathbf{K}_0^T \mathbf{S}_\epsilon^{-1} \mathbf{K}_\infty)^{-1} \mathbf{K}_0^T \mathbf{S}_\epsilon^{-1} \mathbf{K}_\infty.\end{aligned} \tag{6.44}$$

This only requires that the forward model be linear between $\mathbf{x}$ and $\mathbf{x}_\infty$, so is valid for moderately nonlinear problems. $\mathbf{G}$ and $\mathbf{A}$ for the full iteration, when $\mathbf{K}$ is updated at each step, are of the same form but with $\mathbf{K}_0$ replaced by $\mathbf{K}_\infty$. The error due to keeping $\mathbf{K}$ constant depends on how close these versions of $\mathbf{A}$ and $\mathbf{G}$ are to the correct values:

$$\mathbf{x}_\infty - \hat{\mathbf{x}} = (\mathbf{A}_\infty - \mathbf{A})(\mathbf{x} - \mathbf{x}_a) + (\mathbf{G}_\infty - \mathbf{G})\epsilon. \tag{6.45}$$

Evaluating the size of this error for any particular case requires an evaluation of the weighting function at $\mathbf{x}_\infty$, which could then be used to carry out a final step of the iteration to improve the solution. Whether this would be faster than the full Gauss–Newton method depends on the balance between the cost of evaluation of $\mathbf{K}$ at each step and the cost of the increased number of iterations required by this first order method, and can only be determined by experiment.

6.6 Direct Multiple Regression

If the problem is nearly linear, and a sufficiently large sample of cases is available where remote measurements have been made in coincidence with direct measurements, for example from radiosondes, then a set of multiple regression coefficients can be computed to relate the state to the measurements.

Given an ensemble of pairs of measurements $\mathbf{y}$ and corresponding states $\mathbf{x}$, we wish to find the matrix $\mathbf{G}$ such that the state estimate $\hat{\mathbf{x}} = \bar{\mathbf{x}} + \mathbf{G}(\mathbf{y} - \bar{\mathbf{y}})$ minimises the mean square error in the estimate

$$\mathcal{E}\{(\mathbf{x} - \hat{\mathbf{x}})^T(\mathbf{x} - \hat{\mathbf{x}})\} = \mathcal{E}\{(\mathbf{x} - \bar{\mathbf{x}} - \mathbf{G}[\mathbf{y} - \bar{\mathbf{y}}])^T(\mathbf{x} - \bar{\mathbf{x}} - \mathbf{G}[\mathbf{y} - \bar{\mathbf{y}}])\}. \qquad (6.46)$$

Setting the derivative with respect to $\mathbf{G}$ equal to zero leads immediately to another version of the normal equations, compare section 4.2:

$$\mathbf{G} = \mathcal{E}\{(\mathbf{x} - \bar{\mathbf{x}})(\mathbf{y} - \bar{\mathbf{y}})^T\}(\mathcal{E}\{(\mathbf{y} - \bar{\mathbf{y}})(\mathbf{y} - \bar{\mathbf{y}})^T\})^{-1}. \qquad (6.47)$$

These expected values are covariance matrices, and may be estimated by taking means over the ensembles of data.

This process has been used for temperature sounding (Smith et al., 1970), by searching for coincidences between radiosonde measurements and satellite measurements, and computing regression coefficients. These were updated periodically, so that changes in instrument calibration were automatically allowed for, and the weighting functions did not need be evaluated. In this case it is necessary to deal with clouds specially, as they enter the equations in a very nonlinear way.

If we use only an ensemble of states, and compute the measurements from a linear forward model, $\mathbf{y} - \bar{\mathbf{y}} = \mathbf{K}(\mathbf{x} - \bar{\mathbf{x}}) + \boldsymbol{\epsilon}$, then it is easy to show that

$$\mathcal{E}\{(\mathbf{x} - \bar{\mathbf{x}})(\mathbf{y} - \bar{\mathbf{y}})^T\} = \mathbf{S}_e \mathbf{K}^T \quad \text{and} \quad \mathcal{E}\{(\mathbf{y} - \bar{\mathbf{y}})(\mathbf{y} - \bar{\mathbf{y}})^T\} = \mathbf{K}\mathbf{S}_e\mathbf{K}^T + \mathbf{S}_\epsilon \qquad (6.48)$$

where $\mathbf{S}_e = \mathcal{E}\{(\mathbf{x} - \bar{\mathbf{x}})(\mathbf{x} - \bar{\mathbf{x}})^T\}$ is the covariance of the ensemble of states. Thus

$$\mathbf{G} = \mathbf{S}_e \mathbf{K}^T (\mathbf{K}\mathbf{S}_e\mathbf{K}^T + \mathbf{S}_\epsilon)^{-1}, \qquad (6.49)$$

which is identical to the maximum *a posteriori* solution. Thus linear MAP solution can be regarded as a simulation of a multiple regression method where coefficients are found which best relate the state to the measurement.

If the forward model is truly linear, and good quality *in situ* measurements are available, then the regression method has the major advantage that it provides a convenient way of implementing the MAP solution without having to characterise the instrument with great care, and without having to implement an accurate forward model. However, few remote measurements are truly linear, and the result will be only as accurate as the independent measurements.

6.7 Linear Relaxation

A linear relaxation method is of the form

$$\mathbf{x}_{i+1} = \mathbf{x}_i + \mathbf{G}[\mathbf{y} - \mathbf{F}(\mathbf{x}_i)] \tag{6.50}$$

where the matrix $\mathbf{G}$ is chosen so that if $\mathbf{y} \neq \mathbf{F}(\mathbf{x}_i)$ then something is added to $\mathbf{x}_i$ in the right sense to reduce the difference. In the simplest form where the state is a profile and the weighting functions are peaked at different altitudes, as in our standard example, the state vector might be chosen to be the value of the profile at the peaks, with some specified interpolation method. Then $\mathbf{G}$ might be a diagonal matrix, so that if the measurement in one channel is incorrect, the profile is modified at the altitude corresponding to the peak of that channel. The values of the elements G_{ii} must be chosen so that the correction is of an appropriate amplitude. A more sophisticated choice would be to use a matrix that is an approximate inverse of $\mathbf{K}$, for example $\mathbf{G} = (\mathbf{K}_0^T \mathbf{K}_0 + \gamma \mathbf{I}_n)^{-1} \mathbf{K}_0^T$, based on the Twomey–Tikhonov method.

It should be noted that the solution will be of the form $\mathbf{x}_0 + \mathbf{G}\mathbf{a}$ where $\mathbf{a} = \sum_i \mathbf{y} - \mathbf{F}(\mathbf{x}_i)$, i.e. the difference from the first guess will be a linear combination of the columns of $\mathbf{G}$. This can be used to provide a constraint on the retrieved profile.

The convergence analysis of the linear relaxation is straightforward. If $\mathbf{G}$ has been chosen appropriately $\mathbf{x}_i$ will converge to a solution as the iteration proceeds. Writing the solution as $\hat{\mathbf{x}} = \mathbf{x}_\infty$, it must satisfy

$$\hat{\mathbf{x}} = \hat{\mathbf{x}} + \mathbf{G}[\mathbf{y} - \mathbf{F}(\hat{\mathbf{x}})]. \tag{6.51}$$

Note that this implies that $\mathbf{y} = \mathbf{F}(\hat{\mathbf{x}})$, an exact solution, provided that $\mathbf{G}$ fully spans measurement space. Subtract Eq. (6.51) from Eq. (6.50) to obtain the difference of $\mathbf{x}_{i+1}$ from the final solution as

$$(\mathbf{x}_{i+1} - \hat{\mathbf{x}}) = (\mathbf{x}_i - \hat{\mathbf{x}}) - \mathbf{G}[\mathbf{F}(\mathbf{x}_i) - \mathbf{F}(\hat{\mathbf{x}})]. \tag{6.52}$$

Expanding $\mathbf{F}(\mathbf{x})$ about $\hat{\mathbf{x}}$, e.g. Eq. (5.11), and substituting for $\mathbf{F}(\mathbf{x}_i)$ gives

$$\begin{aligned}(\mathbf{x}_{i+1} - \hat{\mathbf{x}}) &= (\mathbf{x}_i - \hat{\mathbf{x}}) - \mathbf{G}\hat{\mathbf{K}}(\mathbf{x}_i - \hat{\mathbf{x}}) + O(\mathbf{x}_i - \hat{\mathbf{x}})^2 & (6.53)\\ &= (\mathbf{I}_n - \mathbf{G}\hat{\mathbf{K}})(\mathbf{x}_i - \hat{\mathbf{x}}) + O(\mathbf{x}_i - \hat{\mathbf{x}})^2. & (6.54)\end{aligned}$$

If $\mathbf{G}$ is exactly orthogonal to $\hat{\mathbf{K}}$ (any exact linear solution) then $\mathbf{I}_n - \mathbf{G}\hat{\mathbf{K}} = \mathbf{O}$ and convergence is second order, i.e. the linear component of the error is fitted exactly on each iteration, and the remaining error is proportional to the square of that from the previous stage. If this is not the case, convergence is first order and we can examine its progress by decomposing $\mathbf{I}_n - \mathbf{G}\hat{\mathbf{K}}$ into eigenvectors:

$$(\mathbf{I}_n - \mathbf{G}\hat{\mathbf{K}}) = \mathbf{R}\mathbf{\Lambda}\mathbf{L}^T, \tag{6.55}$$

giving

$$\mathbf{x}_{i+1} - \hat{\mathbf{x}} = \mathbf{R}\mathbf{\Lambda}\mathbf{L}^T(\mathbf{x}_i - \hat{\mathbf{x}}) + O(\mathbf{x}_i - \hat{\mathbf{x}})^2 \tag{6.56}$$

or
$$\mathbf{L}^T(\mathbf{x}_{i+1} - \hat{\mathbf{x}}) = \mathbf{\Lambda}\mathbf{L}^T(\mathbf{x}_i - \hat{\mathbf{x}}) + O(\mathbf{x}_i - \hat{\mathbf{x}})^2. \tag{6.57}$$

By changing the variable to $\mathbf{z}_i = \mathbf{L}^T(\mathbf{x}_i - \hat{\mathbf{x}})$ or $\mathbf{x}_i - \hat{\mathbf{x}} = \mathbf{R}\mathbf{z}_i$ we find that the coefficients $\mathbf{z}_i$ converge independently:

$$\mathbf{z}_{i+1} = \mathbf{\Lambda}\mathbf{z}_i + O(\mathbf{z}_i^2) \quad \text{or} \quad z_{j,i+1} = \lambda_j z_{j,i} + O(z_{j,i}^2). \tag{6.58}$$

Thus $(\mathbf{x}_i - \hat{\mathbf{x}})$ can be expressed in terms of components such that each iteration reduces the coefficient of the jth component by a factor λ_j.

As an example consider a linear relaxation solution of the standard example. Let the state vector consist of values of the profile at the peaks of the weighting functions, interpolated linearly between these levels, and extrapolated linearly at the top and bottom at the same gradient as the layers below and above respectively. The eigenvalues for this case are given in table 6.1, together with the number of iterations required to reduce the initial error in the coefficient by a factor of ten. The corresponding eigenvectors are shown in Fig. 6.3(a). It can be seen that the large scale structures correspond to small eigenvalues, which should therefore converge relatively quickly, while the eigenvalues close to unity corresponding to small scale structure which will converge slowly. This allows the iteration to be stopped before unrealistic fine structure grows and dominates.

Table 6.1 Eigenvalues corresponding to the eigenvectors in Fig. 6.3(a), and number of iterations required to reduce error by a factor of 10. The negative eigenvalue corresponds to an error component that alternates in sign as the iteration proceeds.

Number	Eigenvalue	Iterations
1	0.0371	0.7
2	0.3426	2.2
3	-0.6033	4.6
4	0.6413	5.2
5	0.8444	13.6
6	0.9454	41
7	0.9847	149
8	0.9965	666

A simulation of the convergence is shown in Fig. 6.3(b). In this case the true profile is the U. S. Standard Atmosphere, and the first guess is a constant 250 K. No noise has been added. Iteration numbers 0 to 4 and 100 are shown. Even at iteration 100 we would not expect vectors 7 and 8 to have converged, and at iteration 4, only vectors 1 to 3 should have converged. Notice the alternating nature of the convergence at the top, due to the negative eigenvalue of vector 3. The retrieval would not be expected to converge to the true profile eventually, because the true state does not have its nodes at the same altitudes as the retrieval, so it cannot be

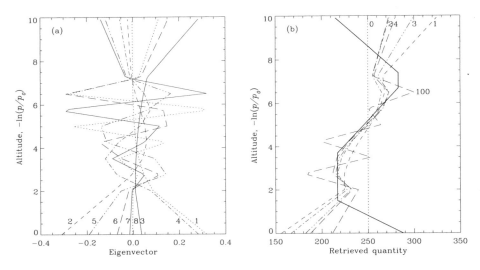

Fig. 6.3 (a) Eigenvector analysis for the convergence of the linear relaxation retrieval. The annotation corresponds to the eigenvalue number in table 6.1. (b) Convergence of a linear relaxation retrieval for the standard case. The true profile is the solid line, and the iterations are labelled with iteration number.

represented exactly. The oscillatory nature of iteration 100 is a consequence of this together with the ill-conditioning of the problem.

6.8 Nonlinear Relaxation

A nonlinear relaxation method follows the same general idea as the linear relaxation, except that the iteration equation is nonlinear. The simplest form, popularised by Chahine (1968, 1970), uses the representation described above for the simple linear relation, where the profile is represented by a linear interpolation between values located at the peaks of a set of weighting functions. At iteration i the value at each level j is modified by multiplying by the ratio of the corresponding measured to computed signal:

$$x_j^{i+1} = x_j^i \frac{y_j}{F_j(\mathbf{x}^i)}. \tag{6.59}$$

For well peaked weighting functions, one would expect this to converge to a profile which produces the right computed signals.

Convergence can be analysed by taking logarithms:

$$\ln x_j^{i+1} = \ln x_j^i + [\ln y_j - \ln F_j(\mathbf{x}^i)]. \tag{6.60}$$

We can see that this is the same as the linear relaxation, with a transformed state vector whose elements are logarithms of the original, a similarly transformed measurement vector and a relaxation matrix $\mathbf{G}$ which is the unit matrix.

The weighting function corresponding to the transformed problem has elements

$$K'_{jk} = \frac{x_k}{F_j(\mathbf{x})} \frac{\partial F_j(\mathbf{x})}{\partial x_k}. \tag{6.61}$$

As in the case of linear relaxation, the convergence will depend on the eigenvalues of $\mathbf{I} - \mathbf{K}'$, and the process will converge to an exact solution ($\mathbf{G} = \mathbf{I}$ does span the whole of state space in this case) unless stopped at a suitable point. As poorly determined fine structure generally converges more slowly than the well determined broad structure, stopping the convergence early has the effect of producing a reasonable stable solution, as in the case of linear relaxation.

A nonlinear relaxation in which the state vector is not constrained by the peaks of the weighting functions has been developed by Twomey et al. (1977). Any reasonable vertical grid can be used. The iteration is

$$x_j^{i+1} = x_j^i \frac{y_k}{F_k(\mathbf{x}^i)} K_{kj} + x_j^i [1 - K_{kj}] \tag{6.62}$$

where j is the height index, k is the channel index, i is the iteration number and K_{kj} is the weighting function normalised so that the peak value is unity. The iteration is applied sequentially over all channels, and then repeated until convergence. One cycle of the inner loop can be rearranged as

$$x_j^{i+m} = x_j^i \prod_{k=1}^{m} \left\{ 1 + [y_k - F_k(\mathbf{x}^i)] \frac{K_{kj}}{F_k(\mathbf{x}^i)} \right\}. \tag{6.63}$$

The effect is to modify the profile at all levels in response to a difference between the measurement and the forward model, using the weighting function as a weight. This would be rather similar to using $\mathbf{G} = \alpha \mathbf{K}^T$ in the linear relaxation, where α is a scaling constant to obtain the appropriate sensitivity.

We can analyse the convergence by taking logarithms:

$$\ln x_j^{i+m} = \ln x_j^i + \sum_k \ln \left\{ 1 + [y_k - F_k(\mathbf{x}^i)] \frac{K_{kj}}{F_k(\mathbf{x}^i)} \right\}. \tag{6.64}$$

Close to the solution we can expand the terms inside the summation:

$$\ln x_j^{i+m} = \ln x_j^i + \sum_k [y_k - F_k(\mathbf{x}^i)] \frac{K_{kj}}{F_k(\mathbf{x}^i)}, \tag{6.65}$$

which is a linear relaxation with

$$G_{jk} = \frac{K_{kj}}{F_k(\mathbf{x}^i)} \tag{6.66}$$

showing that this is equivalent to using a diagonal matrix for α with $1/F_k(\mathbf{x}^i)$ on the diagonal. Convergence will be determined as for the linear relaxation by the eigenvalues and vectors of $\mathbf{I} - \mathbf{GK}$.

6.9 Maximum Entropy

In Chapter 2 we discussed the use of maximum entropy as a tool for estimating probability density functions given limited information about them. That in itself is an inverse problem, and the maximum entropy principle has been used for a wide variety of inverse problems, not necessarily only those involving *pdf*s. Strictly, it is only an appropriate constraint when the quantity being retrieved is a probability density, or could reasonably be interpreted as a probability density, which is hardly ever the case with atmospheric inverse problems. Nevertheless if the state is known to be a positive quantity then maximum entropy provides an *ad hoc* positivity constraint, but so does the rather simpler process of retrieving the logarithm of the state with an optimal estimator. One area in which MaxEnt or MEM (as it has come to be known) has performed particularly well is in improving noisy images.

The exact retrieval can be formulated with a maximum entropy constraint as follows. We use a state vector $\mathbf{x}$ which is normalised so that the elements sum to unity, and minimise its 'configurational entropy' $S = \sum_i x_i \ln x_i$ subject to the constraints $\mathbf{y} = \mathbf{F}(\mathbf{x})$ and $\sum_i x_i = 1$:

$$\frac{\partial}{\partial x_k}\left(\sum_i x_i \ln x_i + \sum_j \lambda_j [F_j(\mathbf{x}) - y_j] + \mu \sum_i x_i\right) = 0 \qquad (6.67)$$

where λ_j and μ are Lagrangian multipliers. This gives

$$\ln x_k + 1 + \sum_j \lambda_j K_{jk} + \mu = 0 \qquad (6.68)$$

where K_{jk} is $\partial F_j/\partial x_k$. The maximum entropy exact solution is therefore

$$x_k = \exp\left[-1 - \mu - \sum_j \lambda_j K_{jk}\right] \qquad (6.69)$$

where μ and λ_j are chosen to satisfy the constraints. The unit area constraint can be satisfied immediately by putting

$$x_k = \exp\left[-\sum_j \lambda_j K_{jk}\right] \Big/ \sum_k \exp\left[-\sum_j \lambda_j K_{jk}\right] \qquad (6.70)$$

but the requirement that the retrieval satisfy the measurements exactly gives

$$y_l = F_l\left(\exp\left[-\sum_j \lambda_j K_{jk}\right] \Big/ \sum_k \exp\left[-\sum_j \lambda_j K_{jk}\right]\right) \qquad (6.71)$$

which has to be solved iteratively for λ_j, for example by Newton's method.

This solution has some resemblance to others, in that it involves a linear combination of the weighting functions. $\sum_j \lambda_j K_{jk}$ is in the row space of $\mathbf{K}$, but as the exponential is a nonlinear function, the maximum entropy solution will have components in the null space.

To allow for measurement error in **y**, we should not really attempt to find an exact solution. Instead we might try to minimise a cost function that includes entropy and measurement error:

$$\frac{\partial}{\partial x_k}\left(\sum_i x_i \ln x_i + \lambda \sum_{jl}[F_j(\mathbf{x}) - y_j]S_{jl}^{-1}[F_l(\mathbf{x}) - y_l] + \mu \sum_i x_i\right) = 0 \quad (6.72)$$

where λ is not now a Lagrangian multiplier, but a trade-off parameter, allocating a fraction of the cost to the χ^2 term. This χ^2 cannot be used as a Lagrangian constraint, because it does not have an exact value, only an expected value of p, the rank of the problem. The maximisation give the following implicit equation for **x**:

$$\ln x_k + 1 + \mu + \lambda K_{jk}S_{jl}^{-1}(F_l(\mathbf{x}) - y_l) = 0 \quad (6.73)$$

where μ can be eliminated by the same method as for the exact solution, and the equation must now be solved iteratively for **x**, for a range of values of λ, the value being chosen that gives $\chi^2 \simeq p$.

6.10 Onion Peeling

It is sometimes possible to make use of a particular feature of a problem to simplify the retrieval method. A possibility of this kind arises in the case of limb sounding, where a method known as 'onion peeling' has become popular (Russell and Drayson, 1972). This is simply a matter of sequentially retrieving from the top of the atmosphere, assuming at each stage that the retrieval for the levels above is correct and does not need updating. To do this the state vector must have a level corresponding directly to each measurement level, so that the weighting function matrix is square. At each level the geometry of the measurement must ensure that the signal depends only on the state vector at that level and above, so that the weighting function matrix is upper triangular (for a state vector whose last element corresponds to the top). Thus at each stage only a scalar parameter is retrieved. In the linear case this is equivalent to equations with a triangular matrix, solved by back substitution.

A little care is needed to ensure that the conditions for an 'onion peelable' problem are satisfied, For example a state vector comprising temperature on an absolute height grid, with a reference pressure at the bottom would not be suitable because the hydrostatic equation would cause the signal at tangent height z to depend on the temperature below as well as above z.

The method has the advantage of speed and simplicity, particularly in the nonlinear case where only the diagonal of the weighting function matrix need be computed. The process is simply to solve the single parameter inverse problem $y_i = F_i(x_i, x_{i+1}, \ldots x_n)$ for x_i, sequentially for $i = n \ldots 1$, taking n to be the top level. Traditionally the equation is solved exactly at each level, although problems usually arise in the top few levels where the signal is small and relative errors are large. Some *ad hoc* treatment may be applied for the instability in the top few levels, but it is quite possible to use an optimal estimator at each level if some *a priori* estimate of **x** is available (Connor and Rodgers, 1988).

Conceptually we can compare this with a sequential estimator in which the measurements are incorporated from the top down. If the incorporation of signal y_i by a sequential optimal estimator improves the error on x_i, but makes little difference to the errors on $x_{i+1}, \ldots x_n$, then we would do just as well with onion peeling. This is likely to happen if most of the information at each level comes from that level, and less from higher levels.

The 'exact onion peel' approach give the exact solution to the problem for both the linear and nonlinear cases, but the 'optimal onion peel' does not give the same result as the full optimal estimate, as the signal at level i cannot influence the retrieval at levels above.

The error analysis, unfortunately, does not retain the efficiency advantage of the onion peeling method if done properly. In the linear exact case the retrieval is $\hat{\mathbf{x}} = \mathbf{K}^{-1}\mathbf{y}$, so the correct retrieval error covariance is $\hat{\mathbf{S}} = \mathbf{K}^{-1}\mathbf{S}_\epsilon(\mathbf{K}^{-1})^T$. It is tempting to onion peel the errors as follows. The forward model and the exact retrieval for level i are

$$y_i = K_{ii}x_i + \sum_{j=i+1}^{n} K_{ij}x_j + \epsilon_i \qquad (6.74)$$

$$\hat{x}_i = (y_i - \sum_{j=i+1}^{n} K_{ij}x_j)/K_{ii} \qquad (6.75)$$

so the error variance might be expected to be given by

$$\sigma_{x_i}^2 = (\sigma_{\epsilon_i}^2 + \sum_{j=i+1}^{n} K_{ij}^2 \sigma_{x_j}^2)/K_{ii}^2 \qquad (6.76)$$

but unfortunately the off-diagonal elements are not computed, and this ignores their effect. Furthermore, in the nonlinear case the advantage of only needing to compute the diagonal of **K** is lost.

Chapter 7

The Kalman Filter

Often measurements are made sequentially, in such a way that the state being measured varies smoothly between successive measurements, or is at least correlated in some way between measurements. Typical examples are when measurements are made from a satellite at fixed (or even varying) intervals along a nadir or limb measurement track, or from the surface at frequent time intervals, and there is little difference between the atmospheric states between successive measurements. Another example, on a larger scale, is when global measurements are made daily, and the underlying atmosphere is evolving steadily according to the equations of motion.

In such cases the previous measurement can provide prior information about the state at the current time, provided the evolution of the state in time (and/or position) can be modelled. This provides the basis for the Kalman filter (Kalman, 1960), which is used to estimate discrete time series or states which are continuously evolving, and are governed by linear differential operators. Only the discrete version will be discussed here. The Kalman filter has found applications in a very wide variety of areas, ranging from the prediction of satellite orbits to the control of oil refineries.

The general formulation of the problem for discrete times $t = 1, 2, \ldots$ comprises an evolution or prediction equation and a measurement equation:

$$\mathbf{x}_t = \mathcal{M}_t(\mathbf{x}_{t-1}) + \boldsymbol{\xi}_t \tag{7.1}$$

$$\mathbf{y}_t = \mathbf{F}_t(\mathbf{x}_t) + \boldsymbol{\epsilon}_t \tag{7.2}$$

where $\mathbf{x}_t$ is the state at time t, $\mathcal{M}_t$ is a known model of the evolution operator, the *dynamic model* or *system model*, that transforms $\mathbf{x}_{t-1}$ into $\mathbf{x}_t$, $\boldsymbol{\xi}_t$ is a random vector, the *process noise*, which may express stochastic terms or unmodelled variations in the state. This stochastic prediction equation may be as simple as persistence, $\mathcal{M} = 1$, or as complex as an atmospheric general circulation model. The second equation is the measurement model, relating the measurement $\mathbf{y}_t$ with its experimental error $\boldsymbol{\epsilon}_t$ to the state, by way of a known forward model $\mathbf{F}$, in the usual way. The statistics of $\boldsymbol{\xi}$ and $\boldsymbol{\epsilon}$ are assumed known. The problem to be solved is: given a sequence of values of $\mathbf{y}_t$, make a sequence of best estimates of $\mathbf{x}_t$.

7.1 The Basic Linear Filter

The basic Kalman filter operates on the linear version of the problem:

$$\mathbf{x}_t = \mathbf{M}_t \mathbf{x}_{t-1} + \boldsymbol{\xi}_t \tag{7.3}$$

$$\mathbf{y}_t = \mathbf{K}_t \mathbf{x}_t + \boldsymbol{\epsilon}_t \tag{7.4}$$

where $\mathbf{M}$ and $\mathbf{K}$ are known matrices, which may be time varying.

The filter operates sequentially in t (which we may think of as time, though it could be space). At time $t - 1$ an estimate of $\mathbf{x}_{t-1}$ has been made, namely $\hat{\mathbf{x}}_{t-1}$, with error covariance $\hat{\mathbf{S}}_{t-1}$. The stochastic prediction equation, Eq. (7.3) is used to construct a prior estimate $\mathbf{x}_{at}$ and its covariance $\mathbf{S}_{at}$ at time t:

$$\mathbf{x}_{at} = \mathbf{M}_t \hat{\mathbf{x}}_{t-1} \tag{7.5}$$

$$\mathbf{S}_{at} = \mathbf{M}_t \hat{\mathbf{S}}_{t-1} \mathbf{M}_t^T + \mathbf{S}_{\xi t} \tag{7.6}$$

where $\mathbf{S}_{\xi t}$ is the covariance of the prediction error. This is then combined with the measurement at time t using the optimal estimation equations, e.g. Eq. (4.6) and (4.7), to produce an updated estimate of the state:

$$\mathbf{G}_t = \mathbf{S}_{at} \mathbf{K}_t^T (\mathbf{K}_t \mathbf{S}_{at} \mathbf{K}_t^T + \mathbf{S}_\epsilon)^{-1} \tag{7.7}$$

$$\hat{\mathbf{x}}_t = \mathbf{x}_{at} + \mathbf{G}_t (\mathbf{y}_t - \mathbf{K}_t \mathbf{x}_{at}) \tag{7.8}$$

$$\hat{\mathbf{S}}_t = \mathbf{S}_{at} - \mathbf{G}_t \mathbf{K}_t \mathbf{S}_{at} \tag{7.9}$$

where $\mathbf{G}_t$ is the Kalman gain matrix, functionally identical to that of the maximum *a posteriori* estimator. Some prior estimate is required at $t = 0$, but this can often use a covariance matrix with elements sufficiently large as to be noncommittal. The quality and the spacing of the data does not have to be uniform. This formalism provides the best estimate that can be made of the state at time t given the measurements up to and including time t.

Example: a scalar random walk

A simple example of a times series which may be Kalman filtered is the random walk with one parameter. The prediction model is

$$x_t = x_{t-1} + \xi_t \tag{7.10}$$

where the variance of ξ_t is σ_ξ^2, and the measurement model corresponds to direct observation of x_t:

$$y_t = x_t + \epsilon_t \tag{7.11}$$

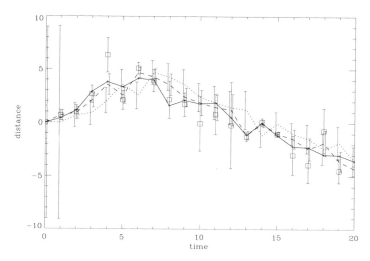

Fig. 7.1 A simulation of Kalman filtering a random walk. Solid: the actual walk; squares with error bars: the measurement; dotted and left hand error bars: prior estimate; dashed and right hand error bars: filtered estimate.

where the measurement noise variance is σ_ϵ^2. Assume that measurements are made at times $t = 0, 1, \ldots$ and that there are no missing or bad data. For the update cycle at time t, the prior estimate x_t^a and its variance σ_{at}^2 are, following Eqs. (7.5) and (7.6):

$$\begin{aligned} x_t^a &= \hat{x}_{t-1} \\ \sigma_{at}^2 &= \hat{\sigma}_{t-1}^2 + \sigma_\xi^2. \end{aligned} \qquad (7.12)$$

From Eq. (7.7) the Kalman gain is

$$G_t = \frac{\hat{\sigma}_{t-1}^2 + \sigma_\xi^2}{\hat{\sigma}_{t-1}^2 + \sigma_\xi^2 + \sigma_\epsilon^2}. \qquad (7.13)$$

The updated state estimate is, from Eq. (7.8),

$$\hat{x}_t = x_t^a + G_t(y_t - x_t^a) = (1 - G_t)x_t^a + G_t y_t \qquad (7.14)$$

with variance from Eq. (7.9):

$$\hat{\sigma}_t^2 = \sigma_{at}^2 - G_t \sigma_{at}^2 = (1 - G_t)(\hat{\sigma}_{t-1}^2 + \sigma_\xi^2) = \frac{\sigma_\epsilon^2(\hat{\sigma}_{t-1}^2 + \sigma_\xi^2)}{\hat{\sigma}_{t-1}^2 + \sigma_\xi^2 + \sigma_\epsilon^2}. \qquad (7.15)$$

A simulated example is shown in Fig. 7.1, where the 'true' time series, is indicated by the solid line, is a random walk with $\sigma_\xi^2 = 1$. The measurements are indicated by the empty squares with error bars. The measurement variance, σ_ϵ^2, varies from point to point and has been simulated as the modulus of a Gaussian random variable with mean zero and standard deviation 2, so that there are some good and some poor measurements. The Kalman filtered estimate, $\hat{x}$, is given by the dashed line, with error bars which are slightly offset to the right for clarity. The prior estimate, x_a, is given by the dotted line; its value is the same as $\hat{x}$ from the previous time step, and its variance (plotted as error bars offset to the left) is obtained by adding unity to $\hat{\sigma}^2$ from the previous time step. The filtered estimate is a weighted combination of the measurement and the prior estimate. In some cases the measurement is more accurate and has greater weight, and in others the prior estimate is more accurate. It is clear that the filtered estimate is significantly closer to the true state than the raw measurements.

⇒ **Exercise** 7.1: In the case where σ_ξ^2 and σ_ϵ^2 are both constant, examine the behaviour of $\hat{\sigma}_t^2$ at large times.

7.2 The Kalman Smoother

The Kalman filter is particularly appropriate for real time processing, where a best estimate of some current quantity, such as a spacecraft ephemeris, is needed immediately, given all of the measurements made to date. Another application is in retrospective processing, where the best estimate of some quantity is needed given data from times both before and after the reference time. In this case an extension known as the Kalman smoother can be used. Given the whole time series of measurements, filters can be run both forwards and backwards in time. The backward filter may require a modified evolution operator, and will provide the best estimate based on future measurements. The two estimates can be combined with appropriate weights to give an estimate based on all the data, with the proviso that any measurements made at the reference time are only used once, i.e. the forward estimate is combined with the backward prior estimate, or vice-versa.

In detail the process is as follows. Run the filter described in section 7.1, storing $\mathbf{x}'_{at}$ and $\mathbf{S}'_{at}$ at each time where a filtered state is required, where the prime now indicates the forward filter. The required times need not be the same as the times where measurements are made, but Eqns. (7.5) and (7.6) should be evaluated for them. Run the filter backwards, starting with some noncommittal prior estimate at time T, the end of the time series, and carrying out the following steps:

$$\begin{aligned}
\mathbf{x}''_{a,t} &= \mathbf{M}''_t \hat{\mathbf{x}}''_{t+1} \\
\mathbf{S}''_{a,t} &= \mathbf{M}''_t \hat{\mathbf{S}}''_{t+1} \mathbf{M}''^T_t + \mathbf{S}''_{\xi t} \\
\mathbf{G}''_t &= \mathbf{S}''_{at} \mathbf{K}^T_t (\mathbf{K}_t \mathbf{S}''_{at} \mathbf{K}^T_t + \mathbf{S}_\epsilon)^{-1} \\
\mathbf{x}''_t &= \mathbf{x}''_{at} + \mathbf{G}''_t (\mathbf{y}_t - \mathbf{K}_t \mathbf{x}''_{at}) \\
\mathbf{S}''_t &= \mathbf{S}''_{at} - \mathbf{G}''_t \mathbf{K}_t \mathbf{S}''_{at}
\end{aligned} \quad (7.16)$$

where the double prime indicates the backward estimate. At times when a filtered estimate is required, combine the saved $\mathbf{x}'_{at}$ optimally with the backward filtered $\mathbf{x}''_t$

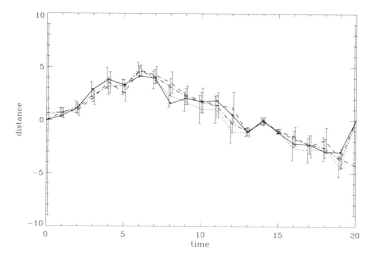

Fig. 7.2 A simulation of Kalman smoothing a random walk. Solid: the actual walk; dashed and right error bars: forward estimate; dotted and left error bars: backward estimate; dash-dot and central error bars: combined estimate.

according to

$$\begin{aligned}\hat{\mathbf{S}}_t &= (\mathbf{S}'_{at}{}^{-1} + \mathbf{S}''_t{}^{-1})^{-1} \\ \hat{\mathbf{x}}_t &= \hat{\mathbf{S}}_t(\mathbf{S}'_{at}{}^{-1}\mathbf{x}'_{at} + \mathbf{S}''_t{}^{-1}\mathbf{x}''_t).\end{aligned} \qquad (7.17)$$

A simulation of a scalar Kalman smoother is shown in Fig. 7.2. The case used is identical to that of Fig. 7.1, but the curves shown are the forward estimate (dashed, error bars to the right), the backward estimate (dotted, error bars to the left) and the combined, or smoothed, estimate (dash-dot, central error bars). The combined error bars are smaller than either the forward or the backward, and the combined estimate lies between the other two.

⇒ **Exercise** 7.2: Find an expression for the backward model $\mathbf{M}''$, in terms of the forward model $\mathbf{M}$, assuming that it is independent of t.

7.3 The Extended Filter.

The extended Kalman filter is required when the prediction equation and/or the forward model is non-linear. It uses the full nonlinear prediction Eq. (7.1) to calculate the prior state estimate:

$$\mathbf{x}_{at} = \mathcal{M}_t(\hat{\mathbf{x}}_{t-1}) \qquad (7.18)$$

but linearises it about $\hat{\mathbf{x}}_{t-1}$ to obtain the prior state estimate covariance $\mathbf{S}_{at}$:

$$\mathbf{M}_t = \frac{\partial \mathcal{M}_t(\hat{\mathbf{x}}_{t-1})}{\partial \mathbf{x}} \quad (7.19)$$

$$\mathbf{S}_{at} = \mathbf{M}_t \hat{\mathbf{S}}_{t-1} \mathbf{M}_t^T + \mathbf{S}_{\xi t}. \quad (7.20)$$

The forward model is linearised at the measurement time about the prior estimate to obtain $\mathbf{K}_t$ for calculating the gain matrix:

$$\mathbf{K}_t = \frac{\partial \mathbf{F}(\mathbf{x}_{at})}{\partial \mathbf{x}} \quad (7.21)$$

$$\mathbf{G}_t = \mathbf{S}_{at} \mathbf{K}_t^T (\mathbf{K}_t \mathbf{S}_{at} \mathbf{K}_t^T + \mathbf{S}_\epsilon)^{-1} \quad (7.22)$$

but the full nonlinear forward model is used to update $\mathbf{x}_{at}$:

$$\hat{\mathbf{x}}_t = \mathbf{x}_{at} + \mathbf{G}_t[\mathbf{y}_t - \mathbf{F}(\mathbf{x}_{at})] \quad (7.23)$$

$$\hat{\mathbf{S}}_t = \hat{\mathbf{S}}_{at} - \mathbf{G}_t \mathbf{K}_t \mathbf{S}_{at}. \quad (7.24)$$

It may be necessary to carry out some iteration of the last four equations if the linearisation of the forward model at $\mathbf{x}_{at}$ is not good enough to be used at $\hat{\mathbf{x}}_t$. If this is a possibility, then $\mathbf{F}(\hat{\mathbf{x}}_t)$ should be evaluated for comparison with $\mathbf{y}_t$.

7.4 Characterisation and Error Analysis

The analysis described for the general retrieval in Chapter 3 can be applied to the Kalman filter. Using the equations for the basic linear filter, substitute (7.4) and (7.5) in (7.8) to express $\hat{\mathbf{x}}_t$ in terms of $\mathbf{x}_t$ and $\hat{\mathbf{x}}_{t-1}$:

$$\begin{aligned}\hat{\mathbf{x}}_t &= \mathbf{G}_t \mathbf{K}_t \mathbf{x}_t + \mathbf{G}_t \boldsymbol{\epsilon}_t + (\mathbf{I} - \mathbf{G}_t \mathbf{K}_t) \mathbf{M}_t \hat{\mathbf{x}}_{t-1} \\ &= \mathbf{A}_t \mathbf{x}_t + \mathbf{B}_t \hat{\mathbf{x}}_{t-1} + \mathbf{G}_t \boldsymbol{\epsilon}_t \end{aligned} \quad (7.25)$$

where $\mathbf{A}_t = \mathbf{G}_t \mathbf{K}_t$ and $\mathbf{B}_t = (\mathbf{I} - \mathbf{A}_t)\mathbf{M}_t$. $\mathbf{A}_t$ is an averaging kernel for the state at time t, and should be similar to a unit matrix, so that $\mathbf{B}_t$ should be 'small' in some sense, depending on the details of the problem. Substituting Eq. (7.25) into itself recursively shows that $\hat{\mathbf{x}}_t$ is a smoothed version of the time series of $\mathbf{x}_t$'s plus a time smoothing of the measurement errors:

$$\begin{aligned}\hat{\mathbf{x}}_t = &\; \mathbf{A}_t \mathbf{x}_t + \mathbf{G}_t \boldsymbol{\epsilon}_t \\ &+ \mathbf{B}_t(\mathbf{A}_{t-1}\mathbf{x}_{t-1} + \mathbf{G}_{t-1}\boldsymbol{\epsilon}_{t-1}) \\ &+ \mathbf{B}_t \mathbf{B}_{t-1}(\mathbf{A}_{t-2}\mathbf{x}_{t-2} + \mathbf{G}_{t-2}\boldsymbol{\epsilon}_{t-2}) \\ &+ \ldots \end{aligned} \quad (7.26)$$

The time scale of the smoothing depends on the magnitude of the $\mathbf{B}_t$ terms. The set of matrix weights $\mathbf{A}_t$, $\mathbf{B}_t \mathbf{A}_{t-1}$, $\mathbf{B}_t \mathbf{B}_{t-1} \mathbf{A}_{t-2}, \ldots$ form a two dimensional averaging kernel, showing how the estimated state $\hat{\mathbf{x}}_t$ is related to the true state at t and all preceding times. In the case of Kalman smoothing this two dimensional averaging kernel extends into both future and past times. Eq. (7.26) can be rearranged to show the smoothing and retrieval noise components of the error separately:

$$\begin{aligned}\hat{\mathbf{x}}_t - \mathbf{x}_t &= (\mathbf{A}_t - \mathbf{I})\mathbf{x}_t + \mathbf{B}_t\mathbf{A}_{t-1}\mathbf{x}_{t-1} + \mathbf{B}_t\mathbf{B}_{t-1}\mathbf{A}_{t-2}\mathbf{x}_{t-2} + \ldots \\ &\quad + \mathbf{G}_t\boldsymbol{\epsilon}_t + \mathbf{B}_t\mathbf{G}_{t-1}\boldsymbol{\epsilon}_{t-1} + \mathbf{B}_t\mathbf{B}_{t-1}\mathbf{G}_{t-2}\boldsymbol{\epsilon}_{t-2} + \ldots\end{aligned} \quad (7.27)$$

but this does not lead to usefully simple expressions for the smoothing error and retrieval noise covariances.

7.5 Validation

For a Kalman filter to operate correctly, the matrices $\mathbf{M}_t$, $\mathbf{S}_{\xi t}$, $\mathbf{K}_t$ and $\mathbf{S}_{\epsilon t}$ must all be 'correct'. To determine whether the filter is producing reasonable results, the statistical behaviour of any or all of the following quantities can be examined:

$$\begin{aligned}\mathbf{y}_t - \mathbf{y}_{at} &= \mathbf{y}_t - \mathbf{K}_t\mathbf{x}_{at} \\ \hat{\mathbf{x}}_t - \mathbf{x}_{at} &= \mathbf{G}_t(\mathbf{y}_t - \mathbf{y}_{at}) \\ \hat{\mathbf{y}}_t - \mathbf{y}_{at} &= \mathbf{K}_t(\hat{\mathbf{x}}_t - \mathbf{x}_{at}) \\ \mathbf{y}_t - \hat{\mathbf{y}}_t &= \mathbf{y}_t - \mathbf{K}_t\hat{\mathbf{x}}_t.\end{aligned} \quad (7.28)$$

Differences involving $\mathbf{x}_t$ itself cannot be used unless independent direct measurements are available. These quantities are all linearly related to one another, so only one need be used. If the measurement equation or forward model is correct then

$$\mathbf{y}_t - \mathbf{y}_{at} = \mathbf{K}_t(\mathbf{x}_t - \mathbf{x}_{at}) + \boldsymbol{\epsilon}_t \quad (7.29)$$

and its covariance should be given by:

$$\mathbf{S}_{ay} = \mathbf{K}_t\mathbf{S}_{at}\mathbf{K}_t^T + \mathbf{S}_\epsilon = \mathbf{K}_t(\mathbf{M}_t\hat{\mathbf{S}}_{t-1}\mathbf{M}_t^T + \mathbf{S}_{\xi t})\mathbf{K}_t^T + \mathbf{S}_\epsilon. \quad (7.30)$$

Thus $\mathbf{S}_{ay}$ can be evaluated from a real data set, and compared with this theoretical expression. Another useful diagnostic is

$$\chi^2 = (\mathbf{y}_t - \mathbf{y}_{at})^T \mathbf{S}_{ay}^{-1}(\mathbf{y}_t - \mathbf{y}_{at}) \quad (7.31)$$

which should produce a χ^2 distribution with m degrees of freedom. These diagnostics will not identify the source of a problem directly, but can be used to indicate when all is well.

If an ensemble of states is available, another test can be constructed from the evolution equation. From Eq. (7.5) the following expression for the ensemble covariance can be derived:

$$\mathbf{S}_e = \mathbf{M}\mathbf{S}_e\mathbf{M}^T + \mathbf{S}_\xi \quad (7.32)$$

provided $\mathbf{M}$ is independent of t. This can be used to determine whether $\mathbf{S}_e$, $\mathbf{M}$, and $\mathbf{S}_\xi$ are internally consistent, or at least to estimate $\mathbf{S}_\xi$ from the other two. If a directly measured time series is available, then $\mathbf{M}$ and $\mathbf{S}_\xi$ may be estimated using multiple regression.

Chapter 8
Global Data Assimilation

8.1 Assimilation as a Inverse Problem

Remote measurements of the atmosphere from satellites are primarily used to study the global atmosphere, with two important applications being research and operational weather forecasting. For many purposes we need to analyse the data to produce global two, three of four dimensional fields of the measured quantities. Particularly for weather forecasting, the accuracy of the final product depends critically on the accuracy with which the global field represent the true state of the atmosphere.

The global analysis problem is of the same kind as the profile retrieval problem, but enormously larger. The state vector is the state of the whole atmosphere, rather than a single profile, and the measurements are known functions of the state. It is required to find the best state estimate consistent with the measurements and with any *a priori* information that may be available. The extra complication is that the state is evolving with time, but the evolution equations are known, in principle. The Kalman filter is designed for exactly such circumstances, but global analysis is such a large problem that Kalman filtering is still a hope for the future. At present, faster and simpler methods are commonly used.

The atmospheric state may be defined by a set of model variables such as temperature, pressure, wind velocities and humidity on a latitude, longitude, height and time grid, although many equivalent ways of representing the state, such as spherical harmonics, are possible. Until recently many analysis methods could only assimilate measurements of the model variables themselves, so that satellite data had to be retrieved to produce profiles of model variables before they can be used. However the direct assimilation of remote measurements without an intermediate retrieval is now being developed as one of the goals of modern assimilation methods.

Historically, measurements were made at the surface and from radiosondes at the agreed times of noon and midnight universal time, sometimes with extra measurements at 0600 and 1800 UT. Global analyses were made at these synoptic times, from which forecasts were run. Consequently many analysis techniques were designed around synoptic data. With the advent of satellite data, which provides continuous measurements, many techniques could not take full advantage of the new data source, resorting to methods such as inserting the data at the nearest synoptic time. Dealing properly with asynoptic data is one of the requirements of a successful assimilation method.

This chapter is not intended as a comprehensive treatment of assimilation methods. An excellent textbook on the subject is *Atmospheric Data Analysis* by R. Daley, (1991), and more recent reviews of the subject can be found in a special issue of the *Journal of the Meteorological Society of Japan*, vol. 75, no. 1B, 1997, with many useful papers, particularly the reviews by Talagrand, Courtier, Cohn and Daley. Here I will try to show how assimilation is related to retrieval methods, and will discuss how best to incorporate indirect measurements into global analyses.

8.2 Methods for Data Assimilation

Meteorological data analysis began with the forecaster subjectively drawing contours on charts of synoptic observations, with the aid of the forecast from the previous day to guide his intuition. With the advent of numerical forecasting models it became clear that objective numerical methods were needed to set up the initial conditions. It was quickly realised that this was not a straightforward process, not least because the atmosphere (and hence many forecast models) sustains high wave-speed inertio-gravity waves, and any errors in the analysis are likely to give rise to erroneously large amplitudes unless they are specifically suppressed. Early methods, such as that suggested by Bergthorsson and Döös (1955) and Cressman (1959) were essentially based on a linear relaxation from a background field, usually the forecast for the analysis time, towards the observations, but as computer power has increased, more attention has been paid to optimal methods, and modern methods are based on estimation theory.

8.2.1 *Successive correction methods*

Bergthorsson and Döös proposed a single step method of correcting a background field on the basis of observations. The principles they used were soon developed into the successive correction method. The process can be described in terms of the notation we have been using for retrieval theory by noting the following equivalences:

(i) The state vector, $\mathbf{x}$, is a set of model variables on the model grid.
(ii) A background field, $\mathbf{x}_a$, is the result of a numerical weather prediction model, often blended with climatology.
(iii) The measurement $\mathbf{y}$ is a set of direct measurements of one or more of the quantities in the state vector (not an indirect measurement) not necessarily (and not normally) located on the model grid.

(iv) The forward model, $\mathbf{F}(\mathbf{x})$, is an interpolation in the model grid to the measurement location.

The analysis proceeds basically as follows, though there are many variants. The observed data and corresponding background values at the measurement locations are $\mathbf{y}$ and $\mathbf{y}_a = \mathbf{F}(\mathbf{x}_a)$. The analysed value at grid point i is a weighted mean of the background value and the measured values within a radius of influence:

$$\hat{x}_i - x_{ai} = \sum_k W_{ik}(y_k - y_{ak}) \qquad (8.1)$$

where the sum is over all observations within some radius r and

$$W_{ik} = \frac{\sigma_a^2 w_{ki}}{\sigma_a^2 w_{ki} + \sigma_k^2} \qquad (8.2)$$

σ_a^2 is the variance of the background field, σ_k^2 is the variance of measurement y_k, and w_{ki} is a weight which depends only on the distance from the observation to the grid point, being unity at the grid point and zero at the radius of influence.

Bergthorsson and Döös proposed a single step of the analysis, and derived the form of the weight function statistically. Later variants are iterative, using weights which change with iteration number, and often with weight functions which are algebraic functions of radius.

8.2.2 *Optimal interpolation*

Eliassen (1954) and Gandin (1963) considered the statistical basis of the successive correction analysis process and as a result gave rise to the *optimal interpolation* approach. This was the first attempt to carry out assimilation using the principles of estimation theory. It is based on updating a prior state with new measurements with weights which depend more correctly on estimates of their respective accuracies. As such it may be regarded as an approximation to the Kalman filter, with the following features and modifications:

(i) The state vector, $\mathbf{x}$, is the complete set of model variables.
(ii) The prior state at time t, $\mathbf{x}_t^a$, is the result of a numerical weather prediction model.
(iii) The prior covariance at time t, $\mathbf{S}_t^a$, is not evolved from the previous estimate. Instead a precomputed function is used which is representative of the forecast error of the model. Horizontal and vertical cross-covariances are ignored or approximated.
(iv) The measurement $\mathbf{y}$ is a set of direct measurements of one or more of the quantities in the state vector (not an indirect measurement) not necessarily (and not normally) located on the model grid.
(v) The measurement error covariance, $\mathbf{S}_\epsilon$, is diagonal.

(vi) The forward model, $\mathbf{F}(\mathbf{x})$, is an interpolation in the model grid to the measurement location. Thus $\mathbf{K}$ is an interpolation operator.

(vii) The quantity $\mathbf{KS}_t^a \mathbf{K}^T$ is not evaluated explicitly. Rather it is represented by an interpolation of the forecast error to the measurement location.

(viii) The analysis error covariance $\hat{\mathbf{S}}_t$ is not evaluated, as it is not needed to compute $\mathbf{S}_{t+1}^a$ in the next cycle.

Most of the modifications have the effect of eliminating the computationally expensive parts of the Kalman filter, but they also eliminate the associated information content. Nevertheless, optimal interpolation proved to be a significant step forward in data analysis, and methods based on its principles became almost universally used by numerical weather prediction centres. They have only recently begun to be supplanted by more advanced methods.

Analyses resulting from successive correction or optimal estimation generally contain non-meteorological structure, that is are not states which could easily have been produced by integrating the model equations from a reasonable prior state. The symptom is usually unrealistically large amplitude inertio-gravity waves if the model is subsequently integrated forward from the analysed state. Consequently the fields need further adjustment in a process known as 'initialisation'. The details of initialisation methods are outside the scope of this book, and will not be described. What is really required is some means of assimilating data so that the model itself is used as a constraint, and analysed fields are always possible model fields. This can be provided by the use of adjoint methods.

8.2.3 Adjoint methods

The general assimilation method will attempt to minimise some cost function, J, in just the same way as the MAP profile retrieval:

$$J = (\mathbf{x} - \mathbf{x}_a)^T \mathbf{S}_a^{-1} (\mathbf{x} - \mathbf{x}_a) + [\mathbf{y} - \mathbf{F}(\mathbf{x})]^T \mathbf{S}_\epsilon^{-1} [\mathbf{y} - \mathbf{F}(\mathbf{x})] \qquad (8.3)$$

where $\mathbf{F}$ now describes both the physics of the measurement and the evolution of the atmosphere. For profile retrieval a Gauss–Newton descent algorithm is commonly used:

$$\mathbf{x}_{n+1} = \mathbf{x}_n - [\nabla_\mathbf{x}(\nabla_\mathbf{x} J)]^{-1} \nabla_\mathbf{x} J \qquad (8.4)$$

where $\nabla_\mathbf{x}(\nabla_\mathbf{x} J)$ is approximated by the form $(\mathbf{K}^T \mathbf{S}_\epsilon^{-1} \mathbf{K} + \mathbf{S}_a^{-1})$, and $\mathbf{K} = \nabla_\mathbf{x} \mathbf{F}(\mathbf{x})$. For the assimilation problem $\nabla_\mathbf{x}(\nabla_\mathbf{x} J)$ can be very expensive to evaluate and invert for large state vectors.

A cheaper approach is to use steepest descent:

$$\mathbf{x}_{n+1} = \mathbf{x}_n - \gamma_n \nabla_{\mathbf{x}_n} J \qquad (8.5)$$

where γ_n is the step size, to be chosen. The gradient is given by:

$$\nabla_{\mathbf{x}_n} J = \mathbf{S}_a^{-1}(\mathbf{x} - \bar{\mathbf{x}}) - \mathbf{K}^T \mathbf{S}_\epsilon^{-1}(\mathbf{y} - \mathbf{F}(\mathbf{x})) \quad (8.6)$$

Convergence is only first order, but the descent direction is much more quickly evaluated. The adjoint method is a steepest descent method, in which:

(i) The measured data $\mathbf{y}$ is least squares fitted to the predictions of a dynamical model in four dimensions over a time interval $t = 0 \ldots T$.
(ii) The state vector is the state at $t = 0$ only.
(iii) Model errors are ignored. It is assumed correct and its time evolution from the state at $t = 0$ is used as a strong constraint.
(iv) Normally no prior state estimate is used.

Thus we take measurements $\mathbf{y}_t = \mathbf{F}_t(\mathbf{x}_t) + \epsilon_t$ of the state $\mathbf{x}_t$ at times $t = 0 \ldots T$, which may be of different kinds at each time. The state evolves according to the operator $\mathcal{M}_t$ without the error term $\boldsymbol{\xi}_t$ that is used in the Kalman filter:

$$\mathbf{x}_t = \mathcal{M}_t \mathbf{x}_{t-1} \quad (8.7)$$

Thus the quantity to be retrieved, $\mathbf{x}_0$ at $t = 0$, determines all the $\mathbf{x}_t$ according to the dynamical model. The forward model is expressed as a function of the state as:

$$\mathbf{y}_t = \mathbf{F}_t(\mathbf{x}_t) + \epsilon_t = \mathbf{F}_t(\mathcal{M}_t \ldots \mathcal{M}_1 \mathbf{x}_0) + \epsilon_t \quad (8.8)$$

We require the state $\mathbf{x}_0$ which minimises

$$J = \sum_{t=0}^{T} (\mathbf{y}_t - \mathbf{F}_t(\mathbf{x}_t))^T \mathbf{S}_{\epsilon,t}^{-1} (\mathbf{y}_t - \mathbf{F}_t(\mathbf{x}_t)) \quad (8.9)$$

for which we evaluate the gradient

$$\nabla_{\mathbf{x}_0} J = -\sum_{t=0}^{T} (\partial \mathbf{F}_t(\mathbf{x}_t)/\partial \mathbf{x}_0)^T \mathbf{S}_{\epsilon,t}^{-1} (\mathbf{y}_t - \mathbf{F}_t(\mathbf{x}_t)) = 0 \quad (8.10)$$

so it can be used in the steepest descent algorithm. Putting $\mathbf{K}_t = \partial \mathbf{F}_t(\mathbf{x}_t)/\partial \mathbf{x}_t$ and $\mathbf{M}_t = \partial \mathbf{x}_t/\partial \mathbf{x}_{t-1}$, a linearised version of the operator $\mathcal{M}_t$, we can write

$$\partial \mathbf{F}_t(\mathbf{x}_t)/\partial \mathbf{x}_0 = \mathbf{K}_t \mathbf{M}_t \ldots \mathbf{M}_0 \quad (8.11)$$

where $\mathbf{M}_0 = \mathbf{I}$, a unit matrix, has been included to simplify the algebra. Hence

$$\nabla_{\mathbf{x}_0} J = -\sum_{t=0}^{T} (\mathbf{M}_t \ldots \mathbf{M}_0)^T \mathbf{K}_t^T \mathbf{S}_{\epsilon,t}^{-1} (\mathbf{y}_t - \mathbf{F}_t(\mathbf{x}_t)) = 0 \quad (8.12)$$

A straightforward evaluation of this would involve the forward integration of the dynamical model, updating $\mathbf{M}_t \ldots \mathbf{M}_1$ at each time step, and whenever there is data the calculation of the vector $\mathbf{K}_t^T \mathbf{S}_{\epsilon,t}^{-1}(\mathbf{y}_t - \mathbf{F}_t(\mathbf{x}_t))$ to be multiplied by $\mathbf{M}_t \ldots \mathbf{M}_1$. The expensive part of this operation is the large matrix multiplication involved in updating $\mathbf{M}_t \ldots \mathbf{M}_1$. Fortunately the sum can be evaluated much more efficiently by working backwards. Consider the following sequence:

$$\boldsymbol{\lambda}_T = \mathbf{K}_T^T \mathbf{S}_{\epsilon,T}^{-1}[\mathbf{y}_T - \mathbf{F}_T(\mathbf{x}_T)] \tag{8.13}$$

then repeat

$$\boldsymbol{\lambda}_t = \mathbf{M}_{t+1}^T \boldsymbol{\lambda}_{t+1} + \mathbf{K}_t^T \mathbf{S}_{\epsilon,t}^{-1}[\mathbf{y}_t - \mathbf{F}_t(\mathbf{x}_t)] \tag{8.14}$$

backwards from $t = T - 1$ to $t = 0$. This results in $\boldsymbol{\lambda}_0 = \nabla_{\mathbf{x}_0} J$, with only a matrix by vector multiply at each step.

This result can also be obtained by considering the minimisation of J with respect to all of the $\mathbf{x}_t$, $t = 0 \ldots T$, treated as independent variables, but with Lagrangian constraints to satisfy Eq. (8.7) (Daley, 1991).

Eq. (8.14) is called the *adjoint equation* for the problem. It integrates $\boldsymbol{\lambda}_t$ backwards using the adjoint, or transpose, of a linearisation of the model describing the evolution of $\mathbf{x}_t$, with a forcing term dependent on the difference between the measurements and the forward model. A forward integration of the model followed by a backward integration of the adjoint equation provides the derivative of the cost function for the initial state chosen. This can then be used in a steepest descent algorithm to improve the estimate of the initial state.

The adjoint method proceeds as follows:

(i) Start at a first guess $\mathbf{x}_0$
(ii) Integrate the dynamics forwards in time to T, saving $\mathbf{x}_t$ at each time step
(iii) Integrate the adjoint model backwards in time, using the saved values $\mathbf{x}_t$, to obtain $\boldsymbol{\lambda}_0$. A special adjoint, or *tangent linear* model is written which linearises $\mathcal{M}_t$ and computes $\mathbf{M}_t^T \boldsymbol{\lambda}_t$ on the fly, rather than evaluating the full $n \times n$ matrix $\mathbf{M}_t$.
(iv) Update $\mathbf{x}_0$ using $\boldsymbol{\lambda}_0$ for $\nabla_{\mathbf{x}_0} J$ in a steepest descent algorithm.
(v) Repeat until converged.

8.2.4 Kalman filtering

The Kalman filter would appear ideal in principle for assimilation, but unfortunately the problem is often so large that to carry it out as described in Chapter 7 would be prohibitively expensive. A typical general circulation or forecasting model has a spatial resolution of 1–4°horizontally and 30–60 levels vertically, giving 10^6 to 10^7 grid points. At each grid point the atmosphere is defined by about 5 variables. Thus a typical state vector might have 10^7 or more elements. The state covariance matrix would then require 10^{14} or more elements just to store it, unless some kind of short cut were used.

The time evolution of the state is a large computation, but is one that is carried out in any case when integrating the equations of motion. The main difficulty in applying the full Kalman filter is in the computation of the time evolution of the covariance of the prior state:

$$\mathbf{S}_t^a = \mathbf{M}_{t-1}\hat{\mathbf{S}}_{t-1}\mathbf{M}_{t-1}^T + \mathbf{S}_M \tag{8.15}$$

where $\mathbf{M}$ may be thought of as either a matrix or as an operator represented by a single time step of the equations of motion. Kalman filtering has not been applied without approximation to the problem of the global atmosphere, although it has been tested in simple cases such as limited areas and a single layer model, for example by Cohn and Parrish (1991).

Various approximations to the Kalman filter have been proposed, with mixed success. Examples of approximations that have been tried are:

- Covariance modelling, such as is used in optimal interpolation.
- Dynamics simplification for evolving forecast error covariances.
- Reduced resolution model for evolving forecast error covariances.
- Local approximation: only evolve covariance elements for nearby points.
- Limited filtering: fixed gain matrix and asymptotic covariance.
- Monte Carlo: forecast an ensemble.

The filter operates by combining information from previous time steps, as represented by $\mathbf{S}_t^a$, with information measured at the current time step. Often most of the information is contained in the prior covariance, so that any approximation is likely to reduce information. Todling and Cohn (1994) have compared a range of suboptimal approximations to the Kalman filter, and have concluded that the most important aspect for a good sub-optimal scheme is the accurate representation of the *a priori* covariance. They found that a dynamically balanced forecast error covariance is essential for successful performance, and the use of 'initialisation' to compensate for data–model imbalance sometimes results in poor performance.

8.3 Preparation of Indirect Measurements for Assimilation

The use of remote sounding data in assimilations has generally been a matter of assimilating retrievals, although recently much work has been done on assimilating radiances directly (e.g. Eyre *et al.* 1993; Andersson *et al.* 1994; Rizzi and Matricardi 1998; Derber and Wu 1998; Courtier *et al.* 1998; Rabier *et al.* 1998; Andersson *et al.* 1998). However retrievals and radiances are not the only alternatives, any form in which the total information content of the measurement is preserved can lead to possible approaches, for example those discussed by Joiner and DaSilva (1998).

The assimilation of retrievals involves several problems. The retrieval may have used a forecast for its *a priori* state, so that its error can be correlated with the that of the initial state of the model into which it is being assimilated. For practical reasons of data quantity, the error covariance of each retrieval is not usually considered, rather some precomputed typical covariance (often diagonal) is used instead. Retrievals may be spatially correlated as a result of using the same *a priori* state, or using nearby previous retrievals as *a priori*, or due to the use of a cloud-clearing algorithm. All of these can lead to the under-use of the information content of the indirect measurement, and to unaccounted influence of the *a priori* on the assimilation.

The assimilation of radiances brings its own set of difficulties. Generally the assimilator is not involved in the design of the instrument, and may not fully appreciate all of the implications of the indirect measurement, particularly its systematic error sources. The specification of the measurement covariance is more satisfactory for radiances, but the presence of clouds usually means that an *ad hoc* cloud-clearing scheme is used rather than the explicit retrieval of cloud distribution, with the consequence that the measurement error covariance varies with the cloud state. Furthermore, the next generation of operational remote sounders, such as AIRS (Aumann and Pagano, 1994) and IASI (Simeoni *et al.*, 1997), will have very large numbers of channels, many more than the degrees of freedom for signal, and the problem of radiance assimilation will become much larger if some form of data compression is not used.

We should therefore think in terms of identifying the most effective way of interfacing remote sounding technology (the 'data supplier') to data assimilation (the 'assimilator'). Considerations to be borne in mind when developing such an interface include:

(i) *Retrieval system validation*

The data supplier will undoubtedly expect to carry out retrievals as part of the instrument validation process. The primary aim of validation can be stated simply as ensuring that the forward model of the measurement is adequate, and this should include tests on the reasonableness of retrievals using the forward model. He should not rely on assimilation of his own data for this purpose, and probably should not even use it without great caution, because its relation to the raw measurements is not simple. Nevertheless, assimilations of other data sources is a useful independent source for comparison.

(ii) *Efficiency of transferring and storing the data*

The amount of data transferred to the assimilator should be as small as convenient, consistent with retaining its full information content.

(iii) *Information content of the data*

The data should represent the measurements with no significant loss of information, and should not add any apparent information (e.g. *a priori*) that does not appear in the original measurements. A description of the information content (by error analysis and characterisation) should be part of the data.

(iv) *Instrument forward model*

The assimilator should preferably not need to understand the full details of the instrument, or maintain the code of a detailed forward model of the measurement. The data itself should describe this as far as possible.

As an interface we can use any linear transformation of the measurements, as long as the error description and characterisation are similarly transformed, and the linearisation is valid within the range of the errors. The transformation may be of the original data or of a retrieval. The aim is that each measurement $\mathbf{y}_i$ is to be described in the generic form

$$\mathbf{y}_i = \mathbf{K}_i(\mathbf{x} - \mathbf{x}_i) + \epsilon \tag{8.16}$$

where the supplier provides a measurement $\mathbf{y}_i$, a corresponding linearisation point $\mathbf{x}_i$ and weighting function matrix $\mathbf{K}_i$, and the covariance of the error $\mathbf{S}_\epsilon$, perhaps separated into random and systematic components. The question is how to do this efficiently (e.g. only diagonal covariances) while including all of the information content of the measurement.

8.3.1 Choice of profile representation

The first practical consideration is in the choice of state vector. A data supplier may have several customers who use different internal representations in their assimilation models, and different from that which the supplier uses for his own retrieval and validation process. The supplier should use a state vector which is capable of representing all of the information content of the measurement, and which can be obtained from the different assimilators' state vectors using e.g. interpolation operators $\mathbf{W}$.

8.3.2 Linearised measurements

One option for an interface is that of a linearisation of the measurements, transformed in such a way that the amount of data to be conveyed is minimum. We can transform the data in such a way that the error covariance is a unit matrix (or at least diagonal) so that the components can be assimilated independently, and the inversion of the error covariance matrix is trivial. The first step is to scale by $\mathbf{S}_\epsilon^{-\frac{1}{2}}$ so that the measurement error covariance is a unit matrix:

$$\begin{align}
\mathbf{y}' &= \mathbf{S}_\epsilon^{-\frac{1}{2}}[\mathbf{y} - \mathbf{F}(\mathbf{x}_0, \mathbf{b})] \tag{8.17} \\
&= \mathbf{S}_\epsilon^{-\frac{1}{2}}\mathbf{K}(\mathbf{x} - \mathbf{x}_0) + \mathbf{S}_\epsilon^{-\frac{1}{2}}\epsilon \tag{8.18} \\
&= \mathbf{K}'(\mathbf{x} - \mathbf{x}_0) + \epsilon' \tag{8.19}
\end{align}$$

where $\mathbf{x}_0$ is the linearisation point, which could conveniently be the data provider's own retrieval, in which case the elements of $\mathbf{y}'$ would be of order unity. The data supplier could provide $\mathbf{y}'$, $\mathbf{K}'$ and $\mathbf{x}_0$ for each measurement, but for the apparently overdetermined case $m > n$, this is more data than is needed. We can reduce the dimensionality of $\mathbf{K}'$ and $\mathbf{y}'$ to at most n by various techniques, of which singular value decomposition is probably the most convenient, and at the same time we can eliminate the need to provide $\mathbf{x}_0$. Express $\mathbf{K}'$ as $\mathbf{U}\mathbf{\Lambda}\mathbf{V}^T$ and write:

$$\mathbf{y}' + \mathbf{K}'\mathbf{x}_0 = \mathbf{K}'\mathbf{x} + \boldsymbol{\epsilon}' = \mathbf{U}\mathbf{\Lambda}\mathbf{V}^T\mathbf{x} + \boldsymbol{\epsilon}' \qquad (8.20)$$

and define $\mathbf{y}''$ by

$$\mathbf{y}'' = \mathbf{U}^T(\mathbf{y}' + \mathbf{K}'\mathbf{x}_0) = \mathbf{\Lambda}\mathbf{V}^T\mathbf{x} + \boldsymbol{\epsilon}''. \qquad (8.21)$$

The covariance of $\boldsymbol{\epsilon}'' = \mathbf{U}^T\boldsymbol{\epsilon}'$ is still a unit matrix. As only p elements of $\mathbf{\Lambda}$ are non-zero, where p is the rank of $\mathbf{K}'$, $\mathbf{y}''$ need only have p elements. The data supplier now provides p elements of $\mathbf{y}''$ and the $p \times n$ matrix $\mathbf{K}'' = \mathbf{\Lambda}\mathbf{V}^T$. It is likely that elements corresponding to values of λ_i rather less than unity can also be omitted, as these are in the near-null space.

This approach leaves the measurement forward model to the data provider, without increasing the data to be transferred and stored too much. The amount of data transferred depends on the information content of the measurement, rather than the number of channels or state vector elements.

The data supplier:

(i) Carries out a formal error analysis to obtain $\mathbf{S}_\epsilon$.
(ii) Carries out his own retrieval, obtaining a value for $\mathbf{x}_0$. As part of this process he should compute $\mathbf{K}$ and $\mathbf{y} - \mathbf{F}(\mathbf{x}_0, \mathbf{b})$.
(iii) Finds the SVD of $\mathbf{S}_\epsilon^{-\frac{1}{2}}\mathbf{K}$.
(iv) Provides the assimilator with $\mathbf{y}'' = \mathbf{U}^T\mathbf{S}_\epsilon^{-\frac{1}{2}}[\mathbf{y} - \mathbf{F}(\mathbf{x}_0, \mathbf{b}) + \mathbf{K}\mathbf{x}_0]$, and $\mathbf{K}''$.

The bulk of the data will be in $\mathbf{K}''$, but this is essential as it is the transformed tangent linear model for the measurement, and releases the assimilator from understanding the forward model.

The assimilator's task is now straightforward, and linear. The linear pseudo-measurements $\mathbf{y}''$ are assimilated using the weighting functions $\mathbf{K}''$ and unit error covariance. This approach does not require the assimilator to understand the instrument and its error analysis by providing a linearisation of both in a convenient form. Apart from the SVD calculation, the data provider will be doing all of the calculations himself in any case.

8.3.3 *Systematic errors*

Systematic errors are correlated between successive and hence geographically close measurements, and random errors are uncorrelated. The usual approach in retrieval and in assimilation is to lump them both together, and treat them as random errors. In terms of retrievals, an individual retrieval may still be a correct MAP or ML estimate, but if retrievals are combined, as in constructing assimilations or climatologies, then the error estimation of the combination is likely to be wrong. When averaging profiles, random errors are reduced an $N^{-\frac{1}{2}}$, but systematic errors are not. As far as possible the two types of errors should be treated separately, with systematic errors providing a lower limit below which errors cannot be reduced by averaging. With modern remote sounders, it is common to find that systematic errors dominate.

The linearised instrument model can be written as

$$\mathbf{y} = \mathbf{F}(\mathbf{x}_0, \mathbf{b}) + \mathbf{K}(\mathbf{x} - \mathbf{x}_0) + \boldsymbol{\epsilon}_r + \boldsymbol{\epsilon}_s \quad (8.22)$$

where $\boldsymbol{\epsilon}_r$ and $\boldsymbol{\epsilon}_s$ are the random and systematic errors respectively. To reduce this to the same kind of simple form as above, we will need more steps. We first reduce the measurement to $\mathbf{y}''$ with p elements, as above, but after first scaling by $(\mathbf{S}_{\epsilon_r} + \mathbf{S}_{\epsilon_s})^{-\frac{1}{2}}$. The scaling is not essential, but convenient to reduce everything to non-dimensionality, and approximately unit size. This will give a representation as before, with p elements:

$$\mathbf{y}'' = \mathbf{U}^T(\mathbf{y}' + \mathbf{K}'\mathbf{x}_0)' = \mathbf{\Lambda}\mathbf{V}^T\mathbf{x} + \boldsymbol{\epsilon}''_r + \boldsymbol{\epsilon}''_s, \quad (8.23)$$

where $\mathbf{S}_{\epsilon''_r}$ and $\mathbf{S}_{\epsilon''_s}$ are not unit matrices, but their sum is. In order to avoid having to specify one of the matrices in full, we can carry out a further process of diagonalisation by scaling the result by the eigenvectors of $\mathbf{S}_{\epsilon''_r}$.

⇒ **Exercise** 8.1: Follow through the suggested algebra and find an expression for the final transformed linearised instrument model.

Thus information about a linearised measurement, together with separate diagonal random and systematic error descriptions can be constructed for use in an assimilation.

8.3.4 Transformation of a characterised retrieval

As an alternative we may consider using a retrieval and its characterisation as discussed in chapter 3 to construct an interface to assimilation. The retrieval is related to the state by

$$\hat{\mathbf{x}} = \mathbf{x}_a + \mathbf{A}(\mathbf{x} - \mathbf{x}_a) + \boldsymbol{\epsilon}_x = \mathbf{A}\mathbf{x} + (\mathbf{I} - \mathbf{A})\mathbf{x}_a + \boldsymbol{\epsilon}_x \quad (8.24)$$

Using $\hat{\mathbf{x}}$ with covariance $\hat{\mathbf{S}}$ for assimilation is clearly inappropriate, because it contains contributions from $\mathbf{x}_a$ which may have been derived from the numerical prediction itself. The proper thing to do is to regard $\hat{\mathbf{x}} - \mathbf{x}_a$ as a measurement, using $\mathbf{x}_a$ as no more than a linearisation point, with weighting function $\mathbf{A}$ and error covariance $\mathbf{S}_x$. Alternatively we could take $\hat{\mathbf{x}} - (\mathbf{I} - \mathbf{A})\mathbf{x}_a$ to be a measurement of $\mathbf{A}\mathbf{x}$ with the same error covariance. Both of these have the effect of eliminating the *a priori* from the information provided to the assimilator. To minimise the data quantity the approach described in section 8.3.2 for radiances can be used, as the equations are of exactly the same form.

Joiner and da Silva (1998) have proposed two methods closely related to this approach, but using different ways of eliminating the null space part of the retrieval to reduce data quantity. The first method they call 'null-space filtering', and projects the retrieval onto the null-space component of either $\mathbf{I} - \mathbf{A}$ or $(\mathbf{I} - \mathbf{A})\mathbf{S}_a(\mathbf{I} - \mathbf{A})$, defined by means of the singular vectors or eigenvectors respectively that have negligible singular- or eigen-values. These both eliminate most, but not all of the *a priori* component without needing to know $\mathbf{x}_a$. The second method is 'partial eigen-decomposition', in which the retrieval is expressed in terms of the leading eigenvectors of $\mathbf{K}^T\mathbf{S}_\epsilon^{-1}\mathbf{K}$. This is essentially the same as the use of linearised measurements suggested in section 8.3.2, as these eigenvectors are the same as the singular vectors of $\mathbf{K}' = \mathbf{S}_\epsilon^{-\frac{1}{2}}\mathbf{K}$.

Chapter 9

Numerical Methods for Forward Models and Jacobians

The heart of a successful and accurate retrieval method is the forward model. If it does not accurately represent the physics of the measurement then there is little hope that the retrieval will be satisfactory. Thus the forward model must include all of the relevant physics, but it must also be numerically efficient when there are large amounts of data to be processed. At the same time as designing the forward model, we must also design the model which computes the weighting functions, or Jacobian of the forward model. The weighting function evaluation is usually the most important part of the process in terms of computing speed, and needs especial care.

In this chapter I have gathered together some approaches to the commonest calculations that are needed for forward models. The emphasis is of course on radiative transfer, but there are related practical topics such as integration of the hydrostatic equation. The chapter is not intended as a comprehensive treatment of radiative transfer in planetary atmospheres, for that the reader should consult for example Goody and Yung (1989), Andrews, Holton and Leovy, (1987), Liou (1992) or Thomas and Stamnes (1999).

9.1 The Equation of Radiative Transfer

The monochromatic radiance emerging from an atmospheric path can be written as:

$$L(\nu, z) = L(\nu, 0)\tau(\nu, 0, z) + \int_0^z J(z') \frac{\mathrm{d}}{\mathrm{d}z'} \tau(\nu, z', z) \, \mathrm{d}z', \qquad (9.1)$$

where z is any convenient distance coordinate along the path, $L(\nu, 0)$ is the radiance at wavenumber ν incident at one end of the path, $L(\nu, z)$ is the radiance emerging at the other, $\tau(\nu, z', z)$ is the transmittance of the path from z' to z, and $J(z)$ is the source function which depends on both thermal emission and scattering. In local thermodynamic equilibrium (LTE) it is given by

$$J(z') = [1 - \varpi(\nu)]B[\nu, T(z')] + \frac{\varpi(\nu)}{4\pi} \int_{4\pi} L(\nu, z, \mathbf{\Omega}')P(\nu, \mathbf{\Omega}, \mathbf{\Omega}') \, d\mathbf{\Omega}', \qquad (9.2)$$

where $B[\nu, T(z')]$ is the Planck function at temperature $T(z')$, $\varpi(\nu)$ is the single scattering albedo, the ratio of the scattering coefficient to the total extinction coefficient:

$$\varpi(\nu) = \sigma(\nu)/[\kappa(\nu) + \sigma(\nu)], \qquad (9.3)$$

$L(\nu, z, \mathbf{\Omega}')$ is the radiance incident at z from direction $\mathbf{\Omega}'$, and $P(\nu, \mathbf{\Omega}, \mathbf{\Omega}')$ is the phase function, describing the efficiency of scattering from $\mathbf{\Omega}'$ into $\mathbf{\Omega}$.

Most remote measurements are described by variants of Eqs. (9.1) and (9.2). Measurements of absorption of solar (or other) radiation by atmospheric gases is described by the first term on the right hand side of (9.1), where $L(\nu, 0)$ is the solar (or other) radiance, and the transmittance is a function of the absorber distribution. Measurements of thermal emission require both terms in Eq. (9.1), but only the first term of Eq. (9.2), although in the case of limb measurements $L(\nu, 0)$ is the emission of cold space, and will be negligible except in the microwave region. Measurements of e.g. UV backscattering require only the second term of Eq. (9.2), and thermal emission may be ignored. For the moment the scattering term will be ignored for simplicity.

Most instruments have a finite spectral bandpass, so that the quantity measured in channel l with normalised spectral response is $F_l(\nu)$ is:

$$L_l(z) = \int F_l(\nu)L(\nu, 0)\tau(\nu, 0, z) \, d\nu + \int F_l(\nu) \int_0^z J[\nu, T(z')] \frac{d}{dz'}\tau(\nu, z', z) \, dz' \, d\nu \qquad (9.4)$$

where $\int F_l(\nu) \, d\nu = 1$. The absorption, as expressed by the transmittance function, may be due to molecular vibration-rotation transitions, or to a range of possible continuum processes. Transmittance is related to the absorption coefficient $\kappa_i(\nu, z'')$ by:

$$\tau(\nu, z', z) = \exp\Big[-\int_{z'}^{z} \sum_i \kappa_i(\nu, z'')\rho_i(z'') \, dz''\Big], \qquad (9.5)$$

where i refers to the i-th absorber and $\rho_i(z'')$ is its density. The absorption coefficient may be a sum over a large number of spectral lines:

$$\kappa_i(\nu, z'') = \sum_j k_{ij}[T(z'')]f_{ij}[\nu, p(z''), T(z'')], \qquad (9.6)$$

where the strength k_{ij} of the j-th line of the i-th absorber is temperature dependent, and its normalised shape $f_{ij}[\nu, p(z''), T(z'')]$ may also depend on pressure p. There is also the phenomenon of 'line mixing' which sometimes must be accounted for

Fig. 9.1 The level and layer numbering convention used here.

(e.g. Rosencrantz, 1975, Strow, 1988). In this case the interaction of the spectral lines of a molecular band is more complex than simple addition.

Clearly a forward model based on this physics must be either very time consuming or be approximated in various ways. A model which evaluates the monochromatic radiance explicitly according to the above prescription is called a 'line-by-line' model. Some well known line-by-line codes which are generally available FASCODE (Clough, et al., 1985), GENLN2 (Edwards, 1992) and LINEPACK (Gordley, 1994).

Information about line strengths and widths for atmospheric molecules can be found in the HITRAN data base, described by Rothman et al., (1996), and available from http://www.hitran.com. The physics of molecular vibration-rotation bands is discussed in many texts, for example Banwell and McCash (1994).

9.2 The Radiative Transfer Integration

To carry out the outer integration, with respect to z' in Eq. (9.1), we require a set of quadrature points. The path will be divided into layers (we will number them from 1 to n), with path parameters such as pressure, temperature and mixing ratios specified either at the layer boundaries ('levels', numbered from 0 to n, so that layer i lies between levels $i-1$ and i, see Fig. 9.1) or as layer averages. Some interpolation rule must be specified so that the path parameters are defined at all points in the path. This set of levels for the forward model is not necessarily the same as the grid defined for the state vector.

The outer integral can be written in several equivalent forms:

$$L(z) = L(0)\tau(0,z) + \int_0^z B(z') \frac{d\tau(z',z)}{dz'} \, dz' \tag{9.7}$$

$$= L(0)\tau(0,z) + \int_{\tau(0,z)}^1 B(\tau) \, d\tau \tag{9.8}$$

$$= [L(0) - B(0)]\tau(0,z) + B(z) - \int_0^z \tau(z',z) \frac{dB(z')}{dz'} \, dz' \tag{9.9}$$

$$= [L(0) - B(0)]\tau(0,z) + B(z) - \int_{B(0)}^{B(z)} \tau[z(B'),z] \, dB' \tag{9.10}$$

⇒ **Exercise 9.1**: Derive the third form from the first

of which a numerical version of the second form is popular. In this, a set of levels is defined, and the transmittance computed from each level to the end of the path. The Planck function is taken to be a constant, $\bar{B}_i$, within each layer, so that the radiance is given by a sum over layers:

$$L_n = L_0 \tau_0 + \sum_1^n \bar{B}_i (\tau_i - \tau_{i-1}) \tag{9.11}$$

where $\tau_i = \tau(z_i, z_n)$. This is simply the histogram rule for numerical integration applied to Eq. (9.8). Expressions can be derived based on other quadratures, such as the trapezium rule, Simpson's rule or Gaussian quadrature. Another possibility is to fit a cubic spline to $B(\tau)$ using the levels as nodes, and integrate the spline. A form similar to Eq. (9.10) but actually derived from Eq. (9.11) by rearranging terms is:

$$L_n = (L_0 - \bar{B}_1)\tau_0 + \bar{B}_n + \sum_1^{n-1} \tau_i (\bar{B}_i - \bar{B}_{i-1}) \tag{9.12}$$

A recursive evaluation, based using Eq. (9.11) successively for each layer, is to repeat for $i = 1 \ldots n$:

$$L_i = L_{i-1} \tilde{\tau}_i + \bar{B}_i (1 - \tilde{\tau}_i) = \bar{B}_i + (L_{i-1} - \bar{B}_i) \tilde{\tau}_i \tag{9.13}$$

where $\tilde{\tau}_i$ is the effective transmittance of the layer:

$$\tilde{\tau}_i = \frac{\tau(0, z_i)}{\tau(0, z_{i-1})}. \tag{9.14}$$

In the monochromatic case where $\tau = \exp(-\chi)$, this is the same as the layer transmittance, but not if τ is the mean over some spectral region. The term $\bar{B}_i(1 - \tilde{\tau}_i)$, which corresponds to $\int B(\tau) \, d\tau$, can be improved if some reasonable functional form for $B(\tau)$ is assumed in each layer. The first suggestion might be to assume that $B(\tau)$ is linear in τ, giving the trapezium rule integral. However this is no better than the histogram rule, the lowest order error term in τ being a factor of two worse. A better suggestion for the monochromatic case is to assume that B is linear in optical thickness χ, on the basis that temperature is linear in height, e.g. Wiscombe (1976). In this case the integral over layer i becomes

$$\int B(\tau) \, d\tau = \int_0^{\chi_i} \left[B_i + (B_{i-1} - B_i) \frac{\chi}{\chi_i} \right] e^{-\chi} \, d\chi \tag{9.15}$$

$$= B_i + \frac{B_{i-1} - B_i}{\chi_i} \left(1 - e^{-\chi_i}\right) - (B_{i-1} - B_i) e^{-\chi_i} \tag{9.16}$$

where χ_i is the optical thickness of the layer and χ is measured from level i. In the optically thin limit, this gives $\chi_i (B_{i-1} + B_i)/2$, as expected, and in the optically thick limit, when $\exp(-\chi_i) \to 0$ it gives $B_i - (B_{i-1} - B_i)/\chi_i$, the Planck function at one optical depth into the path, again as expected. In the general case it can be rearranged to be of the form (Clough et al., 1992)

$$\int B(\tau)\,\mathrm{d}\tau = B(\chi^*)(1 - \tilde{\tau}_i) \tag{9.17}$$

where $\chi^* = 1 - \chi/(e^\chi - 1)$, i.e. the effective Planck function of the layer is the actual Planck function at optical depth χ^* from level i. Its limits are, as expected, unity in the thick limit and $\chi/2$ in the thin limit. Care is needed to evaluate it numerically as $\chi \to 0$. The 'linear in χ' approximation requires somewhat more calculation, but produces a more accurate result (Clough et al., 1992), and may allow fewer and thicker layers to be used.

9.3 Derivatives of Forward Models: Analytic Jacobians

Virtually all retrieval methods make use of derivatives of the forward model, so we must consider ways to evaluate them efficiently. For a complicated model it may appear that the simplest way is to evaluate the forward model for a reference state vector, and re-evaluate it successively for the same state vector but with each element in turn perturbed by some small amount. This process is straightforward to program, but takes a long time to evaluate relative to the original forward model, and may be either subject to rounding errors if the perturbation is too small, or to nonlinearity if too large.

If the retrieval model is to be used for large quantities of data, for example from satellites, a better approach is to write computer code for the derivative of the forward model, even though that may be tedious and difficult. It is usually possible to make enough savings in processing time that the effort is worth it. A simple example of how savings arise can be seen by considering a forward model which is the transmittance of an atmospheric path of n layers and a state vector which contains the absorber amount x_i in each layer. In this case the transmittance is evaluated as

$$\tau = \exp\left(-\sum_{i=1}^{n} \kappa_i x_i\right) \tag{9.18}$$

where the absorption coefficient in layer i is κ_i. Its derivative with respect to x_k is simply

$$\frac{\partial \tau}{\partial x_k} = \kappa_k \exp\left(-\sum_{i=1}^{n} \kappa_i x_i\right) \tag{9.19}$$

which can be evaluated at the same time as τ at the cost of a single multiplication per layer, rather than n complete re-evaluations of τ with a perturbed values of each x_i.

It is possible to approach the coding of analytic Jacobians in a systematic manner. The method is simply an application of the chain rule of differentiation to the computer code of the forward model. In order to get second order convergence, it is important that the derivative evaluated is that of the forward model as actually implemented, rather than a separate implementation of the derivative of the original forward model algebra. I will illustrate this with is slightly more complicated example than the simply the transmittance, namely the thermal emission from the same atmospheric path, where the state vector includes both layer temperatures T_i and absorber mixing ratios q_i.

The emitted radiance is computed by dividing the path into homogeneous segments, and evaluating the radiative transfer equation discretely. Thus we can write the functional dependence of the radiance as

$$L = L(\boldsymbol{\chi}, \mathbf{b}) \tag{9.20}$$

where $\boldsymbol{\chi}$ and $\mathbf{b}$ are vectors of the optical depth and the Planck function respectively in each path segment, and L represents for example Eq. (9.11) with $\tau = \exp(-\sum \chi_i)$.

The derivative of L with respect to the state vector $\mathbf{x} = (\mathbf{T}, \mathbf{q})$ can be written

$$\frac{\partial L}{\partial \mathbf{x}} = \left(\frac{\partial L}{\partial \boldsymbol{\chi}}\right)_{\mathbf{b}} \left(\frac{\partial \boldsymbol{\chi}}{\partial \mathbf{x}}\right) + \left(\frac{\partial L}{\partial \mathbf{b}}\right)_{\chi} \left(\frac{\partial \mathbf{b}}{\partial \mathbf{x}}\right) \tag{9.21}$$

The partials of L on the right hand side of this expression are easily evaluated at the same time as L, but those with respect to $\mathbf{x}$ need a little more attention. The term $\partial \mathbf{b}/\partial \mathbf{x}$ is in principle a matrix, but in practice only the elements $\partial B_k/\partial T_k$ are non zero. They can be easily calculated at the same time as B_k. If $B(T)$ is evaluated from the explicit expression for the Planck function, then some of the components such as $\exp(h\nu/k_B T)$ can be reused. If it is obtained by interpolation in a table in order to save time, then the gradient can be also obtained from the interpolation. The term $\partial \boldsymbol{\chi}/\partial \mathbf{x}$ is also a matrix. Elements of the optical thickness vector are related to the absorption coefficients and the absorber amounts u_i by an expression such as:

$$\chi_i = \sum_{k=1}^{i} \kappa_k u_k \tag{9.22}$$

giving

$$\left(\frac{\partial \chi_i}{\partial \mathbf{x}}\right) = \sum_k u_k \left(\frac{\partial \kappa_k}{\partial \mathbf{x}}\right) + \kappa_k \left(\frac{\partial u_k}{\partial \mathbf{x}}\right) \tag{9.23}$$

A more complicated expression for χ_i could clearly be accommodated, for example one involving level values and an integration rather than segment means.

For elements x_j of $\mathbf{x}$ which are mixing ratios, $\partial \kappa_k/\partial x_j = 0$, and $\partial u_k/\partial x_j = u_{ak}$, the total amount of air in the segment. For elements which are temperatures,

$\partial \kappa_k / \partial x_j$ is zero unless x_j refers to the same segment as κ_k, when it is found from the temperature dependence of the absorption coefficient, easily evaluated at the same time as κ_k by the interpolation procedure.

The only slightly complicated term for elements which are temperatures is $\partial u_k / \partial x_j$, the derivative of absorber amount in a layer with respect to temperature. If the model grid is on pressure levels, then these terms are zero, but if on height levels then the hydrostatic and gas equations are involved, but the algebra is straightforward. If refraction is involved, however, it may become quite complicated!

In all of the calculations described, opportunities are found for saving time, because many terms that would be evaluated in a perturbation method are simply not needed. It is typically found that a combined forward model and analytic Jacobian calculation takes about two or three times as long as a forward model alone, which is a major improvement on the factor of n required by the perturbation method. However the perturbation method is still needed as a check on the coding of the analytic method. The general principles illustrated here can be applied to almost any forward model.

9.4 Ray Tracing

The atmosphere is a refractive medium, so when computing the signal to be expected at an instrument we must consider whether straight line propagation is an adequate approximation or whether refraction must be taken into account. Of course in the case of radio occultation the refraction is itself the target of the measurement, but in almost any other occultation or limb emission measurement, refraction is likely to by important. The only exception is at very high altitudes where the density and hence the refractivity is small.

The International Association of Geodesy in 1999 adopted the following formula, based on Ciddor (1996), for the variation of refractive index with temperature T in Kelvin, and pressure p and water vapour partial pressure e in mb:

$$n = 1 + \left(N_g \frac{p}{p_0} \frac{T_0}{T} - \frac{11.27 e}{T} \right) \times 10^{-6} \qquad (9.24)$$

where $p_0 = 1013.25$ mb, $T_0 = 273.15$ K and N_g, the refractivity (defined as $10^6 \times (n-1)$) of dry air at STP, is given by

$$N_g = 287.6155 + 4.8866 \lambda^{-2} + 0.068 \lambda^{-4} \qquad (9.25)$$

at wavelength λ in micrometers. For more detail, see Ciddor (1996). We can see that $n - 1 \simeq 3 \times 10^{-4}$ at surface pressure, and has a significant dependence on water vapour in the troposphere. Given the distribution of refractivity, we wish to be able to calculate the trajectory of a ray from or to an instrument, and perhaps to integrate the equation of radiative transfer along it.

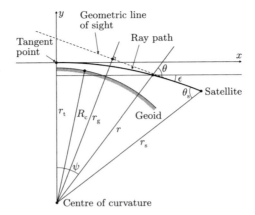

Fig. 9.2 The coordinate system for analysing the propagation of a ray through the atmosphere.

Consider the propagation of the ray shown in Fig. 9.2. It easy to see that the curvature of a ray is proportional to the component of the gradient of refractive index at right angles to the ray:

$$\frac{d\epsilon}{ds} = \frac{1}{n}\left(\frac{\partial n}{\partial t}\right)_s \qquad (9.26)$$

where ϵ is the direction of propagation, s, t are coordinates along and perpendicular to the ray respectively.

⇒ **Exercise** 9.2: Explain why this is the case.

In Cartesian coordinates this can be written:

$$\frac{d\epsilon}{ds} = -\frac{\sin\epsilon}{n}\left(\frac{\partial n}{\partial y}\right)_x + \frac{\cos\epsilon}{n}\left(\frac{\partial n}{\partial x}\right)_y \qquad (9.27)$$

where ϵ is measured from the x-axis so that $dy/dx = \tan\epsilon$.

9.4.1 Choosing a coordinate system

The Earth is not spherical, therefore we must either use a complicated coordinate system, or make an approximation. For this purpose it is convenient use a reference ellipsoid to represent the sea-level geoid, and choose a coordinate system relative to that. Ignore the effect of horizontal gradients of refractive index at right angles to the direction of propagation, so that all of the refraction takes place in a vertical plane through the line of sight. We approximate the local shape of the geoid by a circle with the same curvature as the intersection of the ellipsoid with that plane, with a radius of curvature R_c. This is different from the local radius of the Earth, R_e. At the poles, which are flattened, it would be greater, while at the equator it is equal to R_e for an east-west line of sight, $\phi = \pm 90°$, and less than R_e in all other directions.

The radius of curvature of an ellipsoid at geodetic latitude ϕ, in a direction with azimuth angle α to the meridian plane is given by

$$R_c = \left(R_{NS}^{-1} \cos^2 \alpha + R_{EW}^{-1} \sin^2 \alpha \right)^{-1} \tag{9.28}$$

where the N-S and E-W radii of curvature are

$$R_{NS} = R_q^2 R_p^2 (R_q^2 \cos^2 \phi + R_p^2 \sin^2 \phi)^{-\frac{3}{2}} \tag{9.29}$$

$$R_{EW} = R_q^2 (R_q^2 \cos^2 \phi + R_p^2 \sin^2 \phi)^{-\frac{1}{2}} \tag{9.30}$$

and the equatorial and polar radii are, for the WGS-84 reference ellipsoid, $R_q = 6378.138$ km and $R_p = 6356.752$ km respectively.

9.4.2 Ray tracing in radial coordinates

For ray tracing in the atmosphere, radial coordinates are more appropriate, especially if the atmosphere is horizontally homogeneous. Consider a ray propagating in a plane in the coordinate system defined by Fig. 9.2. An element of the path lies at radius r from the origin, at an angle ψ to a reference direction, and the ray is propagating at angle θ to the radius vector. The element of distance along the ray is ds. We can describe the refraction in terms of the radial equivalent of equation Eq. (9.27), together with two equations relating the coordinates:

$$\frac{d(\theta + \psi)}{ds} = -\frac{\sin \theta}{n} \left(\frac{\partial n}{\partial r} \right)_\psi + \frac{\cos \theta}{nr} \left(\frac{\partial n}{\partial \psi} \right)_r \tag{9.31}$$

$$\frac{dr}{ds} = \cos \theta \tag{9.32}$$

$$\frac{d\psi}{ds} = \frac{\sin \theta}{r} \tag{9.33}$$

where n is a function of r and ψ. We can use Eqs. (9.31) and (9.33) to obtain

$$\frac{d\theta}{ds} = -\sin \theta \left[\frac{1}{r} + \frac{1}{n} \left(\frac{\partial n}{\partial r} \right)_\psi \right] + \frac{\cos \theta}{nr} \left(\frac{\partial n}{\partial \psi} \right)_r \tag{9.34}$$

In principle we can integrate Eqs. (9.32), (9.33) and (9.34) numerically using any standard technique such as Runge–Kutta to give $r(s)$, $\psi(s)$ and $\theta(s)$. We can of course use any of s, r, ψ or θ, or various combinations of them, as the independent variable.

9.4.3 Horizontally homogeneous case

The differential equations of section 9.4.2 cannot be solved explicitly in the general case, but there are simple situations in which solutions can be found. In the case when the atmosphere is horizontally homogeneous, or circularly symmetric in the plane of the ray, the ψ derivative is zero, and as n is a function of r only, $\partial n/\partial r = \mathrm{d}n/\mathrm{d}r$. In this case Eq. (9.34) can be written:

$$\frac{\mathrm{d}\theta}{\mathrm{d}s} = -\sin\theta \left[\frac{1}{r} + \frac{1}{n}\frac{\mathrm{d}n}{\mathrm{d}r}\right] \quad (9.35)$$

using Eq. (9.32) this becomes

$$\frac{\cos\theta}{\sin\theta}\frac{\mathrm{d}\theta}{\mathrm{d}r} = -\left[\frac{1}{r} + \frac{1}{n}\frac{\mathrm{d}n}{\mathrm{d}r}\right] \quad (9.36)$$

which integrates explicitly to give Snell's Law in circular symmetry:

$$nr\sin\theta = r_g \quad (9.37)$$

where r_g is a constant, the geometric tangent point or *impact parameter*. It can be expressed in terms of $nr\sin\theta$ at the tangent point or at the satellite:

$$r_g = n(r_t)r_t = r_s \sin\theta_s \quad (9.38)$$

where r_t is the radius at the tangent point where $\theta = 0$ and r_s is the radius at the instrument, where $\theta = \theta_s$ and $n = 1$. The impact parameter is the radius at the tangent point in the absence of refraction for a view at zenith angle θ_s from the instrument. We can now use this relationship between r and θ to eliminate θ in Eq. (9.32) and integrate to obtain for distance along the ray

$$\int \mathrm{d}s = \int \frac{n(r)r}{(n^2r^2 - r_g^2)^{\frac{1}{2}}} \mathrm{d}r \quad (9.39)$$

and, using Eq. (9.33), for the horizontal position of an element of the path

$$\int \mathrm{d}\psi = \int \frac{r_g}{r(n^2r^2 - r_g^2)^{\frac{1}{2}}} \mathrm{d}r \quad (9.40)$$

The direction of propagation ϵ is $\theta + \psi - \pi/2$. Using Eq. (9.31) to give $\mathrm{d}\epsilon/\mathrm{d}s$ together with Eq. (9.39) we get

$$\int \mathrm{d}\epsilon = -\int \frac{1}{n}\frac{\mathrm{d}n}{\mathrm{d}r}\frac{r_g}{(n^2r^2 - r_g^2)^{\frac{1}{2}}} \mathrm{d}r \quad (9.41)$$

$$= -\int \frac{\mathrm{d}\ln n}{\mathrm{d}(nr)}\frac{r_g}{(n^2r^2 - r_g^2)^{\frac{1}{2}}} \mathrm{d}(nr). \quad (9.42)$$

The forward model for radio occultation is the total bending angle, integrated over the whole path. The second of these two forms shows that it is an Abel transform, (e.g. Arfken, 1995) which can be explicitly inverted to provide an explicit 'exact' inverse method for the radio occultation problem.

These integrations may in principle be carried out numerically using any standard quadrature method, integrating from the spacecraft, the tangent point or the occulted source as required by the problem. Unfortunately the integrand has a singularity at the tangent point which, while not giving an infinity, does cause problems for a numerical method if not properly treated.

A simple approach is that described by Kniezys et al., (1983), in which the variable of integration is changed to $x = r\cos\theta = n^{-1}(n^2r^2 - r_g^2)^{\frac{1}{2}}$, a function of r only. This gives

$$\frac{dx}{ds} = \cos\theta \frac{dr}{ds} - r\sin\theta \frac{d\theta}{ds} \quad (9.43)$$

Using Eqs. (9.32) and (9.35) we obtain

$$\frac{dx}{ds} = 1 + \frac{r}{n}\frac{dn}{dr}\sin^2\theta \quad (9.44)$$

allowing us to obtain s by numerically integrating

$$\int ds = \int \frac{dx}{1 - \gamma(r)\sin^2\theta} = \int \frac{n^2r^2\, dx}{n^2r^2 - \gamma(r)r_g^2} \quad (9.45)$$

where $\gamma(r) = -(r/n)\,dn/dr$ is the ratio of r to the radius of curvature of a horizontal ray at radius r. This form does not have a singularity at the tangent point, provided $\gamma < 1$, and can be integrated numerically without difficulty. Related forms can also be derived for ψ and ϵ:

$$\int d\psi = \int \frac{nr_g\, dx}{n^2r^2 - \gamma(r)r_g^2} \quad (9.46)$$

$$\int d\epsilon = \int \frac{nr_g\gamma(r)\, dx}{n^2r^2 - \gamma(r)r_g^2} \quad (9.47)$$

9.4.4 The general case

In the general case when there is a horizontal gradient of refractivity as well a vertical gradient, we must integrate the differential equations numerically. This can be carried out in radial coordinates, but it is probably better numerically to integrate Eq. (9.27), together with the geometric equations for $d\epsilon/dx$ and $d\epsilon/dx$, to obtain $\epsilon(s)$, $x(s)$ and $y(s)$ in Cartesian coordinates, as the variations are rather smaller than those of $\theta(s)$, $r(s)$ and $\psi(s)$ using radials. The horizontal gradient may be due to horizontal variations of air density and/or water vapour, and there will be a corresponding horizontal gradient of pressure. Hence there is an extra complication in the definition of the tangent point, which depends on whether we are using geometric or pressure coordinates. The lowest altitude of the ray is not necessarily at the same location as the highest pressure along the ray. Geometric coordinates are simpler for ray tracing, I will them used here.

The following analysis gives an indication of the size of the effect in terms of the change in tangent altitude. In the symmetric case, $nr\sin\theta$ is a constant, so let us see how it varies in the general case:

$$\frac{d(nr\sin\theta)}{ds} = nr\cos\theta\frac{d\theta}{ds} + n\sin\theta\frac{dr}{ds} + r\sin\theta\frac{dn}{ds} \tag{9.48}$$

On substituting from Eqs. (9.32) and (9.34) for $d\theta/ds$ and dr/ds, and using

$$\frac{dn}{ds} = \frac{\partial n}{\partial r}\frac{dr}{ds} + \frac{\partial n}{\partial \psi}\frac{d\psi}{ds} = \left(\frac{\partial n}{\partial r}\right)_\psi \cos\theta + \left(\frac{\partial n}{\partial \psi}\right)_r \frac{\sin\theta}{r} \tag{9.49}$$

we find that almost everything cancels and we are left with

$$\frac{d(nr\sin\theta)}{ds} = \left(\frac{\partial n}{\partial \psi}\right)_r \tag{9.50}$$

so that on integrating we get

$$nr\sin\theta = n(r_t)r_t + \int_{s_t}^{s}\left(\frac{\partial n}{\partial \psi}\right)_r ds = n(r_t)r_t + N(s), \tag{9.51}$$

thus defining $N(s)$. Putting $s = s_s$ at the satellite we obtain $r_g = n(r_t)r_t + N(s_s)$, giving an indication of how much the tangent point is perturbed by the horizontal inhomogeneity. We cannot evaluate $N(s_s)$ exactly without integrating to find the path, but we can make an estimate. Take $n(r) - 1$ to be exponentially decreasing in height with a scale height H, as it is approximately proportional to air density, take the slope of a constant refractivity surface $(\partial y/\partial x)_n$ to be a constant, $\tan\mu$, and integrate along an unrefracted ray from the tangent point. Then to first order in H/r_t we find that

$$N(s_s) \simeq [n(r_t) - 1]r_t(\pi r_t/2H)^{\frac{1}{2}}\tan\mu \tag{9.52}$$

With a scale height of about 7 km and $n - 1 \simeq 3 \times 10^{-4}$ at the surface, pressure p_0, we obtain for a ray with tangent pressure p_t

$$N(s_s) \simeq 70(p_t/p_0)\tan\mu \text{ km} \tag{9.53}$$

The slope of a pressure or density surface is proportional to the geostrophic wind velocity. As a worst case, a geostrophic wind in mid-troposphere of order $100\,\mathrm{m\,s^{-1}}$, e.g. in the region of an Atlantic low, corresponds to a slope of order 10^{-3}, so the largest error in tangent height due to ignoring horizontal variations would be of order 35 m, or about 0.5% in pressure or density and hence in path absorber amount. This is probably marginal, but may matter for high accuracy work, for example if trying to retrieve the geopotential height of a pressure surface.

9.5 Transmittance Modelling

In order to evaluate the radiative transfer equation we need to be able to model the transmittance of the atmospheric path. If the absorption coefficient is more or less independent of wavenumber, this is straightforward, but for much of the infrared where molecular vibration-rotation bands are important, the calculations are complicated and computationally expensive, as indicated by Eqs. (9.5) and (9.6), because of the large numbers of spectral lines that might be involved, and the fine spectral grid needed to accommodate the detailed spectral structure. For efficiency we must find either fast ways of evaluating the full equations, or sufficiently accurate approximations.

This section will not discuss the extra complications of scattering and departures from local thermodynamic equilibrium. Suitable texts include Goody and Yung (1989), Liou (1992), Thomas and Stamnes (1999) and López-Puertas and Taylor (2000).

9.5.1 Line-by-line modelling

For high accuracy work, for retrievals from high spectral resolution measurements, and for assessment of approximations it is necessary to be able to evaluate the transmittance of a band of spectral lines in detail. This is a topic which could stretch to several chapters, so I will refer the reader to the literature (e.g. Cowling, 1950; Clough et al., 1992; Gordley et al., 1994; Sparks, 1997; Kuntz and Hopfner 1999) . Here I will discuss only a few aspects related to efficiency of calculation and storage.

The main cost of a line-by-line calculation is in the evaluation of the absorption coefficient, which depends primarily on pressure and temperature, although there are a few circumstances when the partial pressures of the individual constituents are important, particularly when water vapour is involved. If absorption coefficient can be pretabulated as a function of pressure p, temperature T and wavenumber, then the transmittance from z_1 to z_2 at wavenumber ν is particularly simple:

$$\tau(\nu, z_1, z_2) = \exp\left\{ -\int_{z_1}^{z_2} \kappa[\nu, p(z), T(z)]\rho(z)q(z)\,\mathrm{d}z \right\} \qquad (9.54)$$

where $\rho(z)$ is the air density at position z and $q(z)$ is the absorber mixing ratio. In the case where the atmosphere is modelled by homogeneous layers, this simplifies to give Eq. (9.18). The primary considerations are now the size of the ν–p–T grid and the interpolation method, as these determine storage requirements, processing time and accuracy. There are trade-offs of storage versus accuracy in selecting the grid spacings, and between accuracy and processing time in selecting an interpolation method for a given spacing.

Strow et al. (1998) have developed a compression algorithm for transmittance tables which allows the storage to be reduced considerably at the expense of processing time in uncompressing the table. The absorption coefficient does not vary with pressure and temperature in arbitrarily different ways at all wavelengths. There are a few typical variations, such as proportionality to pressure in line wings. If the variation can expressed in terms of a few parameters, then the whole p–T dependence need not be stored for each wavenumber. Possible variations are identified statistically by using singular vectors of a matrix which tabulates absorption coefficient (or some simple function of it) against path conditions. Express the absorption coefficient table as a matrix $\mathbf{K}$ whose element $\mathcal{K}_{ij}$ is $\kappa(\nu_i, p_k, T_l)$ and $j = kn_T + l$, thus laying out the pressure and temperature dependence as a vector rather than as a matrix. This could clearly be extended to include partial pressure dependence if required. Consider the singular vector decomposition of $\mathbf{K}$:

$$\mathbf{K} = \mathbf{U}\mathbf{\Lambda}\mathbf{V}^T \tag{9.55}$$

If there are small singular values are small, then this can be approximated by

$$\mathbf{K} \simeq \sum_{\lambda_r > \epsilon} \lambda_r \mathbf{u}_r \mathbf{v}_r^T \tag{9.56}$$

where the sum is over the significant singular vectors only. The right vectors $\mathbf{v}$ comprise the parameterisation of the variability of the transmittance with pressure and temperature, and the left vectors $\mathbf{u}$ are the coefficients required. Strow finds that typically only a few to several tens of $\mathbf{u}$ coefficients are needed at each wavenumber, far fewer than the number of entries in an absorption coefficient table.

9.5.2 Band transmittance

The band transmittance approximation is useful for instruments which are of low spectral resolution relative to the spacing of spectral lines, and the use of a line-by-line or monochromatic approach would be computationally expensive. The thermal radiance L emerging from some optical path and measured by an instrument with a spectral response $f(\nu)$ is of the form:

$$L = \int\int f(\nu) B[\nu, T(z)] \frac{\mathrm{d}\tau(\nu, z)}{\mathrm{d}z} \, \mathrm{d}z \, \mathrm{d}\nu \tag{9.57}$$

where $f(\nu)$ is the instrumental spectral response, normalised so that $\int f(\nu)\, \mathrm{d}\nu = 1$. In the band transmission approximation we replace this by

$$L = \int \bar{B}[T(z)] \frac{\mathrm{d}\bar{\tau}(z)}{\mathrm{d}z} \, \mathrm{d}z \tag{9.58}$$

where $\bar{\tau}(z) = \int f(\nu)\tau(\nu, z)\, \mathrm{d}\nu$ and $\bar{B}(T) = \int f(\nu) B(\nu, T)\, \mathrm{d}\nu$.

⇒ **Exercise** 9.3: Show that the error in this approximation is:

$$\delta L = \int\int f(\nu)[B(\nu, T) - \bar{B}(T)] \frac{\mathrm{d}}{\mathrm{d}z}[\tau(\nu, z) - \bar{\tau}(z)] \, \mathrm{d}\nu\, \mathrm{d}z \tag{9.59}$$

The error is small if the spectral variation of transmittance, or of its height derivative, is uncorrelated with that of the Planck function. The band transmittance approximation is particularly valuable if the mean transmittance can be further approximated in terms of a small number of parameters such as an absorber amount, a pressure and a temperature only, and so can be pretabulated, and does not have to be computed by a double integration over wavenumber and path every time. Band transmittance functions may be explicit algebraic forms, such as the Goody or Malkmus random models (e.g. Goody and Yung, 1989), or may be tabulations based on line-by-line calculations.

When incorporated into an analytic Jacobian scheme, a band transmittance model needs to be easily differentiable with respect to its parameters, algebraically or numerically, and the parameters need to easily differentiable with respect to the relevant state vector elements.

9.5.3 *Inhomogeneous paths*

It is clear that the transmittance of a homogeneous path, defined only by temperature, pressure and absorber amount, will depend only on those three parameters, while the transmittance of a real atmospheric path will depend on the distribution of the parameters along the path. This is dealt with explicitly (and expensively) in a line-by-line model, but there are approximations which enable some savings to be made.

9.5.3.1 *Curtis–Godson approximation*

The C-G approximation (Curtis, 1953, Godson 1953, Goody and Yung, 1989) approximates the inhomogeneous path by a homogeneous one which agrees exactly in the strong Lorentz and the weak limits for constant temperature paths. In the weak limit, the transmittance depends on the absorber amount, and in the strong limit it depends on absorber amount multiplied by pressure. Hence the equivalent homogeneous path is taken to be defined by an amount u^{cg} at a pressure p^{cg} and temperature T^{cg} given by the following integrals along the path:

$$u^{\text{cg}} = \int \rho(x) q(x) \, dx \tag{9.60}$$

$$u^{\text{cg}} p^{\text{cg}} = \int \rho(x) q(x) p(x) \, dx \tag{9.61}$$

$$u^{\text{cg}} T^{\text{cg}} = \int \rho(x) q(x) T(x) \, dx \tag{9.62}$$

where $\rho(x)$ is air density at position x and $q(x)$ is the absorber mixing ratio. Curtis did not consider temperature variation, and Godson used a more complicated formulation for temperature, but the simple mass-weighted mean temperature has become the commonly used method. If the path is divided into layers each containing an amount m_i of air, the integrals become sums which are easily differentiated with respect to mixing ratio q_i and temperature T_i as part of the Jacobian calculation:

$$u^{\text{cg}} = \sum m_i q_i \tag{9.63}$$

$$u^{cg}p^{cg} \sum m_i q_i p_i \qquad (9.64)$$

$$u^{cg}T^{cg} \sum m_i q_i T_i \qquad (9.65)$$

but remember that m_i will depend on T_j through the hydrostatic and gas equations if the layers are specified geometrically, but not if they are specified by pressure levels. The Curtis–Godson approximation would be used with a band model to evaluate transmittance.

9.5.3.2 *Emissivity growth approximation*

The emissivity growth approximation (Weinreb and Neuendorffer,1973, Gordley and Russell, 1981) also treats the inhomogeneous path problem using a transmittance model for a homogeneous path, but does so in a completely different manner. The approach depends on the shape of the absorption spectrum of a path being more or less much independent of the details of the path, so it can be replaced by any homogeneous path which has the same transmittance. It steps incrementally along the inhomogeneous path, appending the effect of each increment of absorber to the transmittance (or the emissivity) computed so far, according to the following algorithm. Let the absorber path so far have spectral mean transmittance τ_i, and the increment of absorber to be appended to it have amount δu_{i+1}, pressure p_{i+1} and temperature T_{i+1}. An effective absorber amount u^*_{i+1} is found (e.g. by inverse interpolation in a table of $\tau(u,p,T)$) which satisfies

$$\tau(u^*_{i+1}, p_{i+1}, T_{i+1}) = \tau_i \qquad (9.66)$$

and the transmittance of the extended path is approximated by

$$\tau_{i+1} = \tau(u^*_{i+1} + \delta u_{i+1}, p_{i+1}, T_{i+1}), \qquad (9.67)$$

i.e. the parameters of the main path are adjusted so that p and T match the incremental path, whilst keeping its transmittance fixed at τ_i. This will be exact if the detailed spectral profile is the same for all paths having the same spectrally averaged transmittance, as is the case in the strong limit. In the weak limit, the spectral profile does not matter, and EGA will be exact here too.

The emissivity growth approximation can be differentiated straightforwardly for evaluating an analytic Jacobian.

9.5.3.3 *McMillin–Fleming method*

The McMillin–Fleming method, now designated 'OPTRAN', has been developed in several papers over twenty years. The latest version of McMillin *et al.* (1995) is described here. It combines the treatment of the inhomogeneous path with a transmittance model in a way that does not require the use of an equivalent homogeneous path. It essentially carries out a multiple regression of a function of the transmittance, computed line-by-line, of an ensemble of atmospheres onto a set of predictors which includes the Curtis–Godson integrals amongst other similar functions.

The transmittance of a path is expressed as a function of the absorber amount u in the path by

$$\tau(u) = \exp\left[-\int_{u'=0}^{u} k(u')\,du'\right] \quad (9.68)$$

thus defining the effective absorption coefficient $k(u)$ as the derivative of $-\ln\tau(u)$. The effective absorption coefficient will clearly depend on the distribution of absorber, pressure and temperature along the path, and it is this dependence which is approximated by means of a linear function of various predictors Z_j. An ensemble of profiles is constructed, and the effective absorption coefficient and the predictors are computed for each, as a function of u. The fit is carried out separately at a set of values u_i using

$$\hat{k}(u_i) = C_{i0} + \sum_{j=1}^{N} C_{ij} Z_{ij} \quad (9.69)$$

to obtain a set of coefficients C_{ij} by multiple regression. The predictors Z_{ij} include such things as the Curtis–Godson $p_i^{cg} = \int p\,du/(\int du)$ and $T_i^{cg} = \int T\,du/(\int du)$, together with other expressions such as $\int Tu\,du/(\int du)^2$.

The transmittance model is evaluated sequentially along a particular path by evaluating u and the Z_j for the path, calculating $\hat{k}(u_i)$ for neighbouring values of u_i from the regression coefficients using Eq. (9.69), and interpolating k to the required value of u before use in a discrete version of Eq. (9.68):

$$\tau_l = \tau_{l-1} \exp[-k(u)(u_l - u_{l-1})] \quad (9.70)$$

This method can also be differentiated straightforwardly for the evaluation of an analytic Jacobian.

9.5.3.4 *Multiple absorbers*

The models described above have all been described in terms of single absorbers. In practice there are several absorbers involved, each with a different and variable distribution. The line-by-line approach has no problem in dealing with several absorbers, as the absorption coefficients simply add. The McMillin–Fleming method can also be formulated so to treat multiple absorbers by including them in the list of predictors in the multiple regression. The emissivity growth and Curtis–Godson approximations cannot easily be modified, however. In principle the band models could be tabulated in terms of more parameters, but the sizes of the tables needed would rapidly become unmanageable. The spectral mean transmittance of two absorbers is given by:

$$\frac{1}{\Delta\nu}\int_{\Delta\nu}\tau_1(\nu)\tau_2(\nu)\,d\nu = \frac{1}{\Delta\nu}\int_{\Delta\nu}(\bar{\tau}_1+\delta\tau_1(\nu))(\bar{\tau}_2+\delta\tau_2(\nu))\,d\nu$$
$$= \bar{\tau}_1\bar{\tau}_2 + \frac{1}{\Delta\nu}\int_{\Delta\nu}\delta\tau_1(\nu)\delta\tau_2(\nu)\,d\nu \qquad (9.71)$$

The simplest approximation is to assume the absorbers are spectrally uncorrelated, i.e.

$$\int_{\Delta\nu}\delta\tau_1(\nu)\delta\tau_2(\nu)\,d\nu = 0 \qquad (9.72)$$

This is the 'multiplication property' for band transmittance. In any particular case it can only be tested numerically. For a moderate number of spectral lines it may be satisfactory, but often it is found that band contours are correlated, and errors arise.

Chapter 10
Construction and Use of Prior Constraints

10.1 Nature of *a Priori*

As we have seen, if a retrieval problem includes a continuous function as an unknown, then some kind of prior constraint must be used to make the problem well-posed. This may take the form of

(i) a discrete representation with fewer parameters than the number of degrees of freedom of the measurement, as is required by e.g. a maximum likelihood method,

(ii) an *ad hoc* constraint such as smoothness, as used in the Twomey–Tikhonov approach,

(iii) prior information about the state, as in the case of a maximum *a posteriori* method.

In addition, for practical purposes the continuous function in (b) and (c) must be replaced by a discrete representation, which should be in terms of rather more parameters than the number of degrees of freedom. Fundamentally, there exists an ensemble of continuous profiles which are consistent with the measurement. Any method which chooses one of this ensemble as 'the retrieval' must use a prior constraint of some kind to do it, whether by imposing a grid with a spacing such that the problem becomes overconstrained, or by using an explicit prior *pdf*. A representation with a finite number of coefficients, such as a grid with a specified interpolation rule, is a prior constraint which states that any structure that cannot be represented on this grid has zero amplitude.

The term *a priori* is used in a variety of ways in the literature. Case (a) may be described as having no *a priori*, or as having an *a priori* defined by the representation. Case (b) may also be described as having no *a priori*, but as being 'regularised' by the constraint, or as having an *a priori* described by the constraint. Case (c) is always described as *a priori*, but there can be confusion as to whether *a priori* comprises just the expected state, or its covariance matrix, or both. Perhaps the most helpful approach is to describe them all as prior constraints, (a) as a 'hard' constraint, (b) and (c) as 'soft' constraints, (b) as *ad hoc* prior constraint, and (c) as *a priori* information, where the term *a priori* includes both a mean state and its covariance, or more generally, a prior *pdf* of the state.

The main aim of a prior constraint is to restrict the components of the solution that are in the null space or the near-null space of the weighting function matrix. Components of the state which are in the row space are likely to be measured well enough that they are unaffected by any *a priori*. The simplest method is to eliminate the null and near-null space from consideration by means of a representation, but this can have the effect of eliminating part of the information that has been measured if the representation has been chosen for convenience rather than retaining information. To retain all of the measured information, a finer representation and an explicit *a priori* are needed. In the absence of real *a priori* we can resort to various *ad hoc* constraints such as smoothness, or make use of partial information about the prior *pdf* and supplement it with help from the maximum entropy principle, as indicated in section 2.7.

Setting up a prior constraint is not straightforward. If a limited representation is thought appropriate, then the number of parameters to be used, and their nature, needs to be considered carefully. Real prior information is often rather vague in nature and not obviously amenable to formalising. Sometimes a large number of independent measurements are available, for example the routine meteorological radiosonde soundings for tropospheric and lower stratospheric temperature and humidity, but often we may only have a few independent measurements, physical constraints, models, or perhaps only intuition when sounding the remoter parts of the atmosphere. For example all we may have is a reasonable expectation that the temperature of the mesosphere is unlikely to vary by more than about $100\,\mathrm{K}$, or that the concentration of some chemical species is positive and less than about 1 ppb. Once a large set of remote measurements has been made, it may be possible to use them to improve on the *a priori* to be used for future retrievals.

The use of a prior constraint is a critical part of solving an ill-posed problem, but its effect on the retrieval itself is often misunderstood or ignored, particularly by the naive data user. Therefore I will discuss how the retrieval relates to the prior constraint, and what precautions the user should take so as not to be misled when using retrieved data.

In this chapter, the profile is primarily specified by a 'full state vector' $\mathbf{x}$ on a grid fine enough that the effects of discretisation are unimportant for practical purposes. Prior constraint by means of a representation is to be expressed in terms of a 'reduced state vector' $\mathbf{z}$ which implies a full state vector $\mathbf{x} = \mathbf{W}\mathbf{z}$, where $\mathbf{W}$ is an interpolation matrix. The averaging kernel matrix relates $\hat{\mathbf{x}}$ to $\mathbf{x}$, rather than $\hat{\mathbf{z}}$ to $\mathbf{z}$, so the effect of the constraint can be seen explicitly and on the same basis for all types of solution.

10.2 Effect of Prior Constraints on a Retrieval

The relation between a retrieval, the true state of the system, the *a priori* and the measurement errors is discussed in Chapter 3. From Eq. (3.12) it can be summarised as:

$$\hat{\mathbf{x}} = \mathbf{x}_a + \mathbf{A}(\mathbf{x} - \mathbf{x}_a) + \mathbf{G}\epsilon_y = (\mathbf{I}_n - \mathbf{A})\mathbf{x}_a + \mathbf{A}\mathbf{x} + \mathbf{G}\epsilon_y. \tag{10.1}$$

The matrix $\mathbf{I}_n - \mathbf{A}$ indicates the fraction of the *a priori* profile that appears in the retrieved profile. Note that in some cases no $\mathbf{x}_a$ appears to be used. It is usually found on closer examination that in these cases $\mathbf{x}_a = 0$ or a coarse representation has been used in which $\mathbf{A} = \mathbf{I}$. To interpret this equation correctly, the state vector must be a full state vector on a fine grid.

If a reduced representation with fewer levels is being used, then it must be transformed back to the full state vector in order to analyse it. In the linear approximation the constrained forward model becomes $\mathbf{y} = \mathbf{KWz} + \epsilon_y$, and for a retrieval of the form $\hat{\mathbf{z}} - \mathbf{z}_a = \mathbf{G}_z(\mathbf{y} - \mathbf{y}_a)$, the characterisation in terms of $\mathbf{z}$ is

$$\hat{\mathbf{z}} = \mathbf{z}_a + \mathbf{G}_z \mathbf{KW}(\mathbf{z} - \mathbf{z}_a) + \mathbf{G}_z \epsilon_y \tag{10.2}$$

so that $\mathbf{A}_z = \mathbf{G}_z \mathbf{KW}$. For a ML retrieval, this is found to be a unit matrix, because $\mathbf{G}_z = [(\mathbf{KW})^T \mathbf{KW}]^{-1} (\mathbf{KW})^T$, $\mathbf{z}_a$ drops out of the characterisation, and the result is thought not to contain any *a priori*. However $\mathbf{A}_z$ is not the true averaging kernel matrix. This must relate $\hat{\mathbf{x}}$ and $\mathbf{x}$, and is found from:

$$\hat{\mathbf{x}} = \mathbf{W}\hat{\mathbf{z}} = \mathbf{x}_a + \mathbf{W}\mathbf{G}_z \mathbf{K}(\mathbf{x} - \mathbf{x}_a) + \mathbf{W}\mathbf{G}_z \epsilon_y. \tag{10.3}$$

Thus $\mathbf{x}_a$ always contributes because $\mathbf{A} = \mathbf{W}\mathbf{G}_z \mathbf{K}$ is not a unit matrix. The retrieved state is a weighted mean of the true state and the *a priori* state.

In the answer to exercise 3.1 we saw that the right eigenvectors of $\mathbf{A}$ can be used as another representation of $\mathbf{x}$, $\mathbf{x} = \mathbf{Rr}$, where $\mathbf{r}$ is the new reduced state vector. In this representation the characterisation equation is of the form

$$r_i = \lambda_i r_i + (1 - \lambda_i) r_{ai}, \tag{10.4}$$

plus noise, where λ_i is the i-th eigenvalue of $\mathbf{A}$. In the case of a hard constraint, there will be a set of eigenvectors with unit eigenvalue, corresponding to the column space of $\mathbf{W}$, the remaining eigenvalues all being zero. Components of the profile not representable by $\mathbf{W}$ are determined entirely by the *a priori*. Thus it is possible for there to be components of the atmospheric profile which contribute to the measurements but cannot be represented in the retrieval. These components will be aliassed into retrieval noise and into other components of the retrieval. The representation should be chosen to minimise this problem without inducing too much sensitivity to noise.

10.3 Choice of Prior Constraints

10.3.1 *Retrieval grid*

We need an objective way of choosing grids with an appropriate number and locations of levels for both the ML and MAP cases, i.e. the coarsest grid we can use without appreciable loss of information in the case of a MAP retrieval, or the most appropriate grid to provide a prior constraint in the case of a ML retrieval.

10.3.1.1 *Transformation between grids*

We will first examine how to transform both the state and the covariance matrix between grids, in both directions. Consider a state vector $\mathbf{x}$ on a fine grid of n levels, and a reduced vector $\mathbf{z}$ on the coarse grid of l levels, with

$$\mathbf{x} = \mathbf{W}\mathbf{z} \tag{10.5}$$

where the interpolation matrix $\mathbf{W}$ is $n \times l$. This does not mean that if a given atmosphere is represented by $\mathbf{z}$ then the underlying state is $\mathbf{W}\mathbf{z}$, but rather that we are only considering states of the form $\mathbf{W}\mathbf{z}$ as candidates for retrieved profiles. We also need some way of transforming from a general $\mathbf{x}$ to $\mathbf{z}$. Let this be done by

$$\mathbf{z} = \mathbf{W}^*\mathbf{x} \tag{10.6}$$

where $\mathbf{W}^*$ is a $l \times n$ pseudo inverse of $\mathbf{W}$. Clearly $\mathbf{W}^*$ should satisfy $\mathbf{W}^*\mathbf{W} = \mathbf{I}_l$ so that a full state constructed from $\mathbf{z}$ by interpolation will give the same $\mathbf{z}$ back, but we cannot expect $\mathbf{W}\mathbf{W}^* = \mathbf{I}_n$ as fine structure lost in transforming a general $\mathbf{x}$ to $\mathbf{z}$ cannot be restored. Mathematically, as $\mathbf{W}$ is of rank $l < n$, there are an infinite number of matrices which satisfy $\mathbf{W}^*\mathbf{W} = \mathbf{I}_l$, but none that satisfy $\mathbf{W}\mathbf{W}^* = \mathbf{I}_n$. We therefore choose the best approximation in some way, for example by choosing the $\mathbf{W}^*$ which minimises the difference between the actual state $\mathbf{x}$ and the interpolated fit $\mathbf{W}\mathbf{z} = \mathbf{W}\mathbf{W}^*\mathbf{x}$ in the simple least squares sense, namely

$$\mathbf{W}^* = (\mathbf{W}^T\mathbf{W})^{-1}\mathbf{W}^T. \tag{10.7}$$

We may also wish to consider minimising other norms of $\mathbf{x} - \mathbf{W}\mathbf{z}$, such as $(\mathbf{x} - \mathbf{W}\mathbf{z})^T \mathbf{S}_e^{-1}(\mathbf{x} - \mathbf{W}\mathbf{z})$ which gives the best fit over an ensemble with covariance $\mathbf{S}_e$. In this case

$$\mathbf{W}^* = (\mathbf{W}^T\mathbf{S}_e^{-1}\mathbf{W})^{-1}\mathbf{W}^T\mathbf{S}_e^{-1}. \tag{10.8}$$

Alternatively we can start with $\mathbf{z} = \mathbf{W}^*\mathbf{x}$, for example if we wish to retrieve layer averages rather than level values, where $\mathbf{W}^*$ is defined as an averaging operator, and look for a pseudo-inverse $\mathbf{W}$ that allows us to obtain a reasonable $\mathbf{x}$ consistent with $\mathbf{z}$. From the infinite number of possible states, we can choose the one which is shortest in a least squares sense, by finding $\mathbf{W}$ which minimises e.g. $\mathbf{x}^T\mathbf{x}$ or $\mathbf{x}^T\mathbf{S}_e^{-1}\mathbf{x}$

subject to $\mathbf{z} = \mathbf{W}^*\mathbf{x}$. These lead to

$$\mathbf{W} = \mathbf{W}^{*T}(\mathbf{W}^*\mathbf{W}^{*T})^{-1} \tag{10.9}$$

and

$$\mathbf{W} = \mathbf{S}_e\mathbf{W}^{*T}(\mathbf{W}^*\mathbf{S}_e\mathbf{W}^{*T})^{-1}. \tag{10.10}$$

respectively. It is easy to see by substitution that these satisfy Eq. (10.7) or (10.8).

Retrieving in terms of the reduced representation $\mathbf{z}$ restricts us to a subspace of the full state space, and transforms the forward model in the linear case to

$$\mathbf{y} = \mathbf{KWz} + \boldsymbol{\epsilon} + \boldsymbol{\epsilon}_R \tag{10.11}$$

which now has an additional 'representation error' $\boldsymbol{\epsilon}_R = \mathbf{K}(\mathbf{I} - \mathbf{WW}^*)\mathbf{x}$. We must choose the reduced representation so that this additional error is not significant.

If we intend to compare MAP retrievals on different grids, we must ensure that we compare like with like. Not only must the state be properly transformed, but also the prior covariance. If the prior covariance of $\mathbf{x}$ on the high resolution grid is $\mathbf{S}_a$, then the prior covariance of $\mathbf{z}$ is $\mathbf{S}_{za} = \mathbf{W}^*\mathbf{S}_a\mathbf{W}^{*T}$. It is easy to make the mistake of retrieving at high resolution with e.g. $\mathbf{S}_a = \sigma^2\mathbf{I}_n$, and to compare this with the result of using $\mathbf{S}_{za} = \sigma^2\mathbf{I}_l$. A diagonal covariance with elements of say $100\,\mathrm{K}^2$ on a grid of 1 km spacing is not equivalent to the same variance on a grid of 2 km spacing.

It is trickier going the other way. If we only have an estimate of $\mathbf{S}_{za}$ we cannot assume that $\mathbf{S}_a = \mathbf{WS}_{za}\mathbf{W}^T$, as this will be singular (it is rank l), and will carry the implication that any structure not represented by the interpolation has zero variance. If we really have no other information, we should make an educated guess, for example by changing the zero eigenvalues of the eigenvector decomposition of $\mathbf{WS}_{za}\mathbf{W}^T$ to values conservatively extrapolated from the non-zero eigenvalues.

10.3.1.2 *Choice of grid for maximum likelihood retrieval*

In the case of ML methods, for which there are fewer state vector elements than the rank of the problem, the prior constraint consists entirely of the representation used. The sensitivity of the solution to measurement noise will increase with the number of elements, as less well-determined components of state space are included. Fewer elements reduces noise sensitivity, but increases representation error, so there will be an optimum number of elements which minimises total error. We can balance the gain due to improved noise performance as the number of levels is reduced against the loss due to a poorer quality of profile representation. The ML solution will typically be evaluated as

$$\hat{\mathbf{z}} = [(\mathbf{KW})^T\mathbf{S}_\epsilon^{-1}\mathbf{KW}]^{-1}(\mathbf{KW})^T\mathbf{S}_\epsilon^{-1}\mathbf{y} = \mathbf{G}_{\mathrm{ML}}\mathbf{y}, \tag{10.12}$$

thus defining $\mathbf{G}_{\mathrm{ML}}$, giving $\hat{\mathbf{x}} = \mathbf{W}\mathbf{G}_{\mathrm{ML}}\mathbf{y} = \mathbf{W}\mathbf{G}_{\mathrm{ML}}(\mathbf{K}\mathbf{x} + \boldsymbol{\epsilon})$. The total error, including representation error, will be

$$\hat{\mathbf{x}} - \mathbf{x} = (\mathbf{W}\mathbf{G}_{\mathrm{ML}}\mathbf{K} - \mathbf{I})\mathbf{x} + \mathbf{W}\mathbf{G}_{\mathrm{ML}}\boldsymbol{\epsilon}. \tag{10.13}$$

Note that the first term on the right hand side is the smoothing error, as discussed in section 3.2. To select an optimal vertical grid for the ML method, we need a figure of merit. Appropriate choices are the entropy or the trace of the retrieval covariance, $\hat{\mathbf{S}}_{\mathrm{ML}}$, in the fine grid:

$$\hat{\mathbf{S}}_{\mathrm{ML}} = (\mathbf{W}\mathbf{G}_{\mathrm{ML}}\mathbf{K} - \mathbf{I})\mathbf{S}_e(\mathbf{W}\mathbf{G}_{\mathrm{ML}}\mathbf{K} - \mathbf{I})^T + \mathbf{W}\mathbf{G}_{\mathrm{ML}}\mathbf{S}_\epsilon \mathbf{G}_{\mathrm{ML}}^T \mathbf{W}^T. \tag{10.14}$$

We can then optimise the figure of merit with respect to the grid spacing and level placement. As noted in section 3.2.1, to estimate the smoothing error covariance (the first term) the covariance matrix of a real ensemble of states must be known. We can only select an appropriate grid for a ML method if we have a good idea of the statistics of what might be lost or gained by our choice of grid. The optimisation will have to be done numerically with respect to grid spacing and/or detailed placement.

10.3.1.3 *Choice of grid for maximum* a priori *retrieval*

For the MAP retrieval, a low resolution representation is not needed as a prior constraint, because that function is provided by the statistical *a priori*. For accuracy, we would like to make the vertical grid we use as fine as possible, but for speed and efficiency, we want to reduce the number of levels as far as possible. It may be necessary to use a fine grid for the numerical evaluation of the forward model, but the state vector need not consist of the values of the unknown on the forward model grid. We have discussed efficient representations in terms of eigenvector expansions of covariance matrices in section 5.8.3, but there are often practical reasons for wishing to use a simple representation in terms of layer means or values at discrete levels with a specified interpolation, a common consideration being understanding of the product by the naive data user.

A figure of merit such as information content or degrees of freedom for signal should decrease as the number of levels is reduced, because there are parts of state space no longer accessible. Therefore rather than optimising something, we must choose the smallest number of levels consistent with an acceptable loss of information. Both the number of levels and their locations are quantities we might wish to adjust to best effect.

To evaluate a MAP solution in the reduced state space, we must ensure that we use the correct prior covariance, namely $\mathbf{W}^*\mathbf{S}_a\mathbf{W}^{*T}$. A retrieval in a reduced state space gives

$$\hat{\mathbf{z}} = [(\mathbf{W}^*\mathbf{S}_a\mathbf{W}^{*T})^{-1} + \mathbf{W}^T\mathbf{K}^T\mathbf{S}_\epsilon^{-1}\mathbf{K}\mathbf{W}]^{-1}\mathbf{W}^T\mathbf{K}^T\mathbf{S}_\epsilon^{-1}\mathbf{y} = \mathbf{G}_{\mathrm{MAP}}\mathbf{y}, \tag{10.15}$$

thus defining $\mathbf{G}_{\text{MAP}}$. This corresponds to, in full state space,

$$\hat{\mathbf{x}} = \mathbf{W}[(\mathbf{W}^*\mathbf{S}_a\mathbf{W}^{*T})^{-1} + \mathbf{W}^T\mathbf{K}^T\mathbf{S}_\epsilon^{-1}\mathbf{K}\mathbf{W}]^{-1}\mathbf{W}^T\mathbf{K}^T\mathbf{S}_\epsilon^{-1}\mathbf{y}. \qquad (10.16)$$

The averaging kernel matrix will be $\mathbf{W}\mathbf{G}_{\text{MAP}}\mathbf{K}$, as for the ML case, and the error covariance will be of the same form as Eq. (10.14). A suitable figure of merit such as the entropy or trace can be evaluated as a function of number of grid levels. It will degrade as the number of levels is reduced, but for a given number, the placement can be optimised numerically. It might be expected, on the basis of the interpretation of the diagonal $\mathbf{A}$ as data density, section 3.3, that the optimal level placement might divide the data density in the full state vector into approximately equal portions.

10.3.2 Ad hoc *Soft constraints*

10.3.2.1 *Smoothness constraints*

The most commonly used smoothness constraint is the diagonal matrix, with which the retrieval jointly minimises the squared departure from the measurement plus the squared departure from a prior profile, which may be zero. Other possibilities include the squared first or second differences, as suggested by Twomey (1963), see section 6.4.

As smoothness constraints are quadratic, and lead to equations of the same form as *a priori* information, they can be interpreted in the same terms. Twomey's $\mathbf{H}$ matrix can be interpreted as an inverse covariance or as a Fisher information matrix. If it is of the form $\gamma^{-1}\mathbf{I}$, the corresponding covariance matrix is $\gamma\mathbf{I}$, with the interpretation is that the prior variance of each element of the state vector is γ, and independent. If $\mathbf{H}$ represents the squared second difference, as illustrated in Eq. (6.33), the interpretation is less straightforward because $\mathbf{H}$ is singular. However by looking at the eigenvector decomposition of $\mathbf{H} = \mathbf{L}\mathbf{\Lambda}\mathbf{L}^T$, we can see that it corresponds to error patterns $\lambda^{-\frac{1}{2}}\mathbf{l}$, with some, those corresponding to a constant offset and a linear variation in height, having infinite variance. A practical consequence of $\mathbf{H}$ being singular is that it cannot be used in the m-form of the equations.

It should be remembered in interpreting smoothness constraints that they have the grid spacing built in. If retrievals are to be compared at different grid spacings, then any smoothness constraint must be transformed appropriately.

⇒ *Exercise* 10.1: How should this transformation be carried out?

10.3.2.2 *Markov process*

In section 2.6 I introduced an *ad hoc* constraint based on a Markov description of the profile. This model with two free parameters, a variance σ_a and a length scale h can be very useful for a range of applications. The formulation in terms of a length scale, Eq. (2.83):

$$S_{ij} = \sigma_a^2 \exp\left(-|i-j|\frac{\delta z}{h}\right) \qquad (10.17)$$

provides a straightforward way of comparing retrievals on different grids, although it is not strictly the same as a transformation with $\mathbf{W}$ or $\mathbf{W}^*$. Note that the inverse of this matrix has a particularly simple form.

> ⇒ **Exercise** 10.2: What is the form of the inverse of the Markov covariance matrix? Interpret it in terms of Twomey **H**-matrices.

10.3.3 *Estimating* a priori *from real data*

10.3.3.1 *Estimating* a priori *from independent sources*

The most satisfactory source of *a priori* information is from independent high spatial resolution measurements of ensembles of the kind of state you are measuring, for example, as may be obtained from radiosonde measurements. Such data is often available as climatologies partitioned by, for example, latitude and date.

Mean values and variances (or standard deviations) are usually available, covariances less often, so it may be necessary to obtain an original ensemble of data and construct your own means and covariances. In doing this, you should take precautions to ensure that the prior covariance must be positive definite. A covariance matrix with zero eigenvalues implies that something is known exactly *a priori*. If that knowledge is real then the quantity referred to, the coefficient of that eigenvector, need not appear in the state vector. However the knowledge is usually not real, for example it may have arisen because the original data set had been interpolated from a coarser grid to a finer grid, so that there are elements that are linearly dependent on others. The corresponding fine scale variability is not represented in the data, and the covariance gives a false impression of the accuracy with which it is known. Small or zero eigenvalues should always be checked for reasonableness, and adjusted artificially if it is felt that they do not represent reality. A negative eigenvalue is the equivalent of a negative variance of a set of scalar data, and is likely to be the result of a programming error, or a numerical problem such as a rounding error.

10.3.3.2 *Maximum entropy and the estimation of* a priori

Real prior data may or may not be available, or if it is, it may not be complete, in the sense that it may only apply to a smaller part of state space than is seen by the observing system. The question then arises of constructing a reasonable prior *pdf* which will help the inverse method to produce a sensible solution, while not inducing unwanted artifacts.

We may, for example, have some idea of a mean value and variance of the profile at each level. We have seen in section 2.7 that in the scalar case the maximum entropy *pdf* for this case would be Gaussian, and it is easy to see that this generalises to Gaussian with diagonal covariance. However we often have a better idea than that, and can often make a rough guess at for example the variance of first difference. In this case the Markov process described above is useful, as its effect is to constrain a combination of the profile variance and the variance of the first difference. As well as these quantities, we may also have some idea of the means and variances of other functions of the state such as total column amount of a constituent, or even of the indirectly measure quantities. Can we use the statistics of our own measurements to help establish an *a priori*?

We can in some circumstances use the maximum entropy principle which allows us to make a reasonable estimate of a probability density function consistent with all of our knowledge, one that is least committal about the state but consistent with whatever more or less detailed understanding we may have of the state vector prior to the measurement(s).

The most uncertain or least committal *pdf* $P(\mathbf{x})$ may reasonably be taken to be the one which occupies the largest volume of state space consistent with the information we do have, that is the one whose entropy is largest. If we have information expressed as the expected value of some function of the state, e.g. $\mathcal{E}\{f_i(\mathbf{x})\} = s_i$ then we can find the *pdf* which has maximum entropy S:

$$S = -\int P(\mathbf{x}) \ln P(\mathbf{x}) \, d\mathbf{x} \qquad (10.18)$$

subject to a set of constraints:

$$\frac{d}{dP}\left(-\int P(\mathbf{x}) \ln P(\mathbf{x}) \, d\mathbf{x} + \sum_i \lambda_i [\int P(\mathbf{x}) f_i(\mathbf{x}) \, d\mathbf{x} - s_i] + \mu [\int P(\mathbf{x}) \, d\mathbf{x} - 1]\right)$$
$$= 0. \qquad (10.19)$$

Typical functions that we might know are the mean state, $\mathcal{E}\{\mathbf{x}\} = \bar{\mathbf{x}}$, the variance of each element, $\mathcal{E}\{(x_i - \bar{x}_i)^2\}$, and the variance of a linear function of the elements: $\mathcal{E}\{[\mathbf{k}_i^T(\mathbf{x} - \bar{\mathbf{x}})]^2\}$, e.g. a total column or a linearised measured signal. Differentiating with respect to P gives

$$-\ln P(\mathbf{x}) - 1 + \mu + \sum_i \lambda_i f_i(\mathbf{x}) = 0. \qquad (10.20)$$

This implies that if the functions f_i are linear or quadratic, then $\mathbf{P}(\mathbf{x})$ must be Gaussian. For illustration, consider only quadratic functions. The variance of a single element is a particular case of that of a linear function of the state vector with $\mathbf{k} = (0, \ldots, 0, 1, 0, \ldots, 0)$, so consider only $f(\mathbf{x}) = [\mathbf{k}^T \mathbf{x}]^2 = \mathbf{x}^T \mathbf{k} \mathbf{k}^T \mathbf{x}$:

$$-\ln P(\mathbf{x}) - 1 + \mu + \sum_i \lambda_i \mathbf{x}^T \mathbf{k}_i \mathbf{k}_i^T \mathbf{x} = 0 \qquad (10.21)$$

Thus the maximum entropy *pdf* subject to constraints expressing knowledge of various quadratic statistics is gaussian with inverse covariance, or information matrix,

$$\mathbf{S}^{-1} = \sum_i \lambda_i \mathbf{k}_i \mathbf{k}_i^T. \tag{10.22}$$

To go any further, we must ensure that this is nonsingular, e.g. by including a variance for every element of the state vector. The values of the Lagrangian multipliers are in principle obtained by substituting in the constraints:

$$E\{[\mathbf{k}_j^T \mathbf{x}]^2\} = \mathbf{k}_j^T \mathbf{S} \mathbf{k}_j = s_j \tag{10.23}$$

so that

$$\mathbf{k}_j^T [\sum_i \lambda_i \mathbf{k}_i \mathbf{k}_i^T]^{-1} \mathbf{k}_j = s_j. \tag{10.24}$$

Unfortunately does not have an explicit solution for λ_i in the general case, and must be solved numerically.

10.3.4 Validating and improving a priori with indirect measurements

Once we have obtained an ensemble of indirect measurements we are in the position to check whether the *a priori* we are using is appropriate, and possibly to improve it by comparing the covariance of the ensemble of retrievals or measured signals with that expected from the *a priori*.

It might appear incestuous to use a set of measurements to improve their own *a priori*. This would be the case for a single profile or a small sample but not if done carefully, and for a large sample. A retrieval is biassed by *a priori*, but only within experimental error, if the *a priori* is appropriate. A better *a priori* should reduce any bias, and it should be possible to obtain an improved estimate of $\mathbf{x}_e$ from the measurements because of the $\sqrt{N}$ advantage of a large ensemble. For the retrieval of an individual measurement using a retrieved *a priori*, the first-time-round retrieval will have been used in the *a priori*, but it is diluted by all of the other retrievals. Strictly you might consider retrieving profile n using an *a priori* based on all retrievals except n. However with a large ensemble, this will not make any noticeable difference. Use of a retrieved *a priori* is one way of using information from other measurements in retrieval n, and is not very different philosophically from Kalman filtering.

In the nearly linear case the covariance of an ensemble of measured signals $\mathbf{S}_y$ is related to that of the ensemble of states by $\mathbf{S}_y = \mathbf{K} \mathbf{S}_e \mathbf{K}^T + \mathbf{S}_\epsilon$, so we can check whether $\mathbf{S}_a$ is a good estimate of $\mathbf{S}_e$ by comparing $\mathbf{K} \mathbf{S}_a \mathbf{K}^T + \mathbf{S}_\epsilon$ to $\mathbf{S}_y$. We have information about the components of $\mathbf{x} - \mathbf{x}_a$ which are in the row space of $\mathbf{K}$, and hence can test the corresponding components of $\mathbf{S}_a$ and $\mathbf{x}_a$, and perhaps adjust them, particularly if the original was *ad hoc* and of poor quality.

10.3.4.1 The nearly linear case

Consider an ensemble which has a real mean $\mathbf{x}_e$ and covariance $\mathbf{S}_e$, which are poorly known and are approximated by some $\mathbf{x}_a$ and $\mathbf{S}_a$. In the nearly linear case, all of the measurements have the same weighting function, averaging kernel and contribution function matrices. If all is well, and the statistics of the ensemble are as described by the *a priori* state and covariance, we would expect to find the mean measured signal $\bar{\mathbf{y}}$, the mean retrieval $\bar{\hat{\mathbf{x}}}$, and the covariance $\mathbf{S}_{\hat{\mathbf{x}}}$ of the retrievals about the mean to be given by

$$\bar{\mathbf{y}} = \mathbf{F}(\mathbf{x}_a) + \boldsymbol{\epsilon}_s \qquad (10.25)$$

$$\bar{\hat{\mathbf{x}}} = \mathbf{x}_a + \mathbf{G}\boldsymbol{\epsilon}_s \qquad (10.26)$$

$$\mathbf{S}_{\hat{\mathbf{x}}} = \mathbf{A}\mathbf{S}_a\mathbf{A}^T + \mathbf{G}\mathbf{S}_{\epsilon_r}\mathbf{G}^T \qquad (10.27)$$

in the limit of large sample size, provided $\mathbf{A}$, $\mathbf{G}$ and systematic error $\boldsymbol{\epsilon}_s$ are taken to be constant. However if the ensemble mean and covariance differ from the *a priori*, we expect

$$\bar{\mathbf{y}} = \mathbf{F}(\mathbf{x}_a) + \mathbf{K}(\mathbf{x}_e - \mathbf{x}_a) + \boldsymbol{\epsilon}_s \qquad (10.28)$$

$$\bar{\hat{\mathbf{x}}} = \mathbf{x}_a + \mathbf{A}(\mathbf{x}_e - \mathbf{x}_a) + \mathbf{G}\boldsymbol{\epsilon}_s \qquad (10.29)$$

$$\mathbf{S}_{\hat{\mathbf{x}}} = \mathbf{A}\mathbf{S}_e\mathbf{A}^T + \mathbf{G}\mathbf{S}_{\epsilon_r}\mathbf{G}^T \qquad (10.30)$$

⇒ **Exercise** 10.3: Show this.

Comparisons of $\bar{\mathbf{y}}$ with $\mathbf{F}(\mathbf{x}_a)$, $\bar{\hat{\mathbf{x}}}$ with $\mathbf{x}_a$ and $\mathbf{S}_{\hat{\mathbf{x}}}$ with $\mathbf{A}\mathbf{S}_a\mathbf{A}^T + \mathbf{G}\mathbf{S}_{\epsilon_r}\mathbf{G}^T$ will indicate whether improvements can be made, although it is not easy to separate *a priori* errors from systematic errors. Clearly an improved estimate of $\mathbf{x}_e$ can be obtained from Eq. (10.28) or (10.29), by regarding it as a retrieval problem for $\mathbf{x}_e$, with errors given by the systematic error plus any residual noise due to a finite sample. For example using Eq. (10.29):

$$\hat{\mathbf{x}}_e = \mathbf{G}_e(\bar{\hat{\mathbf{x}}} - \mathbf{x}_a + \mathbf{A}\mathbf{x}_a), \qquad (10.31)$$

where $\mathbf{G}_e$ is the gain matrix for the $\mathbf{x}_e$ retrieval. An estimate of $\mathbf{S}_e$ may similarly be found from Eq. (10.30) as:

$$\hat{\mathbf{S}}_e = \mathbf{G}_e(\mathbf{S}_{\hat{\mathbf{x}}} - \mathbf{G}\mathbf{S}_{\epsilon_r}\mathbf{G}^T)\mathbf{G}_e^T \qquad (10.32)$$

but it is instructive, and allows problems to be identified, to consider transforming the problem to the primed coordinates introduced in section 2.4.1, in which the forward model is $\mathbf{y}' = \boldsymbol{\Lambda}\mathbf{x}' + \boldsymbol{\epsilon}'$ and the noise and prior covariances are both unit matrices.

If the prior covariance $\mathbf{S}_a$ is correct then we expect to obtain for the covariance of $\mathbf{y}'$:

$$\mathbf{S}_{\mathbf{y}'} = \boldsymbol{\Lambda}^2 + \mathbf{I}_m \qquad (10.33)$$

but in practice it will be

$$\begin{aligned}\mathbf{S}_{y'} &= \mathbf{\Lambda}\mathbf{S}_{x'}\mathbf{\Lambda} + \mathbf{I}_m \\ &= \mathbf{\Lambda}\mathbf{V}\mathbf{S}_a^{-\frac{1}{2}}\mathbf{S}_e\mathbf{S}_a^{-\frac{1}{2}}\mathbf{V}^T\mathbf{\Lambda} + \mathbf{I}_m\end{aligned} \qquad (10.34)$$

where $\mathbf{V}$ is the matrix of right singular vectors of $\tilde{\mathbf{K}} = \mathbf{S}_{\epsilon_r}^{-\frac{1}{2}}\mathbf{K}\mathbf{S}_a^{\frac{1}{2}}$. We can then estimate $\mathbf{S}_{x'}$ from:

$$\mathbf{S}_{x'} = \mathbf{\Lambda}^{-1}(\mathbf{S}_{y'} - \mathbf{I}_m)\mathbf{\Lambda}^{-1} \qquad (10.35)$$

which is similar to Eq. (10.32), but in the primed space. This is valid if the λ_i are not too small and the subtraction of unity does not make $\mathbf{S}_{y'} - \mathbf{I}_m$ non positive-definite, which should be the case if $\mathbf{S}_{\epsilon_r}$ is correct, but any components for which λ_i is too small to estimate reliably should be included with the null space, for example by using the equivalent of Eq. (10.32), namely

$$\mathbf{S}_{x'} = \mathbf{\Lambda}(\mathbf{I} + \mathbf{\Lambda}^2)^{-1}(\mathbf{S}_{y'} - \mathbf{I}_m)(\mathbf{I} + \mathbf{\Lambda}^2)^{-1}\mathbf{\Lambda} \qquad (10.36)$$

The next step is to transform back to find an improved estimate of $\mathbf{S}_e$. The transformed state is of order p, the effective rank of $\tilde{\mathbf{K}}$ after zero and small λ_i's have been eliminated. Consider it extended to its full order n by adding to $\mathbf{V}$ all of the eigenvectors with zero eigenvalue to make $\mathbf{V}_n$, a square unitary matrix. The prior covariance of the extended $\mathbf{x}'$ will be $\mathbf{I}_n$. The first p rows and columns of the ensemble covariance of $\mathbf{x}'$ have been estimated by the above process as $\mathbf{S}_{x'}$, the remainder can only be obtained from the prior estimate as unity on the diagonal and zero elsewhere. After extending $\mathbf{S}_{x'}$ to order n in this way, an improved estimate of $\mathbf{S}_e$ can be obtained:

$$\hat{\mathbf{S}}_e = \mathbf{S}_a^{\frac{1}{2}}\mathbf{V}_n^T\mathbf{S}_{x'}\mathbf{V}_n\mathbf{S}_a^{\frac{1}{2}} \qquad (10.37)$$

This process will improve estimates of the ensemble mean and covariance in the part of the row space that is well measured, but not in the null space and the near-null space, thus providing a more realistic constraint on individual retrievals, based on the actual behaviour of the atmosphere. The null space and near-null space components of the retrievals will still be provided by the original *a priori* information.

This process will not make much difference to the estimate of components of $\mathbf{x}'$ which are well determined from the measurements, for which $\lambda_i^2 \gg 1$, or those for which the *a priori* is not much improved, for which $N\lambda^2 \ll 1$. It should help improve components between these two limits by combining the measurement with a better *a priori*.

10.3.4.2 The moderately non-linear case

The nonlinear case is more problematical because the averaging kernel depends on the state, so simple averaging does not help. However we can take the retrieval characterisation for sample i of the ensemble to be a forward model for the ensemble mean:

$$\hat{\mathbf{x}}_i = \mathbf{x}_a + \mathbf{A}_i(\mathbf{x}_e - \mathbf{x}_a) + \boldsymbol{\epsilon}_v + \boldsymbol{\epsilon}_r + \boldsymbol{\epsilon}_s \qquad (10.38)$$

where $\boldsymbol{\epsilon}_v = \mathbf{A}_i(\mathbf{x} - \mathbf{x}_e)$ is the component of measurement error due to atmospheric variability, and sequentially update an estimate of $\mathbf{x}_e$ with a sequence of retrievals $\mathbf{x}_i$. Unfortunately to do this we need an estimate of $\mathbf{S}_v = \mathbf{A}\mathbf{S}_e\mathbf{A}^T$, so the process will need to be iterated.

10.4 Using Retrievals Which Contain *a Priori*

The MAP retrieval is a weighted mean of the actual state and the *a priori* state, where the weights are constant for linear problems, but depend on the state in the case of non-linear problems (section 3.1). For certain types of use of the data this causes complications. For some applications of remote sounding data, we do not necessarily want the best estimate of the state, particularly when further statistical analysis is to be carried out. The prime examples are (1) meteorological data assimilation and (2) estimation of means and covariances, both for climate studies and for retrieval validation. We may want to calculate the best estimate of a mean of an ensemble, which is not the same as the mean of the ensemble of best estimates of the individual profiles because the *a priori* contribution is a systematic error.

We should consider separately the requirements of the naive and the sophisticated data user. Meteorological assimilation, for example, is a matter for the sophisticated user for whom the best way of presenting the data may not be as a retrieval at all, as discussed in Chapter 8. For climate studies we should first consider what kinds of results the data user wants, and then decide what is the best kind of product. Clearly the user should be able to take a properly characterised retrieval and derive whatever he wants from it, but the naive user may find this daunting, and the data supplier should consider providing a range of standard products as well as individual retrievals, such as:

(i) Best estimates of mean profiles for specified space/time regions.
(ii) Error variances (or covariances) of the means
(iii) Variances (or covariances) of profile ensembles about the means
(iv) Best estimates of Fourier coefficients for longitude variation
(v) Best estimates of Fourier coefficients for annual variation

One application of particular interest is in validation and retrieval improvement, to be discussed in Chapter 12. The *a priori* itself should be consistent with the mean and covariance of ensembles of retrievals that are produced.

10.4.1 Taking averages of sets of retrievals

Considering random and systematic errors separately, an individual retrieval from a set is characterised by:

$$\hat{\mathbf{x}} = \mathbf{x}_a + \mathbf{A}(\mathbf{x} - \mathbf{x}_a) + \mathbf{G}\epsilon_r + \mathbf{G}\epsilon_s \qquad (10.39)$$

where ideally the random error ϵ_r has a known covariance $\mathbf{S}_{\epsilon_r}$ and the systematic error ϵ_s is a constant for the set of retrievals, but may be thought of as being taken from an ensemble of possible errors with known covariance $\mathbf{S}_{\epsilon_s}$. Consider a finite sample of s retrievals taken from an ensemble of atmospheres with mean $\mathbf{x}_e$ and covariance $\mathbf{S}_e$. The mean of the sample of retrievals is

$$\bar{\hat{\mathbf{x}}} = \mathbf{x}_a + \mathbf{A}(\bar{\mathbf{x}} - \mathbf{x}_a) + \mathbf{G}\epsilon'_r + \mathbf{G}\epsilon_s \qquad (10.40)$$
$$= \mathbf{x}_a + \mathbf{A}(\mathbf{x}_e - \mathbf{x}_a) + \epsilon'_v + \mathbf{G}\epsilon'_r + \mathbf{G}\epsilon_s \qquad (10.41)$$

where $\bar{\mathbf{x}}$ is the mean of the sample of atmospheres, and $\epsilon'_v = \mathbf{A}(\bar{\mathbf{x}} - \mathbf{x}_e)$ represents the error due to the difference between the mean of the sample and the mean of the ensemble. It is a random vector from an ensemble with covariance about $s^{-1}\mathbf{A}\mathbf{S}_e\mathbf{A}^T$. The term $\mathbf{G}\epsilon'_r$ is the error in the mean due to random error ϵ_r, being a random vector from an ensemble with covariance $s^{-1}\mathbf{G}\mathbf{S}_{\epsilon_r}\mathbf{G}^T$. The systematic error ϵ_s is unchanged by averaging. Thus $\bar{\hat{\mathbf{x}}}$ is a biassed estimate of both $\bar{\mathbf{x}}$ and of $\mathbf{x}_e$, because of the presence of $\mathbf{A}$ and $\mathbf{x}_a$ in the expression. The error is, from Eqs. (10.40) and (10.41):

$$\bar{\hat{\mathbf{x}}} - \bar{\mathbf{x}} = (\mathbf{A} - \mathbf{I})(\bar{\mathbf{x}} - \mathbf{x}_a) + \mathbf{G}\epsilon'_r + \mathbf{G}\epsilon_s \qquad (10.42)$$
$$\bar{\hat{\mathbf{x}}} - \mathbf{x}_e = (\mathbf{A} - \mathbf{I})(\mathbf{x}_e - \mathbf{x}_a) + \epsilon'_v + \mathbf{G}\epsilon'_r + \mathbf{G}\epsilon_s \qquad (10.43)$$

The average of an ensemble of MAP retrievals is not a MAP estimate of the ensemble mean. This is probably the heart of the problem for the naive user. We can do better for example by taking Eq. (10.40) as a measurement equation for $\bar{\mathbf{x}}$, and evaluating the implied MAP estimate. However this does require an estimate of its error covariance $\mathbf{G}(s^{-1}\mathbf{S}_{\epsilon_r} + \mathbf{S}_{\epsilon_s})\mathbf{G}^T$ and its prior.

10.4.2 Removing a priori

The MAP retrieval for a moderately linear problem can be written, from Eqs. (5.9) and (5.13) as

$$\hat{\mathbf{S}} = (\mathbf{S}_a^{-1} + \mathbf{K}_l^T \mathbf{S}_\epsilon^{-1} \mathbf{K}_l)^{-1} \qquad (10.44)$$
$$\hat{\mathbf{x}}_{\text{MAP}} = \hat{\mathbf{S}}\{\mathbf{K}_l^T \mathbf{S}_\epsilon^{-1}[\mathbf{y} - \mathbf{F}(\mathbf{x}_l) + \mathbf{K}_l \mathbf{x}_l] + \mathbf{S}_a^{-1} \mathbf{x}_a\}, \qquad (10.45)$$

where $\mathbf{x}_l$ is a linearisation point close to the solution, while the ML solution is formally obtained by substituting $\mathbf{S}_a^{-1} = \mathbf{O}$:

$$\hat{\mathbf{x}}_{\text{ML}} = (\mathbf{K}_l^T \mathbf{S}_\epsilon^{-1} \mathbf{K}_l)^{-1} \mathbf{K}_l^T \mathbf{S}_\epsilon^{-1}[\mathbf{y} - \mathbf{F}(\mathbf{x}_l) + \mathbf{K}_l \mathbf{x}_l]. \qquad (10.46)$$

Thus it might seem that if we are given a MAP solution and the *a priori* used, we can remove the effect of the *a priori* without rerunning the retrieval to obtain the corresponding ML solution:

$$\hat{\mathbf{x}}_{\mathrm{ML}} = (\hat{\mathbf{S}}^{-1} - \mathbf{S}_a^{-1})^{-1}[\hat{\mathbf{S}}^{-1}\hat{\mathbf{x}}_{\mathrm{MAP}} - \mathbf{S}_a^{-1}\mathbf{x}_a]. \qquad (10.47)$$

or, equivalently, we can calculate a ML solution at the same time as a MAP solution once the iteration has converged and all of the relevant matrices have been computed. Unfortunately this will only work if the problem is overconstrained, i.e. if the problem has been formulated with a representation that provides a hard constraint. The symptom of the process failing will be that $\mathbf{K}_l^T \mathbf{S}_\epsilon^{-1} \mathbf{K}_l = \hat{\mathbf{S}}^{-1} - \mathbf{S}_a^{-1}$ is singular, or $\mathbf{K}$ is of rank less than n. However we can use this idea to change the *a priori* rather than remove it, by

$$\hat{\mathbf{x}}_{\mathrm{new}} = (\hat{\mathbf{S}}^{-1} - \mathbf{S}_a^{-1} + \mathbf{S}_b^{-1})^{-1}[\hat{\mathbf{S}}^{-1}\hat{\mathbf{x}}_{\mathrm{MAP}} - \mathbf{S}_a^{-1}\mathbf{x}_a + \mathbf{S}_b^{-1}\mathbf{x}_b]. \qquad (10.48)$$

where $\mathbf{x}_b$, $\mathbf{S}_b$ is the new *a priori*. Note that the $\mathbf{S}_b$ need not be a probabilistic one, it could be an *ad hoc* smoothness constraint. We can also add *a priori* to an unsatisfactory ML solution this way, as long as its linearisation is valid.

To really remove the effect of *a priori* and convert the retrieval to a ML one, it is necessary to replace the explicit soft prior constraint with a hard one by means of a representation, $\mathbf{x} = \mathbf{W}\mathbf{z}$, which will have to be chosen. The MAP retrieval, after transforming back to $\mathbf{x}$, is

$$\begin{aligned}\hat{\mathbf{x}}_{\mathrm{ML}} &= \mathbf{W}(\mathbf{W}^T \mathbf{K}_l^T \mathbf{S}_\epsilon^{-1} \mathbf{K}_l \mathbf{W})^{-1}\mathbf{W}^T \mathbf{K}_l^T \mathbf{S}_\epsilon^{-1}[\mathbf{y} - \mathbf{F}(\mathbf{x}_l) + \mathbf{K}_l \mathbf{x}_l] & (10.49) \\ &= \mathbf{W}[\mathbf{W}^T(\hat{\mathbf{S}}^{-1} - \mathbf{S}_a^{-1})\mathbf{W}]^{-1}[\hat{\mathbf{S}}^{-1}\hat{\mathbf{x}}_{\mathrm{MAP}} - \mathbf{S}_a^{-1}\mathbf{x}_a] & (10.50)\end{aligned}$$

where the linearisation point has not been transformed with $\mathbf{W}$. The only requirement on $\mathbf{W}$ is that $\mathbf{W}^T \mathbf{K}_l^T \mathbf{S}_\epsilon^{-1} \mathbf{K}_l \mathbf{W}$ be nonsingular. The solution error covariance for the $\mathbf{z}$ representation is $(\mathbf{W}^T \mathbf{K}_l^T \mathbf{S}_\epsilon^{-1} \mathbf{K}_l \mathbf{W})^{-1}$, and for the $\mathbf{x}$ representation it is $\mathbf{W}(\mathbf{W}^T \mathbf{K}_l^T \mathbf{S}_\epsilon^{-1} \mathbf{K}_l \mathbf{W})^{-1}\mathbf{W}^T$ in the row space of $\mathbf{W}$, but is infinite in its null space.

Chapter 11

Designing an Observing System

Design of retrieval methods is an integral part of the overall design of an observing system. A proper retrieval study is likely to indicate improvements in the conceptual instrument design by identifying extra parameters that should be measured, and by acting as a test-bed for optimising design choices such as spectral regions to be measured. Designing an instrument without a detailed retrieval study is likely to lead to problems in interpreting the data. Often much of observing system design is obvious, but it is worth doing a systematic and comprehensive study to obtain the best possible overall result.

Given a target atmospheric quantity to be measured, and a proposed measurement technique, the main tasks are to design and build the instrument, to design and build the operational retrieval system and to validate the resulting data products. Part of designing the instrument is an end-to-end simulation of the instrument and retrieval method to determine the feasibility of the selected measurement technique, to optimise the instrument and observing pattern design, and to set requirements on the knowledge of forward model parameters, including instrument parameters, atmospheric *a priori* data and basic physical parameters such as spectral line data.

The operational retrieval method will not necessarily be the same as the one required for design. It may be more complex in some areas, e.g. the treatment of nonlinearity, and simpler in others, e.g. the built-in diagnostics. Validation in the general sense will include comparisons of the data products and their statistics with the characteristics predicted by the error analysis and diagnostics of the operational retrieval, as well as comparisons with other instruments.

11.1 Design and Optimisation of Instruments

The first step is to select an appropriate instrument type and an observing geometry. For example, ozone has absorption features in the microwave (the pure rotation band), the infrared (vibration-rotation bands at around $10\,\mu$m, and in the ultraviolet and visible (electronic transitions with some vibrational and rotational structure).

The microwave and infrared radiation can be measured in thermal emission or in absorption of solar (or perhaps lunar or stellar) radiation. The UV and visible radiation can be measured in absorption, from the surface or in occultation, or as solar scattering. The viewing geometry for absorption may be directly from the surface or in occultation from a satellite. The viewing geometry for thermal emission or scattering may be from the surface, *in situ* or from space, and if from space it may be quasi-nadir or limb viewing. In the case of scattering there is also the possibility of providing your own source in the form of Lidar. There are many options. How should we determine the characteristics of each so that we can choose the most appropriate for any particular application? Typically it is usually a matter of assessing whether the instrument types for which the experimenter has expertise are appropriate for the problem in hand, and if not to look for another problem (or obtain the expertise!).

Once the measurement requirements have been specified, and the style of instrument and observing geometry to be considered have been selected, the process of design and optimisation should include:

(i) Designing a forward model for the instrument, algebraically
(ii) Identifying the forward model parameters and state vector elements
(iii) Building a numerical forward model and the corresponding derivative model
(iv) Building a numerical retrieval method
(v) Instrumenting the retrieval method with diagnostics
(vi) Selecting *a priori* information, both atmospheric and instrumental
(vii) Determining whether the proposed measurements are adequate to determine the proposed target parameters
(viii) Optimising the retrieval characteristics with respect to the instrument design.

The end-to-end simulation can also be used to optimise the instrument design. In the simplest form, the instrument design parameters may be varied manually, with the aid of an experimenters intuition, but it is also possible to optimise some overall quality indicator such as Shannon information content automatically. The simulation should also be used to specify the accuracy with which instrument parameters should be determined in the laboratory, and to identify physics parameters and atmospheric *a priori* which may need further attention.

11.1.1 *Forward model construction*

The parameters of the forward model should include everything that may affect the signal detected by the instrument in the three main categories:

(i) Instrument parameters such as spectral response, temperature dependences, detector noise, etc.

(ii) Atmospheric parameters such as temperature and constituent distributions, including the target quantities.
(iii) Physics parameters such as spectral data for the relevant gases.

The signal should at this stage be regarded as the raw output of the instrument in engineering units before any effects of the calibration have been applied. The calibration parameters are, of course, forward model parameters. Initial values and variances should be set up for all of the parameters, together with covariance matrices where appropriate. One of the aims of the instrument design and optimisation is to determine which forward model parameters should be targets for retrieval, and which ones are known well enough *a priori* to be regarded as constants. The forward model may be used for this purpose by examining the sensitivity of the measurements with respect to all of its parameters. An initial division can be made on the basis of a comparison of the sensitivity of the computed signal to changes of the parameter within its error bars, although the final division should depend on the sensitivity of the retrieval.

11.1.2 *Retrieval method and diagnostics*

A retrieval method will be needed, equipped with a range of diagnostics to determine whether the accuracy and resolution of the measurement of the target quantity satisfies the original requirements. The retrieval method for design purposes need not be as comprehensive as an operational one, and many simplifications are possible. Its prime purpose is to compute system characteristics rather than retrievals as such, although the computation of retrievals is a useful test that it is working correctly. As such, a linear retrieval will normally be adequate, linearised about the state for which the signals are to be computed. The linearisation can be carried out straightforwardly by perturbing the forward model numerically. This is a relatively slow method, but it is fairly flexible, and straightforward to code.

The retrieval method should be capable of retrieving all forward model parameters that may significantly affect the signal. *A priori* information about all of the forward model parameters will be needed, together with a conservative description of the prior state of the atmosphere. The simulation should be able to produce the following diagnostics of the observing system:

(i) Sensitivity of the measured signal to the target quantities and the other forward model parameters. This includes the weighting function matrix.
(ii) Sensitivity of the retrieval to the target quantities and the other forward model parameters. This includes the averaging kernel matrix.
(iii) The contributions to the signal covariance of uncertainties in each of the independent forward model parameters. It is of course important to distinguish the contributions of random and systematic sources of error.
(iv) The contributions to the total retrieval error covariance of measurement

noise and the uncertainties in each of the independent forward model parameters.
(v) If you have a source of realistic atmospheric *a priori*, evaluate the smoothing error.
(vi) The Shannon information content and the degrees of freedom for signal.

These diagnostics will enable you to determine whether the observing system is capable of producing results to the resolution, accuracy and precision needed, and help you to identify which sources of error need further attention.

11.1.3 *Optimisation*

An initial estimate of the feasibility of an observing system will depend on a comparison between the total error variance of the retrieval and the original requirement, and on whether the averaging kernels can provide the spatial resolution required. If the requirements are not satisfied, it may be possible to improve matters by optimising instrument design parameters. Even if they are satisfied, it may be worth optimising the instrument beyond the requirements if it can be done at little cost.

Any observing system can be described in terms of a multitude of design parameters. Optimising the performance of the system is a matter of adjusting the parameters for best performance, within constraints set by both the physics of the measurement and practical requirements such as cost and weight. Design parameters may include such things as spectral resolution, number and placement of channels, size and design of input optics, scan patterns, spatial resolution, integration times and frequency of measurement, etc. Many of the design choices interact, and the effect of design choice on the quality of the eventual product is often far from obvious. Often the only way to objectively optimise the design is to simulate the observing system numerically, select some quality characteristic, and to optimise that with respect to the design parameters.

In order to carry out a successful and meaningful optimisation the quality parameter should be a single scalar. It is not in general possible to simultaneously minimise several different parameters, as each one would have its minimum at a different place. If several quantities are separately important then a weighted mean will have to be minimised, and the different requirements traded off against each other by varying the weights until a satisfactory compromise is reached. The quantities that make up a suitable overall quality parameter might include such things as the retrieval accuracy, the information content, the degrees of freedom for signal, and the spatial resolution, depending on the scientific requirements of the measurement. Note that some of these items depend on the retrieval method chosen, so that it is important to remember that the retrieval method is part of the observing system, and also needs optimising.

If maximum accuracy or minimum solution variance is needed, the obvious approach is to minimise the total solution variance, $\sum_i \hat{S}_{ii}$, i.e. trace($\hat{\mathbf{S}}$). However a moment's thought shows that this is not appropriate, as variances are not a uniform set of quantities, and may even be in different units. Therefore some scaling of the solution covariance is needed, so that the elements are dimensionless and comparable. The simplest quantity to scale with is based on the prior covariance. The corresponding scaled state vector would be dimensionless: $\mathbf{z} = \mathbf{S}_a^{-\frac{1}{2}}\mathbf{x}$, with a unit prior covariance. The solution (posterior) covariance of $\mathbf{z}$ is

$$\hat{\mathbf{S}}_z = \mathbf{S}_a^{-\frac{1}{2}}\hat{\mathbf{S}}\mathbf{S}_a^{-\frac{1}{2}}. \tag{11.1}$$

The trace of this matrix is a reasonable quantity to minimise, as an appropriately scaled retrieval covariance:

$$\sum_i \hat{S}_{z,ii} = \text{trace}(\mathbf{S}_a^{-\frac{1}{2}}\hat{\mathbf{S}}\mathbf{S}_a^{-\frac{1}{2}}) = \text{trace}(\hat{\mathbf{S}}\mathbf{S}_a^{-1}). \tag{11.2}$$

From Eqs. (2.79) and (2.80), we can see that

$$\text{trace}(\hat{\mathbf{S}}\mathbf{S}_a^{-1}) = \text{trace}(\mathbf{I}_n - \mathbf{A}) = n - d_s. \tag{11.3}$$

so that minimising the trace is the same as maximising the degrees of freedom for signal. Alternatively, the information content of the retrieval, Eq. (2.73), could be optimised by maximising the determinant of $\hat{\mathbf{S}}\mathbf{S}_a^{-1}$. The same result should be obtained if the information content of the measurement alone is maximised, provided an information-preserving retrieval is used, e.g. an optimal estimator.

11.1.4 *Specifying requirements for the accuracy of parameters*

Sensitivity studies of retrieval characteristics with respect to the instrument parameters are needed to determine the accuracy with which the parameters must be known and hence the accuracy with which they must be measured, both in the laboratory as part of the instrument characterisation and in the field as part of the routine ongoing instrument calibration.

Similarly sensitivity studies with respect to atmospheric and physical forward model parameters such as spectral data may indicate that further laboratory work is needed to improve their accuracy.

These studies are part of the design process, and should be carried out before the instrument is built, as they may have consequences for the design, for example the internal radiometric calibration system.

11.2 Design of the Operational Retrieval Method

A retrieval method for routine processing of the data from an instrument has different design requirements from one used for studying the feasibility of the method or for optimising the instrument design. It must be able to deal correctly with the full nonlinearity of the measurement, including possible extremes of the atmospheric state and instrument behaviour, and should be able to detect and report unexpected behaviour of the instrument and the atmosphere. It does not have to be able to retrieve all of the possible forward model parameters, and does not need so wide a range of diagnostics, but it should be capable of providing a routine description of the characteristics and accuracy of the retrieval. It should be designed with the eventual use of the data in mind, i.e. it should be built to suit the customer.

11.2.1 Forward model construction

The forward model for retrieval should be capable of reproducing the measurements within measurement error but also within the limitations set by the available computing power. Operational forward model parameters are constants, and this usually allows for some quantities to be precomputed, so that the calculations can be carried out more efficiently. For example, atmospheric transmittance models can be fixed, as their dependence on the instrument spectral response can be built in.

However the computation of the Jacobians of the forward model must now be carried out efficiently, and this normally means that the algebraic derivative of the forward model must be coded, rather than using numerical perturbation methods.

11.2.2 State vector choice

The retrieval method for design and optimisation will have been used to determine which of the forward model parameters are sufficiently poorly known that they need to be retrieved—which should of course include the original target quantities! The next question is to decide in which form they should be retrieved. There are often many possible equivalent representations of the state, some of which may be numerically more efficient than others. A typical example is the temperature profile which may be defined on a height or a pressure grid, and in some circumstances it may be more sensible to retrieve e.g. the Planck function profile instead. Considerations include linearity of the forward model, which has consequences for the number of iterations required by the retrieval, and simplicity of formulation, which may have consequences for the coding and running time for the forward model and its derivative code.

A profile is a continuous function, and a measurement is of a finite number of quantities, very often fewer independent quantities than you might expect. This should be kept in mind when choosing a profile representation. The number of degrees of freedom for signal, d_s is a useful measure of the minimum number of parameters required. Choosing a representation is usually a matter of choosing an altitude or pressure grid, together with an interpolation procedure, although other approaches such as finite-order polynomials and fourier series are options. The primary considerations are the choice of the number of levels or parameters, and the form of the representation. The number of levels should be enough to represent all of the information that may be provided by both the measurements and the *a priori*, if used. The interpolation method should represent the normal behaviour of the profiles, if known, reasonably well. If the number of levels is too small, generally smaller than quantities such as the rank of $\mathbf{K}$, the rank of $\tilde{\mathbf{K}}$, or the degrees of freedom for signal, d_s, then all of the information in the measurements cannot be represented in the retrieval, and some will be lost. If the number of levels is larger than needed then the problem is underconstrained and *a priori* is necessary. In this case the only penalties are the extra computing time needed, and the difficulty of explaining the nature of the retrieval to the naive data user.

11.2.3 Choice of vertical grid coordinate

The vertical profile of temperature, pressure and gas concentrations must be represented as a finite set of parameters, some or all of which may be used as elements of the state vector. These will normally be specified on some grid, with some specified interpolation rule. It should be noted that there are two grids involved, which may or may not be the same. The fundamental one is the grid defined for the state vector, but it also may be necessary to define a finer scale grid in order to carry out the numerical calculations to the required accuracy. Choice of state vector grid spacing has been discussed in Chapter 10.

There are usually two constraints on the profile, namely the ideal gas equation and the hydrostatic equation:

$$p = \rho RT/M \tag{11.4}$$
$$\mathrm{d}p = -\rho g\, \mathrm{d}z \tag{11.5}$$

where p is pressure, R is the molar gas constant, T is temperature, M is the molar mass of air, g is gravity and $\mathrm{d}z$ is a height increment. Note that g will depend on height and latitude, and M will also vary with height, both in the troposphere where water mixing ratios can be significant and at high altitudes where O_2 is dissociated. For very high accuracy work, it may be necessary to use a non-ideal gas equation.

Consequently there are several different ways in which the vertical distribution of the mass of the air can be described. The basic parameters are temperature, pressure, density and height, and any two of these in conjunction with the gas equation and the hydrostatic equation will determine the other two, except that in some cases a reference level may be required to provide an absolute height scale, as one of the constraints is a differential. Other parameters such as potential temperature may also be used, provided they are related to the basic parameters in a known way. In practice only three vertical coordinates are normally considered, namely pressure (or equivalently, $\ln p$), absolute height relative to the geoid, or height relative to a pressure level, the choice depending on which is most convenient for evaluating the forward model. Thus the mass distribution may be described as temperature, density or height on a pressure scale, or as temperature, density or pressure on a height scale.

⇒ **Exercise** 11.1: Investigate the relationships between these different possibilities, and determine which ones need a reference height or pressure.

The profile may be discretised in terms of a set of values on discrete levels or layers, regularly or irregularly spaced in height or in pressure, together with an interpolation rule between the values. Some possibilities are:

(i) A set of pressure levels, separated by constant temperature layers. Between the levels density is proportional to pressure. One level may be given an absolute height.

(ii) A set of pressure levels with temperature given at the levels, and interpolated linearly in $\ln p$. Between the levels density is not a simple function of pressure.

(iii) Density at a set of height levels, with $\ln \rho$ interpolated linearly in height. Pressure and temperature are not simple functions of height.

(iv) Temperature at a set of heights, interpolated linearly, and with pressure given at one level. Pressure and density are not simple functions of height.

It is clearly more satisfying if the representation is hydrostatically consistent in detail, but this may lead to complicated algebra, and it may not lead to significant error if narrow non-hydrostatic layers are used.

11.2.3.1 *Choice of parameters describing constituents*

The quantity of a target species may be expressed as a mixing ratio, a gas density, a layer amount, or in the case of water vapour as specific humidity. We can also transform these quantities, for example by using the logarithm of mixing ratio. All are equally good as descriptions of the profile, but have different advantages and disadvantages in terms of retrieval. Some relevant considerations are:

Linearity of the forward model as a function of the parameter. This is important for the speed of convergence of the retrieval and the validity of the characterisation and error analysis. For weak absorbers anything proportional to the absorber amount in the path (mixing ratio, density or layer amount) is suitable, but for strong absorbers some numerical experimentation is needed.

A wide range of magnitude between state vector elements can produce numerical problems by provoking ill-conditioning in systems of equations to be solved. This can arise if layer amounts or gas densities are used, because of the wide range of air density with altitude. It is easily dealt with by scaling the state vector with some standard values, e.g. retrieving the ratio of the state vector to the *a priori*, or by a scale-reducing (but nonlinear) transformation such as the logarithm or fourth root.

Gaussian distribution of the a priori. This is needed for linearity of the retrieval problem and hence speed of convergence. Parameters proportional to absorber amount are likely to be non-Gaussian if variances are not small compared with the amount, because the amount is required to be positive. A log-normal distribution of amount may be more suitable, i.e. a Gaussian distribution when the state vector is the logarithm of mixing ratio, density or layer amount.

Positivity of the retrieved gas amount. We know *a priori* that mixing ratios are positive. We can use a transformation that enforces positivity, or simply be aware that the formal error bounds of a retrieval encompass zero.

Bias when computing statistics of retrieved amounts is possible if the state vector is a nonlinear function of amount, and error is significant, i.e. when looking for small effects in noise. This happens because when transformed back to amount, the distribution of error becomes non-Gaussian, and its mean is not the same as its median. A simple average is then misleading. A better approach is to average the transformed quantity.

There is no straightforward choice, suitable for all problems. You will have to experiment with the possibilities, and decide which set of advantages and disadvantages is most appropriate for the problem in hand.

11.2.4 A priori *information*

A priori information is central to the Bayesian view of the inverse problem and is a critical part of regularising an ill-posed problem. Selection and construction of *a priori* are discussed in detail in Chapter 10.

11.2.5 *Retrieval method*

The counsel of perfection would be to use e.g. the optimal estimator expressed in Eq. (5.9), perhaps using some variant of the Levenberg-Marquardt method to improve convergence. In the case of Gaussian noise the full covariance $\mathbf{S}_\epsilon$ would be used, and in the case of a Gaussian prior the full covariance $\mathbf{S}_a$ would be used. However there are circumstances when efficiency must be considered over optimality. The non-Gaussian case is often one of these, and it is almost universal to assume Gaussian statistics regardless. In the case of a large measurement vector, the calculation (effectively) of $\mathbf{S}_\epsilon^{-1}$ may be prohibitive, and may have to be approximated. It goes without saying that the effect of such approximations must be tested, and the trade-off between processing time and accuracy considered.

11.2.6 *Diagnostics*

A retrieval should produce a routine error analysis and characterisation so that the user is aware of the nature of the retrieved data. This is needed so that the user of the measurements knows exactly what the retrievals mean, and so that they are not over-interpreted. The minimum set comprises a statement of the resolution, and a statement of the precision and accuracy. Often a representative set will suffice, but if the problem is nonlinear the characteristics will be state dependent, and each retrieved state should include its own characterisation. A useful set is the averaging kernel matrix, the retrieval noise covariance, and the systematic error covariance. Unfortunately this set of matrices occupies rather more space than a retrieval itself!

The retriever will require a more extensive set of diagnostics in order to determine whether the operational retrievals are satisfactory. These will be discussed in Chapter 12.

Chapter 12

Testing and Validating an Observing System

Once the instrument has been built, its characteristics determined in the laboratory, and it is producing real data from the atmosphere, the results must be validated to confirm that the retrieved quantities are as accurate as is possible, and that their relationship to the true state is properly understood. The purpose of validation in this general sense is to confirm that the theoretical characterisation and error analysis actually represent the properties of the real data. Discrepancies at this stage may be a result of the instrument not performing to specification, or of the retrieval method (usually the forward model) not properly representing the instrument. The testing and validation process includes such things as

(i) A full error analysis and characterisation of the observing system.
(ii) Internal consistency checks, for example the comparison of measurements (signals and retrievals) for different instrument settings if appropriate, or comparison of measurements taken at the same place and closely spaced in time, such as on ascending and descending sections of the orbit. These can reveal errors in the understanding of how the instrument works, and its representation in the forward model, and can give an independent estimate of the measurement precision.
(iii) Comparison of the measured signals with independent direct measurements of the profile, if you are lucky enough to have some. This is a very straightforward way to check out the performance of the instrument and your forward model. It is less important to compare retrievals with direct measurements (though it is usually a good idea to do it for political reasons). It is the quality of the forward model that determines the quality of the result, and the inverse method itself can be adequately validated by simulation. Agreement should be within the combined error bounds of the direct measurement and the instrument being validated. But remember that there is no such thing as 'ground truth', if discrepancies are found, both instruments should be investigated.
(iv) Comparison of the measured signals with *a priori* using a χ^2 test. This can

be used to look at distribution of the signals and of χ^2, to test some aspects of the correctness of the forward model, knowledge of the measurement error and the prior *pdf*, its mean and its covariance. It is a poor substitute for comparison with direct measurements, because the accuracy of the *a priori* is usually much poorer than almost any direct measurement.

(v) Retrieval validity tests such as a χ^2 comparison of the signal with the retrieved signal; examination of retrievals that fail to converge properly or that converge outside reasonable limits set by the χ^2 test, and determination of the reasons. This can reveal problems with the numerical method chosen to solve the inverse problem, or with the forward model.

(vi) Comparison of the retrievals with the *a priori*, looking for bias and unexpected variance about mean.

(vii) Comparison of retrievals with other remote sounding instruments.

(viii) Comparison with models, but taking due note of possible modelling errors, and remembering that one of the purposes of the measurements is to validate the models.

12.1 Error Analysis and Characterisation

A full error analysis and characterisation is needed as a basis for any comparisons to be made, and to understand the nature of the retrieval. The following at least, as described in Chapter 3, will be found useful:

(i) *Averaging kernels.* Confirm that the resolution is what you expect or need, and that the areas of the averaging kernels are sensible.

(ii) *Noise sensitivity.* Calculate both the gain matrix $\mathbf{G}$ and the contribution to the solution covariance from random error: $\mathbf{S_x} = \mathbf{G}\mathbf{S}_\epsilon\mathbf{G}^T$. The diagonal of $\mathbf{S_x}$ is a basic first estimate of the retrieval noise.

(iii) *Forward model parameter sensitivity.* For all of the forward model parameters, calculate $\mathbf{GK}_b$ and the individual contribution to the overall retrieval error from each class of forward model parameter.

(iv) *Inverse method parameter sensitivity* If there are any inverse method parameters other than the *a priori*, you should evaluate the contribution to retrieval error due to them, and ensure that it is negligible.

(v) *Smoothing error.* If (and only if) you have a good estimate of a true ensemble covariance $\mathbf{S}_e$, you can estimate the error covariance due to the smoothing inherent in the retrieval for that ensemble, namely $(\mathbf{A} - \mathbf{I})\mathbf{S}_e(\mathbf{A} - \mathbf{I})^T$.

(vi) *Retrieval bias.* Does the retrieval reproduce the *a priori* priori in the absence of error?

12.2 The χ^2 Test

The χ^2 test is a way of testing whether a particular random vector belongs to a given Gaussian distribution. If a vector $\mathbf{z}$ is supposed to be a member of a Gaussian ensemble with zero mean and covariance $\mathbf{S}_z$, then the quantity considered is $\chi^2 = \mathbf{z}^T \mathbf{S}_z^{-1} \mathbf{z}$, i.e. twice the exponent in the Gaussian distribution. The χ^2 test asks the question: 'what fraction of members of the Gaussian distribution have a probability density less than (or greater than) that of the vector being tested?' If the fraction, f, is small then the vector is an outlier, and is described as 'significant at the $100f\%$ level'. The test is usually used to look for cases where χ^2 is too large, but it can also be used to detect cases where χ^2 is too small. In either case a significant result may indicate that the vector is suspicious, or the Gaussian distribution assumed is suspicious. In carrying out a χ^2 test it is of course important to ensure that the covariance matrix for the case being considered is appropriate, i.e. that $\mathbf{S}_z$ really is $\mathcal{E}\{\mathbf{z}\mathbf{z}^T\}$.

Using an eigenvector decomposition $\mathbf{S}_z = \mathbf{L}\mathbf{\Lambda}\mathbf{L}^T$, we can see that

$$\chi^2 = \mathbf{z}^T \mathbf{L}\mathbf{\Lambda}^{-1}\mathbf{L}^T \mathbf{z} = \sum_{i=1}^{n} z_i'^2/\lambda_i \qquad (12.1)$$

is the sum of separate terms for n independent normally-distributed random variables $\mathbf{z}' = \mathbf{L}^T \mathbf{z}$, each z_i' having variance λ_i. If any of the eigenvalues are zero, then the covariance matrix is singular, implying that the corresponding z_i' should also be zero. In the non-singular case χ^2 can be thought of as the square of the radius in an n-dimensional space having coordinates $z_i'/\lambda_i^{\frac{1}{2}}$. Its *pdf* will be the fraction of the ensemble between χ and $\chi + \mathrm{d}\chi$:

$$P_n(\chi^2)\,\mathrm{d}\chi \propto \chi^{n-1} \exp\left(-\tfrac{1}{2}\chi^2\right) \mathrm{d}\chi \qquad (12.2)$$

This can be normalised and rearranged to give

$$P_n(\chi^2)\,\mathrm{d}\chi^2 = \frac{1}{\Gamma(n/2)} \left(\tfrac{1}{2}\chi^2\right)^{\frac{n}{2}-1} \exp\left(-\tfrac{1}{2}\chi^2\right), \qquad (12.3)$$

the χ^2 distribution with n degrees of freedom. Significance tables for the distribution, usually tabulations of $\int_{\chi^2}^{\infty} P_n(\chi^2)\,\mathrm{d}\chi^2$, can be found in statistical texts and computer subroutine libraries. The distribution has the property that $\mathcal{E}\{\chi^2\} = n$ and $\mathrm{var}(\chi^2) = 2n$. For large n it tends to a normal distribution.

It is quite possible for the covariance of a quantity being considered for testing to be singular or nearly so. The prime example is $\hat{\mathbf{x}} - \mathbf{x}_a$, when the problem is underconstrained and $\mathbf{K}$ has a null space. In this case components of $\hat{\mathbf{x}}$ corresponding to the null space should have been supplied entirely from $\mathbf{x}_a$, and the corresponding difference should be zero. Hence the covariance is singular and χ^2 cannot be computed directly. There are various ways of ensuring that we only consider non-null-space components. We can construct a pseudo-inverse with the aid of the eigenvector expansion, namely $\mathbf{S}^* = \mathbf{L}'\mathbf{\Lambda}'^{-1}\mathbf{L}'^T$ where only the p nonzero (or non-small) eigenvalues are retained in $\mathbf{\Lambda}'$. The χ^2 calculated with this inverse would have p degrees of freedom. Alternatively we can expand $\hat{\mathbf{x}} - \mathbf{x}_a$ in terms of eigenvectors, $\mathbf{z}' = \mathbf{L}^T(\hat{\mathbf{x}} - \mathbf{x}_a)$. The elements of $\mathbf{z}'$ are independent and can be tested individually using the normal distribution (which is the same as χ^2 with one degree of freedom) using $\chi^2 = z_i^2/\lambda_i$. The z_i corresponding to zero eigenvalues should of course be zero, but may not be exactly so due to numerical errors.

It can be useful to look at the actual distribution of χ^2 for a large ensemble of vectors. This can give more information about the source of problems, perhaps indicating that the assumed distribution is basically correct, but there is a small population of outliers (faulty measurements) that do not belong, or that there is something wrong with the assumed distribution. In considering individual cases you should remember that even if all is well, a fraction f of a large ensemble is likely to be in the 'significant' category.

12.3 Quantities to be Compared and Tested

There are several different quantities that can be compared, with different comparisons answering different questions. Three quantities depend on the measurement, namely the measurement $\mathbf{y}$ itself, the retrieval, $\hat{\mathbf{x}}$, and the retrieved measurement $\hat{\mathbf{y}} = \mathbf{F}(\hat{\mathbf{x}})$. These may be compared with the prior state, $\mathbf{x}_a$, the prior measurement, $\mathbf{F}(\mathbf{x}_a)$, and other independent (correlative) measurements of the same state, $\mathbf{x}_c$, if any are available. Useful diagnostics are:

(i) The individual differences between the quantities tested and the quantities with which they are being compared;
(ii) χ^2 for the individual differences;
(iii) The mean difference over an ensemble of cases;
(iv) The covariance of differences estimated from an ensemble of cases, compared with the theoretical covariance;
(v) An estimate of the distribution of differences, to compare with a Gaussian distribution;
(vi) The distribution of χ^2 values for an ensemble of measurements to compare with a χ^2 distribution.

12.3.1 *Internal consistency*

If an instrument is designed to make redundant or nearly redundant measurements then there is the possibility of some internal consistency tests. Conceptually if the state could be retrieved using two independent selections of channels, then the two retrievals could be compared as if from two different instruments, using the methods described in section 12.4. Alternatively, if there are independent linear combinations of weighting functions which are the same or similar, then there will also be combinations which give zero or small weighting functions. The corresponding combinations of measurements should then be zero or small. Such combinations can be found by computing the singular vectors and values of the weighting function matrix. Expanding $\mathbf{K}$ as $\mathbf{U}\mathbf{\Lambda}\mathbf{V}^T$ about some reference state $\mathbf{x}_0$ we obtain:

$$\mathbf{y} - \mathbf{F}(\mathbf{x}_0) = \mathbf{U}\mathbf{\Lambda}\mathbf{V}^T(\mathbf{x} - \mathbf{x}_0) + \boldsymbol{\epsilon} \qquad (12.4)$$

or

$$\mathbf{U}^T[\mathbf{y} - \mathbf{F}(\mathbf{x}_0)] = \mathbf{\Lambda}\mathbf{V}^T(\mathbf{x} - \mathbf{x}_0) + \mathbf{U}^T\boldsymbol{\epsilon} \qquad (12.5)$$

If any singular value λ_i is small enough that $\lambda_i \mathbf{v}_i^T(\mathbf{x} - \mathbf{x}_0) \ll \mathbf{u}_i^T \boldsymbol{\epsilon}$, then statistical tests can be carried out on the corresponding element of the transformed measurement, $\mathbf{u}_i^T[\mathbf{y} - \mathbf{F}(\mathbf{x}_0)]$, using the statistics of $\mathbf{u}_i^T \boldsymbol{\epsilon}$ only. More generally we should consider scaling the weighting function matrix to give $\tilde{\mathbf{K}} = \mathbf{S}_\epsilon^{-\frac{1}{2}} \mathbf{K} \mathbf{S}_a^{\frac{1}{2}}$, as introduced in section 2.4. The complement of the conclusion of section 2.4.1 is that the number of independent measurements made to worse than measurement error is the number of singular values of $\tilde{\mathbf{K}} = \tilde{\mathbf{U}}\tilde{\mathbf{\Lambda}}\tilde{\mathbf{V}}^T$ which are less than about unity, where the tilde distinguishes this singular vector decomposition from that of $\mathbf{K}$. These give independent combinations of the measurements which should be zero to within measurement error:

$$\mathbf{y}' = \tilde{\mathbf{U}}^T \mathbf{S}_\epsilon^{-\frac{1}{2}} (\mathbf{y} - \mathbf{y}_a) = \mathbf{\Lambda}\tilde{\mathbf{V}}^T \mathbf{S}_a^{-\frac{1}{2}} (\mathbf{x} - \mathbf{x}_a) + \boldsymbol{\epsilon}' \qquad (12.6)$$

The covariance of $\mathbf{y}'$ is $\mathbf{\Lambda}^2 + \mathbf{I}$, the $\mathbf{I}$ corresponding to the noise term $\boldsymbol{\epsilon}'$. Elements y'_i of $\mathbf{y}'$ or combinations $\tilde{\mathbf{u}}_i^T \mathbf{S}_\epsilon^{-\frac{1}{2}}(\mathbf{y} - \mathbf{y}_a)$ of the measurements which correspond to $\lambda_i \ll 1$ should have unit variance.

The test can be carried out either as a significance test using a single measurement vector, or by comparing an ensemble covariance of the transformed measurement with the transformed noise covariance. Such a test can indicate whether there are problems with the forward model or the measurement error characterisation.

A typical case where this kind of test would be useful is a Fourier Transform Spectrometer, where there are far more channels than degrees of freedom for signal.

Strictly this applies only to nearly linear problems, but by selecting ensembles with small variability, it can be used for other cases.

12.3.2 Does the retrieval agree with the measurement?

In the case of non-linear problems, this test should be a routine part of the numerical solution, rather than something carried out afterwards as part of the validation, because it is needed to check whether the retrieval has converged to a spurious minimum. In the case of linear problems, the test is only needed to check whether the retrieval method is conceptually correct, as it should automatically be satisfied if all is well.

The test is carried out by evaluating the forward model for the retrieval, and comparing with the measurement. The covariance of the difference is not the same as the measurement error covariance, $\mathbf{S}_{\epsilon_y}$, as you might suppose, because the retrieval is correlated with this error. Provided that the problem is linear within the error bounds, we should have

$$\begin{aligned}
\delta\hat{\mathbf{y}} = \hat{\mathbf{y}} - \mathbf{y} &= \mathbf{F}(\hat{\mathbf{x}}) - \mathbf{F}(\mathbf{x}) - \epsilon_y \\
&= \mathbf{K}(\hat{\mathbf{x}} - \mathbf{x}) - \epsilon_y \\
&= \mathbf{K}[(\mathbf{GK} - \mathbf{I})(\mathbf{x} - \mathbf{x}_a) + \mathbf{G}\epsilon_y] - \epsilon_y \\
&= (\mathbf{KG} - \mathbf{I})[\mathbf{K}(\mathbf{x} - \mathbf{x}_a) + \epsilon_y]
\end{aligned} \quad (12.7)$$

using Eq. (3.16). Thus its covariance is

$$\mathbf{S}_{\delta\hat{\mathbf{y}}} = (\mathbf{KG} - \mathbf{I})(\mathbf{KS}_a\mathbf{K}^T + \mathbf{S}_{\epsilon_y})(\mathbf{KG} - \mathbf{I})^T. \quad (12.8)$$

In the case of an optimal retrieval, we can substitute for $\mathbf{G}$ from Eq. (4.41) and obtain

$$\mathbf{S}_{\delta\hat{\mathbf{y}}} = \mathbf{S}_{\epsilon_y}(\mathbf{KS}_a\mathbf{K}^T + \mathbf{S}_{\epsilon_y})^{-1}\mathbf{S}_{\epsilon_y} \quad (12.9)$$

which is clearly always of full rank if $\mathbf{S}_{\epsilon_y}$ is. In this case χ^2 has m degrees of freedom*. If the test indicates significance when a test on $\mathbf{y} - \mathbf{y}_a$ does not (section 12.3.3), then we would expect the cause to be incorrect convergence.

In the case of an instrument with a large number of channels, this test can be expensive in time because of the large matrices involved, even with the aid of Cholesky decomposition. In that case it may be worth approximating $\mathbf{S}_{\delta\hat{\mathbf{y}}}^{-1}$ by $\mathbf{S}_{\epsilon_y}^{-1}$, especially if the latter is diagonal. This will give a test for incorrect convergence which is conservative in the sense of being too loose.

12.3.3 Consistency with the a priori

We can test whether the measurement, the retrieval and the retrieved measurement are consistent with the *a priori*, each test giving different information. A test on the measurement depends on, and so will give information about, the *a priori*, our understanding of the instrument, and our modelling of the measurement. A test on the retrieval will give the same kind of information as the measurement comparison, but will also test the retrieval process.

12.3.3.1 Measured signal and a priori

In the nearly linear approximation, the ensemble covariance of $\mathbf{y} - \mathbf{y}_a$ is expected to be $\mathbf{S}_y = \mathbf{KS}_a\mathbf{K}^T + \mathbf{S}_\epsilon$ and the corresponding χ^2 should have m degrees of freedom. The significance level for individual measurements will indicate cases which need further examination; departure of the actual distribution of χ^2 from that expected

---

*See also Exercise 5.2.

may indicate non-Gaussian *a priori* or noise, which can be further investigated by examining the distribution of $\mathbf{y} - \mathbf{y}_a$ itself. The presence of outliers may indicate intermittent instrument problems or unanticipated atmospheric behaviour (such as the ozone hole!). The individual values of $\mathbf{y} - \mathbf{y}_a$ for significant cases may indicate that some elements of $\mathbf{y}$ are being incorrectly computed; consider using a χ^2 test with one degree of freedom on each element of $\mathbf{y}$ to identify them further. The comparison of the mean difference $\bar{\mathbf{y}} - \mathbf{y}_a$ and covariances may indicate that the ensemble being examined is not consistent with the assumed *a priori*.

12.3.3.2 *Retrieval and* a priori

If the ensemble of data being examined passes the test of comparing the measured signal with the *a priori*, then the next step is to compare an ensemble of retrievals with the *a priori*. We can write:

$$\hat{\mathbf{x}} - \mathbf{x}_a = \mathbf{A}(\mathbf{x} - \mathbf{x}_a) + \mathbf{G}\boldsymbol{\epsilon} = \mathbf{G}[\mathbf{K}(\mathbf{x} - \mathbf{x}_a) + \boldsymbol{\epsilon}] \qquad (12.10)$$

for the difference in the nearly linear approximation in the general case, and in the moderately non-linear approximation in the optimal estimation case (section 5.5), so this is of wider validity than the direct comparison of measurements. The covariance for the general case, is expected to be

$$\mathbf{S}_{\hat{\mathbf{x}}} = \mathbf{G}(\mathbf{K}\mathbf{S}_a\mathbf{K}^T + \mathbf{S}_\epsilon)\mathbf{G}^T \qquad (12.11)$$

and in the optimal case this becomes,

$$\mathbf{S}_{\hat{\mathbf{x}}} = \mathbf{S}_a\mathbf{K}^T(\mathbf{K}\mathbf{S}_a\mathbf{K}^T + \mathbf{S}_\epsilon)^{-1}\mathbf{K}\mathbf{S}_a \qquad (12.12)$$

The number of degrees of freedom will be the same as the rank of $\mathbf{K}$. If $\mathbf{K}$ has rank smaller than n, i.e. if $m < n$ or $\mathbf{K}$ has a null space, $\mathbf{S}_{\hat{\mathbf{x}}}$ will be singular because components of $\hat{\mathbf{x}}$ in the null space are unchanged from their *a priori* values. In this case χ^2 must be calculated in the non-null space of $\mathbf{K}$, best identified in terms of the singular vectors of $\tilde{\mathbf{K}} = \mathbf{S}_\epsilon^{-\frac{1}{2}}\mathbf{K}\mathbf{S}_a^{\frac{1}{2}}$.

⇒ **Exercise** 12.1: Show in detail how χ^2 should be calculated in this case.

If the χ^2 test on $\mathbf{y} - \mathbf{y}_a$ is satisfactory, either for individual measurements or the ensemble, and the corresponding test on $\hat{\mathbf{x}} - \mathbf{x}_a$ is not, then there is a problem with the retrieval process itself.

12.3.3.3 *Comparison of the retrieved signal and the* a priori

Is the retrieved signal consistent with the *a priori*? Like the comparison of $\mathbf{y}$ with the *a priori*, this can only be usefully carried out for the nearly linear case. The retrieved signal is related to the measured signal by

$$\hat{\mathbf{y}} = \mathbf{F}(\hat{\mathbf{x}}) = \mathbf{F}[\mathbf{x}_a + \mathbf{G}(\mathbf{y} - \mathbf{y}_a)] = \mathbf{y}_a + \mathbf{K}\mathbf{G}(\mathbf{y} - \mathbf{y}_a) \qquad (12.13)$$

so that $\hat{\mathbf{y}} - \mathbf{y}_a = \mathbf{KG}(\mathbf{y} - \mathbf{y}_a)$, and its covariance is

$$\mathbf{S}_{\hat{\mathbf{y}}} = \mathcal{E}\{\mathbf{KG}(\mathbf{y} - \mathbf{y}_a)(\mathbf{y} - \mathbf{y}_a)^T\mathbf{G}^T\mathbf{K}^T\} \quad (12.14)$$
$$= \mathbf{KG}(\mathbf{KS}_a\mathbf{K}^T + \mathbf{S}_\epsilon)\mathbf{G}^T\mathbf{K}^T \quad (12.15)$$

in the general case and

$$\mathbf{S}_{\hat{\mathbf{y}}} = \mathbf{KS}_a\mathbf{K}^T(\mathbf{KS}_a\mathbf{K}^T + \mathbf{S}_\epsilon)^{-1}\mathbf{KS}_a\mathbf{K}^T \quad (12.16)$$

in the optimal case. This is of order m and rank $\min(m, n)$, so the calculation of χ^2 will need the same kind of special treatment as the comparison of the retrieval and the *a priori* if $n < m$.

12.4 Intercomparison of Different Instruments

We cannot of course determine whether the retrieval is consistent with the state, the measurement with the forward model for the state, or the state with the *a priori*, because we do not know the true state. We can compare the retrieval(s) with independent measurements, but care is needed because they will also have errors which must be allowed for, and if they are indirect measurements they will also have a characterisation in terms of averaging kernels and error covariance. We can use the same formalism for comparing with direct as with indirect measurements, simply by using $\mathbf{A} = \mathbf{I}$ for a direct measurement, if appropriate.

12.4.1 *Basic requirements for intercomparison*

The purpose of an intercomparison is to determine whether different observing systems agree within their known limitations, i.e. the extent to which the forward models and the error covariances satisfactorily describe the instruments and their errors, and the retrieval methods reproduce the atmosphere. To this end it is necessary that:

(1) The observing systems must both retrieve the same quantity $\mathbf{x}$. The definition of $\mathbf{x}$ includes the parameter retrieved, such as trace gas mixing ratio or concentration, the grid on which it is being retrieved, and any interpolation rule used. However there may also be instrument-specific parameters in the state vectors that are not targets for comparison but need to be retrieved, such as surface reflectivity or instrumental offsets.

(2) An ensemble of states must be defined over which the comparison is to take place. For a theoretical assessment of instrumental capabilities, this can be chosen to be all-encompassing, but need not include states grossly outside the range of atmospheric possibilities. For an actual intercomparison over an ensemble of real atmospheres, it should describe the real ensemble as far as possible, but we should be able to treat cases where it is not well known. It need not be the same as any *a priori* used by either or both retrieval methods, but it could be if appropriate. For

convenience we will assume the comparison ensemble to be described by a Gaussian distribution, with mean $\mathbf{x}_c$ and covariance $\mathbf{S}_c$.

For the characterisation of the observing system, the *a priori* is used primarily as the linearisation point of the retrieval process in terms of its averaging kernel; it is a profile which would be unchanged by measurement and retrieval in the absence of measurement noise. Any intercomparison between different observing systems must allow for possible differences between their *a priori*'s.

A retrieval may be described as optimal. By this term we mean one which is, for example, a MAP or minimum variance method, providing some characteristic of the Bayesian ensemble consistent with the measurement, its noise statistics and some *a priori* state. It is important to note that an optimal retrieval depends on both the measurement and the *a priori*, and an inverse method which is optimal with respect to its own *a priori*, $\mathbf{x}_a$, $\mathbf{S}_a$, will not be optimal with respect to a different ensemble such as $\mathbf{x}_c$, $\mathbf{S}_c$. The intercomparison method must be able to treat observing systems without regard to whether the retrieval methods are optimal in any sense.

12.4.2 Direct comparison of indirect measurements

Let us consider the comparison of an ensemble of pairs of independent retrievals of a state $\mathbf{x}$, each observing system characterised by an averaging kernel and a retrieval noise covariance. In practice we are likely to find that the two systems will have used different state vectors, different prior covariances and different linearisation points. If the state vectors do not contain the same elements, then some transformation or selection of elements should be carried out so that they do. If *a priori*'s differ, then a χ^2 test could be carried out to determine whether they are significantly different. We should also transform the characterisation so that both are linearised about the same mean state $\mathbf{x}_c$, the comparison ensemble mean for preference, if known. We can put:

$$\hat{\mathbf{x}}_i - \mathbf{x}_c + (\mathbf{A}_i - \mathbf{I})(\mathbf{x}_{ai} - \mathbf{x}_c) = \mathbf{A}_i(\mathbf{x} - \mathbf{x}_c) + \boldsymbol{\epsilon}_{xi} \qquad (12.17)$$

so that adding the term $(\mathbf{A}_i - \mathbf{I})(\mathbf{x}_{ai} - \mathbf{x}_c)$ to each retrieval will adjust them for different *a priori*. Remember that even if the retrievals are optimal with respect to their own *a priori*, they will not normally be optimal with respect to the comparison ensemble. For the rest of this section we will assume that this adjustment has been carried out and the linearisation point for the characterisation is $\mathbf{x}_c$.

To carry out a χ^2 test on the difference between the measurements we need the covariance of the difference:

$$\mathbf{S}_{\delta \mathbf{x}} = \mathcal{E}\{(\hat{\mathbf{x}}_1 - \hat{\mathbf{x}}_2)(\hat{\mathbf{x}}_1 - \hat{\mathbf{x}}_2)^T\} \qquad (12.18)$$

Using the averaging kernel representation of the retrieval, $\hat{\mathbf{x}}_i - \mathbf{x}_c = \mathbf{A}_i(\mathbf{x} - \mathbf{x}_c) + \boldsymbol{\epsilon}_{\mathbf{x}_i}$ then we obtain:

$$\mathbf{S}_{\delta \mathbf{x}} = (\mathbf{A}_1 - \mathbf{A}_2)\mathbf{S}_c(\mathbf{A}_1 - \mathbf{A}_2)^T + \mathbf{S}_{\epsilon_1} + \mathbf{S}_{\epsilon_2}^T \qquad (12.19)$$

where $\mathbf{S}_{\epsilon_i}$ is the covariance of $\epsilon_{\mathbf{x}_i}$. There are contributions to this covariance from the retrieval errors of the two measurements, and from the differences in the weighting functions which lead to the difference between the averaging kernels $\mathbf{A}_1 - \mathbf{A}_2$. If there is a joint null space, and some components of the state are not measured by either instrument, then $\mathbf{S}_{\delta\mathbf{x}}$ will be singular. In that case we should use the eigenvector approach described in section 12.2.

State space may be considered as being divided into four subspaces, the joint row space spanned by both sets of weighting functions, the joint null space spanned by neither, and two residual spaces, each spanned by one instrument but not the other. The two instruments can be compared directly in the common row space, with our without the aid of a prior *pdf*, and there is nothing to compared in the common null space. In the residual spaces the two instruments measure different quantities, and by themselves cannot be compared.

12.4.3 *Comparison of linear functions of measurements*

Comparing profiles from different instruments is not comparing like with like, and it is necessary to make an estimate of the contribution to the apparent error from this source, in the form of the term$(\mathbf{A}_1 - \mathbf{A}_2)\mathbf{S}_c(\mathbf{A}_1 - \mathbf{A}_2)^T$ above. An alternative approach is to look for quantities that can be compared properly, with a minimum of error due to this source, that is we look for linear functions of the retrievals that correspond to the same or very similar averaging kernels. Another option is to consider the problem of comparing the direct measurements from two different instruments, both of the form $\mathbf{y} = \mathbf{K}\mathbf{x} + \epsilon$, without carrying out a retrieval. In this section we will describe the development in terms of the comparison of raw measurements, but as both problems are of the same algebraic form, the same approach can be applied equally to the comparison of retrievals. This problem is related to the time series analysis technique known as 'canonical covariance analysis', in which linear functions of two separate time series are sought which have the same time variation (e.g. Bretherton, Smith and Wallace, 1992).

The problem may be stated as follows: given two measurements $\mathbf{y}_i$, $i = 1, 2$, or given ensembles of such pairs of measurements, of linear functions of a state

$$\begin{aligned}\mathbf{y}_1 - \mathbf{y}_{c1} &= \mathbf{K}_1(\mathbf{x} - \mathbf{x}_c) + \epsilon_1 \\ \mathbf{y}_2 - \mathbf{y}_{c2} &= \mathbf{K}_2(\mathbf{x} - \mathbf{x}_c) + \epsilon_2\end{aligned} \quad (12.20)$$

what can be said about whether these assumptions are likely to be correct? The ϵ_i and $\mathbf{x}$ are not known, but are assumed to be normally distributed with known covariances, and the $\mathbf{K}_i$ assumed to be known. In general, $\mathbf{K}_1$ and $\mathbf{K}_2$ will span different parts of state space, so cannot be compared directly. They may even not have any subspace of state space in common. Nevertheless, within the comparison ensemble, there may be components of $\mathbf{x}$ which both instruments measure well enough for a useful comparison to be made. To identify such components, we will look for linear functions of the $\mathbf{y}_i$ with weighting functions which are sufficiently similar that the effect of any differences are less than noise, when applied to states within the comparison ensemble. The variances of the linear functions themselves should be, for the comparison ensemble, large compared with noise.

Let both instruments be described by Eq. (12.20) with $\mathbf{x}_c$, the mean of the comparison ensemble, as the linearisation point. If the available instrument description uses a different linearisation point, it can easily be rewritten in terms of $\mathbf{x} - \mathbf{x}_c$. We require linear functions $\mathbf{l}_i$, $i = 1, 2$, corresponding to the combinations of signals

$$z_i = \mathbf{l}_i^T (\mathbf{y}_i - \mathbf{y}_{ci}), \qquad (12.21)$$

such that $\mathcal{E}\{(z_1 - z_2)^2\}$ is as small as possible. To avoid the trivial solution $z_i = 0$ we will require $\mathcal{E}\{z_i^2\} = 1$, with the consequence that $\mathcal{E}\{(z_1 - z_2)^2\} = 2 - 2\mathcal{E}\{z_1 z_2\}$ so that minimising it is the same as maximising $\mathcal{E}\{z_1 z_2\}$. Substituting Eqs. (12.21) and (12.20) we find that the problem becomes one of maximising

$$\mathcal{E}\{z_1 z_2\} = \mathcal{E}\{\mathbf{l}_1^T (\mathbf{K}_1(\mathbf{x} - \mathbf{x}_c) + \boldsymbol{\epsilon}_1)(\mathbf{K}_2(\mathbf{x} - \mathbf{x}_c) + \boldsymbol{\epsilon}_2)^T \mathbf{l}_2\} = \mathbf{l}_1^T \mathbf{K}_1 \mathbf{S}_c \mathbf{K}_2^T \mathbf{l}_2 \quad (12.22)$$

subject to the unit length constraint

$$\mathcal{E}\{\mathbf{l}_i^T (\mathbf{K}_i(\mathbf{x} - \mathbf{x}_c) + \boldsymbol{\epsilon}_i)(\mathbf{K}_i(\mathbf{x} - \mathbf{x}_c) + \boldsymbol{\epsilon}_i)^T \mathbf{l}_i\} = \mathbf{l}_i^T (\mathbf{K}_i \mathbf{S}_c \mathbf{K}_i^T + \mathbf{S}_{\epsilon_i})\mathbf{l}_i = 1 \quad (12.23)$$

where we have taken $\boldsymbol{\epsilon}_1$, $\boldsymbol{\epsilon}_2$ and $\mathbf{x}$ to be independent. Using the transformation $\mathbf{m}_i = (\mathbf{K}_i \mathbf{S}_c \mathbf{K}_i^T + \mathbf{S}_{\epsilon_i})^{\frac{1}{2}} \mathbf{l}_i$ we find that this is equivalent to maximising

$$\mathcal{E}\{z_1 z_2\} = \mathbf{m}_1^T (\mathbf{K}_1 \mathbf{S}_c \mathbf{K}_1^T + \mathbf{S}_{\epsilon_1})^{-\frac{1}{2}} \mathbf{K}_1 \mathbf{S}_c \mathbf{K}_2^T (\mathbf{K}_2 \mathbf{S}_c \mathbf{K}_2^T + \mathbf{S}_{\epsilon_2})^{-\frac{1}{2}} \mathbf{m}_2 \quad (12.24)$$

subject to $\mathbf{m}_i^T \mathbf{m}_i = 1$. From the form of this equation it is clear that $\mathbf{m}_1$ and $\mathbf{m}_2$ must be left and right singular vectors respectively of

$$(\mathbf{K}_1 \mathbf{S}_c \mathbf{K}_1^T + \mathbf{S}_{\epsilon_1})^{-\frac{1}{2}} \mathbf{K}_1 \mathbf{S}_c \mathbf{K}_2^T (\mathbf{K}_2 \mathbf{S}_c \mathbf{K}_2^T + \mathbf{S}_{\epsilon_2})^{-\frac{1}{2}} \qquad (12.25)$$

and that the corresponding singular value is $\mathcal{E}\{z_1 z_2\}$. The required combined weighting functions are

$$\check{\mathbf{K}}_i = \mathbf{l}_i^T \mathbf{K}_i = \mathbf{m}_i^T (\mathbf{K}_i \mathbf{S}_c \mathbf{K}_i^T + \mathbf{S}_{\epsilon_i})^{-\frac{1}{2}} \mathbf{K}_i \qquad (12.26)$$

The closeness of a singular value λ to unity indicates the closeness of the fit, because $\mathcal{E}\{(z_1 - z_2)^2\} = 2 - 2\lambda$. For any particular singular vector, the variance of $z_1 - z_2$ can be regarded as the sum of two terms, one due to noise, $\mathbf{l}_1^T \mathbf{S}_{\epsilon_1} \mathbf{l}_1 + \mathbf{l}_2^T \mathbf{S}_{\epsilon_2} \mathbf{l}_2$, and the remainder due to the combinations of weighting functions not matching perfectly. Hence if we wish to compare only combinations which agree more or less within noise, then we should consider those for which $2 - 2\lambda$ is not much greater than $\mathbf{l}_1^T \mathbf{S}_{\epsilon_1} \mathbf{l}_1 + \mathbf{l}_2^T \mathbf{S}_{\epsilon_2} \mathbf{l}_2$.

For instruments with large numbers of channels, the matrices to be diagonalised are large, and the evaluation of e.g. $(\mathbf{K}_i \mathbf{S}_c \mathbf{K}_i^T + \mathbf{S}_{\epsilon_i})^{-\frac{1}{2}}$ may be difficult. This can be dealt with by first using a singular vector decomposition of $\tilde{\mathbf{K}}_i = \mathbf{S}_{\epsilon_i}^{-\frac{1}{2}} \mathbf{K}_i \mathbf{S}_c^{\frac{1}{2}}$, which reduces the effective number of channels to not more than the dimension of the state vector.

Appendix A

Algebra of Matrices and Vectors

Much of retrieval theory involves solving linear equations, whether the problem being solved is linear or non-linear, so a thorough understanding of the nature of linear equations and the algebra of matrices an vectors is valuable. This appendix is intended to summarise the main tools that are used in this book. For more detailed background reading and derivations there many textbooks available at all levels.

Matrices are denoted by bold face upper case, e.g. $\mathbf{A}$, column vectors by bold face lower case, e.g. $\mathbf{a}$, the transpose by superscript T, e.g. $\mathbf{A}^T$, so that a row vector is e.g. $\mathbf{b}^T$ and a vector product is e.g. $\mathbf{b}^T \mathbf{a}$. The inverse is written $\mathbf{A}^{-1}$ and the transpose of the inverse may be written $\mathbf{A}^{-T}$.

A.1 Vector Spaces

The general linear problem of the kind encountered in this book is of the form $\mathbf{Kx} = \mathbf{y}$, where $\mathbf{K}$ is an $m \times n$ real matrix, with m rows and n columns, $\mathbf{x}$ is a column vector of n unknowns and $\mathbf{y}$ is a column vector of m measurements:

$$\begin{pmatrix} K_{11} & K_{12} & \ldots & K_{1n} \\ K_{21} & K_{22} & \ldots & K_{2n} \\ \vdots & \vdots & \vdots & \vdots \\ K_{m1} & K_{m2} & \ldots & K_{mn} \end{pmatrix} \begin{pmatrix} x_1 \\ x_2 \\ \vdots \\ x_n \end{pmatrix} = \begin{pmatrix} y_1 \\ y_2 \\ \vdots \\ y_m \end{pmatrix} \quad (A.1)$$

It is convenient, and aids the intuition, to think of linear equations in terms of vector spaces. The state $\mathbf{x}$, for example, may be thought of as a vector or a point in an n-dimensional space which we call *state space*, and similarly the measurement $\mathbf{y}$ may be thought of as a vector or a point in an m-dimensional space called *measurement space*. A coordinate system for a vector space is called a *basis*, and consists on n or m unit vectors which may or may not be orthogonal, but should be linearly independent. Given $\mathbf{x}$, the matrix $\mathbf{K}$ provides a mapping from state space into measurement space. The rows of $\mathbf{K}$, $\mathbf{k}_i$, are vectors of dimension n, and can be thought of as vectors on state space, even though they do not represent states.

The projection (dot product, vector product) of $\mathbf{x}$ onto $\mathbf{k}_i$ which may be written $y_i = \mathbf{k}_i^T \mathbf{x} = \mathbf{x}^T \mathbf{k}_i$ giving the (noise-free) measurement y_i i.e. i-th coordinate of the corresponding vector in measurement space.

Given $\mathbf{y}$ and $\mathbf{K}$ the inverse problem, or the solution of linear equations, is a matter of determining a mapping back from measurement space into state space. Each equation $y_i = \mathbf{k}_i^T \mathbf{x}$ states that the point $\mathbf{x}$ lies in a space of points whose vectors projected onto $\mathbf{k}_i$ give the value y_i. This subspace is of dimension $n-1$ and it lies at right angles to $\mathbf{k}_i$. For example in two dimensions the equation $y = k^1 x_1 + k^2 x_2$ defines a line in (x_1, x_2) space at right angles to the vector (k^1, k^2). Solutions to the set of linear equations lie in the intersection of all of these subspaces, which may be a point, a space of one or more dimensions, or may not exist.

We will have to deal with cases where m may be greater than, less than, or equal to n, and consider the appropriate solution method in each case. The reader may be familiar with the description of the case $m < n$ as underdetermined (underconstrained or ill-posed), and expect the case $m = n$ to be well-determined and $m > n$ to be overdetermined. Unfortunately this simple description is not always correct, as a simple example will illustrate. Consider the 2×2 case:

$$\begin{aligned} x_1 + x_2 &= 1 \\ x_1 + x_2 &= 2 \end{aligned} \tag{A.2}$$

Here we have two equations for two unknowns, but they are both overdetermined and underdetermined. They give us contradictory information about the combination $x_1 + x_2$, and so are overdetermined, and give no information about the linearly independent combination $x_1 - x_2$, and so are underdetermined. This situation, which is normally not so blatantly obvious, is called *mixed-determined*. The following set is also mixed-determined, but less obviously:

$$\begin{aligned} x_1 - 2x_2 + 3x_3 &= 4 \\ 2x_1 - x_2 + 4x_3 &= 3 \\ -x_1 - 4x_2 + x_3 &= 7 \end{aligned} \tag{A.3}$$

because the difference between the first two equations contradicts the difference between the last two. Given these hints, it should be quite clear how to construct linear systems with more equations than unknowns which are underdetermined, well-determined, mixed-determined or overdetermined.

Equations A.3 describe three planes in a 3-dimensional space, each pair of which intersects in a line, but the three lines to not intersect in a point to give a solution, as in the case of a well-determined problem. They are all parallel.

A.2 Eigenvectors and Eigenvalues

The eigenvalue problem associated with an arbitrary square matrix $\mathbf{A}$ of order n is to find *eigenvectors* $\mathbf{r}$ and scalar *eigenvalues* λ which satisfy

$$\mathbf{Ar} = \lambda \mathbf{r} \tag{A.4}$$

If $\mathbf{A}$ is regarded a coordinate transformation, then $\mathbf{r}$ has the same representation in the untransformed and transformed coordinates, apart from a factor λ. This looks like a strange thing to want to know, but eigenvectors turn out to have some extremely useful features. This equation is the same as $(\mathbf{A} - \lambda \mathbf{I})\mathbf{r} = 0$, a homogeneous equation, which can only have a solution other than $\mathbf{r} = 0$ if $\mathbf{A} - \lambda \mathbf{I}$ has rank less than n, i.e. if its determinant is zero. This leads to a polynomial equation of degree n in λ, with n solutions for the eigenvalues. Each λ can be substituted back in Eq. (A.4) and a solution for the corresponding eigenvector found. Eigenvalues may be real or a member of complex conjugate pair. An eigenvector can be scaled by an arbitrary factor, and still satisfy Eq. (A.4), so it is conventional to normalise them so that $\mathbf{r}^T \mathbf{r} = 1$ or, in the case of complex eigenvectors, $\mathbf{r}^\dagger \mathbf{r} = 1$, where $\dagger$ is the Hermitian adjoint, or transpose of the complex conjugate.

The eigenvectors can be assembled into a matrix $\mathbf{R}$:

$$\mathbf{AR} = \mathbf{R}\Lambda \tag{A.5}$$

where the columns of $\mathbf{R}$ are the eigenvectors, and Λ is a diagonal matrix, with the eigenvalues on the diagonal.

We can find the vectors and values of $\mathbf{A}^T$ by transposing (A.5):

$$\mathbf{R}^T \mathbf{A}^T = \Lambda \mathbf{R}^T \tag{A.6}$$

then premultiplying and postmultiplying by $\tilde{\mathbf{L}} = (\mathbf{R}^T)^{-1}$

$$\mathbf{A}^T \tilde{\mathbf{L}} = \tilde{\mathbf{L}}\Lambda \tag{A.7}$$

Thus $\tilde{\mathbf{L}}$ is a matrix whose columns are eigenvectors of $\mathbf{A}^T$, with eigenvalues which are the same as those of the corresponding column of $\mathbf{R}$. However these eigenvectors are not necessarily normalised. Let ν_i be the length of $\tilde{\mathbf{l}}_i$, the i-th column of $\tilde{\mathbf{L}}$, i.e. $\nu_i = (\tilde{\mathbf{l}}_i^T \tilde{\mathbf{l}}_i)^{-1/2}$, and $\mathbf{N}$ be a diagonal matrix containing the ν_i as its diagonal elements. Then $\mathbf{L}$, the normalised matrix of eigenvectors of $\mathbf{A}^T$, is given by $\mathbf{L} = \tilde{\mathbf{L}}\mathbf{N}^{-1}$ and satisfies:

$$\mathbf{A}^T \mathbf{L} = \mathbf{L}\Lambda \tag{A.8}$$

These are called the *left* eigenvectors, because they operate on $\mathbf{A}$ (rather then $\mathbf{A}^T$) on the left: $\mathbf{L}^T \mathbf{A} = \Lambda \mathbf{L}^T$, while $\mathbf{R}$ are the *right* eigenvectors. By postmultiplying Eq. (A.5) by $\mathbf{R}^{-1} = \tilde{\mathbf{L}}^T = \mathbf{N}\mathbf{L}^T$ we can express $\mathbf{A}$ in terms of its eigenvectors as

$$\mathbf{A} = \mathbf{R}\Lambda\mathbf{N}\mathbf{L}^T = \sum_i \lambda_i \nu_i \mathbf{r}_i \mathbf{l}_i^T \tag{A.9}$$

which is described as a 'spectral decomposition' of $\mathbf{A}$.

In the case of a symmetric matrix $\mathbf{S}$, where $\mathbf{S} = \mathbf{S}^T$, we must have $\mathbf{R} = \mathbf{L}$ by symmetry. By premultiplying (A.8) by $\mathbf{L}^T$ and postmultiplying its transpose by $\mathbf{L}$ we see that $\mathbf{L}^T\mathbf{L}\boldsymbol{\Lambda} = \boldsymbol{\Lambda}\mathbf{L}^T\mathbf{L}$, so $\mathbf{L}^T\mathbf{L}$ must be diagonal. As $\mathbf{L}$ is normalised we must have $\mathbf{L}^T\mathbf{L} = \mathbf{L}\mathbf{L}^T = \mathbf{I}$ or $\mathbf{L}^T = \mathbf{L}^{-1}$, i.e. the eigenvectors are *orthonormal*, and $\mathbf{N} = \mathbf{I}$. In this case the eigenvalues are all real. If the matrix is positive definite the eigenvalues are all greater than zero, and similarly for a negative definite matrix.

The following is a summary of useful relations involving eigenvectors:

Asymmetric Matrices

$$\mathbf{AR} = \mathbf{R}\boldsymbol{\Lambda} \quad (A.10)$$
$$\mathbf{L}^T\mathbf{A} = \boldsymbol{\Lambda}\mathbf{L}^T \quad (A.11)$$
$$\mathbf{NL}^T = \mathbf{R}^{-1},\ \mathbf{NR}^T = \mathbf{L}^{-1} \quad (A.12)$$
$$\mathbf{L}^T\mathbf{R} = \mathbf{R}^T\mathbf{L} = \mathbf{N}^{-1} \quad (A.13)$$
$$\mathbf{A} = \mathbf{R}\boldsymbol{\Lambda}\mathbf{NL}^T = \sum_i \lambda_i \nu_i \mathbf{r}_i \mathbf{l}_i^T \quad (A.14)$$
$$\mathbf{A}^{-1} = \mathbf{R}\boldsymbol{\Lambda}^{-1}\mathbf{NL}^T \quad (A.15)$$
$$\mathbf{A}^n = \mathbf{R}\boldsymbol{\Lambda}^n\mathbf{NL}^T \quad (A.16)$$
$$\mathbf{L}^T\mathbf{AR} = \mathbf{N}^{-1}\boldsymbol{\Lambda} \quad (A.17)$$
$$\mathbf{L}^T\mathbf{A}^n\mathbf{R} = \mathbf{N}^{-1}\boldsymbol{\Lambda}^n \quad (A.18)$$
$$\mathbf{L}^T\mathbf{A}^{-1}\mathbf{R} = \mathbf{N}^{-1}\boldsymbol{\Lambda}^{-1} \quad (A.19)$$
$$|\mathbf{A}| = \prod_i \lambda_i \quad (A.20)$$

Symmetric Matrices

$$\mathbf{SL} = \mathbf{L}\boldsymbol{\Lambda} \quad (A.21)$$
$$\mathbf{L}^T\mathbf{S} = \boldsymbol{\Lambda}\mathbf{L}^T \quad (A.22)$$
$$\mathbf{L}^T = \mathbf{L}^{-1} \quad (A.23)$$
$$\mathbf{LL}^T = \mathbf{L}^T\mathbf{L} = \mathbf{I} \quad (A.24)$$
$$\mathbf{S} = \mathbf{L}\boldsymbol{\Lambda}\mathbf{L}^T = \sum_i \lambda_i \mathbf{l}_i \mathbf{l}_i^T \quad (A.25)$$
$$\mathbf{S}^{-1} = \mathbf{L}\boldsymbol{\Lambda}^{-1}\mathbf{L}^T \quad (A.26)$$
$$\mathbf{S}^n = \mathbf{L}\boldsymbol{\Lambda}^n\mathbf{L}^T \quad (A.27)$$
$$\mathbf{L}^T\mathbf{SL} = \boldsymbol{\Lambda} \quad (A.28)$$
$$\mathbf{L}^T\mathbf{S}^n\mathbf{L} = \boldsymbol{\Lambda}^n \quad (A.29)$$
$$\mathbf{L}^T\mathbf{S}^{-1}\mathbf{L} = \boldsymbol{\Lambda}^{-1} \quad (A.30)$$
$$|\mathbf{S}| = \prod_i \lambda_i \quad (A.31)$$

Square roots of matrices

The relation $\mathbf{A}^n = \mathbf{R}\boldsymbol{\Lambda}^n\mathbf{NL}^T$ (where $\mathbf{N} = \mathbf{I}$ for a symmetric matrix) can be used for arbitrary powers of a matrix, in particular the square root such that $\mathbf{A} = \mathbf{A}^{\frac{1}{2}}\mathbf{A}^{\frac{1}{2}}$. This square root of a matrix is not unique, because the diagonal elements of $\boldsymbol{\Lambda}^{\frac{1}{2}}$ in $\mathbf{R}\boldsymbol{\Lambda}^{\frac{1}{2}}\mathbf{NL}^T$ can have either sign, leading to 2^n possibilities.

We only use square roots of covariance matrices in this book. In this case we can see that $\mathbf{S}^{\frac{1}{2}} = \mathbf{L}\boldsymbol{\Lambda}^{\frac{1}{2}}\mathbf{L}^T$ is symmetric. As well as these roots, symmetric matrices can also have non-symmetric roots satisfying $\mathbf{S} = (\mathbf{S}^{\frac{1}{2}})^T\mathbf{S}^{\frac{1}{2}}$, of which the Cholesky decomposition, $\mathbf{S} = \mathbf{T}^T\mathbf{T}$ where $\mathbf{T}$ is upper triangular, is the most useful, see section 5.8.1.1 and Exercise 5.3. There are an infinite number of non-symmetric square roots. If $\mathbf{S}^{\frac{1}{2}}$ is a square root, then clearly so is $\mathbf{X}\mathbf{S}^{\frac{1}{2}}$ where $\mathbf{X}$ is any orthonormal matrix. The inverse symmetric square root is $\mathbf{S}^{-\frac{1}{2}} = \mathbf{L}\boldsymbol{\Lambda}^{-\frac{1}{2}}\mathbf{L}^T$, and the inverse Cholesky decomposition is $\mathbf{S}^{-1} = \mathbf{T}^{-1}\mathbf{T}^{-T}$. The inverse square root $\mathbf{T}^{-1}$ is triangular, and its numerical effect is implemented efficiently by back substitution.

A.3 Principal Axes of a Quadratic Form

Consider the scalar equation:

$$\mathbf{x}^T \mathbf{S} \mathbf{x} = 1 \tag{A.32}$$

where $\mathbf{S}$ is symmetric. This is the equation of a quadratic surface centered on the origin, in n-space. $\mathbf{S}$ might be for example an inverse covariance matrix, when the equation might represent a surface of constant probability density. The normal to the surface is the vector $\nabla_\mathbf{x}(\mathbf{x}^T \mathbf{S} \mathbf{x}) = 2\mathbf{S}\mathbf{x}$, and $\mathbf{x}$ is the radius vector, so

$$\mathbf{S}\mathbf{x} = \lambda \mathbf{x} \tag{A.33}$$

is the problem of finding points where the normal and the radius vector are parallel. These are clearly where the principal axes intersect the surface. At these points, $\mathbf{x}^T \mathbf{S} \mathbf{x} = 1$ too, so $\mathbf{x}^T \lambda \mathbf{x} = 1$ or:

$$\lambda = \frac{1}{\mathbf{x}^T \mathbf{x}} \tag{A.34}$$

Thus the eigenvalues of a symmetric matrix are the reciprocals of the squares of the lengths of the principal axes of the associated ellipsoid. The lengths of the axes are independent of the coordinate system, so will also be invariant under an arbitrary rotation, that is one in which $(distance)^2 = \mathbf{x}^T \mathbf{x}$ is unchanged.

Consider using the eigenvectors of $\mathbf{S}$ to transform the equation for the quadratic surface:

$$\mathbf{x}^T \mathbf{L} \mathbf{\Lambda} \mathbf{L}^T \mathbf{x} = 1 \text{ or } \mathbf{x}'^T \mathbf{\Lambda} \mathbf{x}' = 1 \text{ or } \sum \lambda_i x_i'^2 = 1 \tag{A.35}$$

where $\mathbf{x}' = \mathbf{L}^T \mathbf{x}'$ or $\mathbf{x} = \mathbf{L}\mathbf{x}'$. The result is a quadratic surface in which the principal axes coincide with the coordinate axes.

A.4 Singular Vector Decomposition

The standard eigenvalue problem is meaningless for non-square matrices because $\mathbf{A}\mathbf{r}$ will be of a different dimension from $\mathbf{r}$, and Eq. (A.4) is invalid.

However an eigenvalue problem can be constructed by considering the symmetric problem:

$$\begin{pmatrix} \mathbf{O} & \mathbf{K} \\ \mathbf{K}^T & \mathbf{O} \end{pmatrix} \begin{pmatrix} \mathbf{u} \\ \mathbf{v} \end{pmatrix} = \lambda \begin{pmatrix} \mathbf{u} \\ \mathbf{v} \end{pmatrix} \tag{A.36}$$

where $\mathbf{K}$ is an arbitrary non-square matrix with m rows and n columns and $\mathbf{v}$ is of dimension n and $\mathbf{u}$ is of dimension m. We use the symbol $\mathbf{K}$, because the main application in this book is for the weighting function matrix. The vectors $\mathbf{u}$

and **v** are called 'singular vectors' of **K**, and λ is a singular value. The symmetric eigenvalue problem is equivalent to the 'shifted' eigenvalue problem (Lanczos, 1961):

$$\begin{aligned} \mathbf{Kv} &= \lambda \mathbf{u} \\ \mathbf{K}^T \mathbf{u} &= \lambda \mathbf{v} \end{aligned} \quad (A.37)$$

From Eq. (A.37) we can obtain by substitution:

$$\mathbf{K}^T \mathbf{K} \mathbf{v} = \lambda \mathbf{K}^T \mathbf{u} = \lambda^2 \mathbf{v} \quad (A.38)$$
$$\mathbf{K} \mathbf{K}^T \mathbf{u} = \lambda \mathbf{K} \mathbf{v} = \lambda^2 \mathbf{u} \quad (A.39)$$

showing that **u** and **v** are the eigenvectors of $\mathbf{K}\mathbf{K}^T$ ($m \times m$) and $\mathbf{K}^T\mathbf{K}$ ($n \times n$) respectively, which consequently both must have the same set of eigenvalues, and that the singular value is real. Furthermore, if $(\mathbf{u}, \mathbf{v})$ are a pair of singular vectors with singular value λ, then so is $(\mathbf{u}, -\mathbf{v})$, with singular value $-\lambda$. Thus we need only consider pairs with positive singular values.

A little care is needed in constructing a matrix of singular vectors, because individual **u** and **v** vectors correspond to each other, yet there are potentially different numbers of **v** and **u** vectors. However, if the rank of **K** is p, then there will be p non-zero singular values, and both $\mathbf{K}\mathbf{K}^T$ and $\mathbf{K}^T\mathbf{K}$ will have p non-zero eigenvalues. Thus the surplus eigenvectors will have zero eigenvalues, and can be discarded and we can write:

$$\begin{pmatrix} \mathbf{O} & \mathbf{K} \\ \mathbf{K}^T & \mathbf{O} \end{pmatrix} \begin{pmatrix} \mathbf{U} \\ \mathbf{V} \end{pmatrix} = \begin{pmatrix} \mathbf{U} \\ \mathbf{V} \end{pmatrix} \mathbf{\Lambda} \quad (A.40)$$

where $\mathbf{\Lambda}$ is $p \times p$, $\mathbf{U}$ is $m \times p$, and $\mathbf{V}$ is $n \times p$. There will be $n + m - p$ more eigenvectors of the composite matrix, all with zero eigenvalue. The singular vectors and values have the following properties:

$$\mathbf{KV} = \mathbf{U\Lambda} \quad (A.41)$$
$$\mathbf{K}^T \mathbf{U} = \mathbf{V\Lambda} \quad (A.42)$$
$$\mathbf{U}^T \mathbf{KV} = \mathbf{V}^T \mathbf{K}^T \mathbf{U} = \mathbf{\Lambda} \quad (A.43)$$
$$\mathbf{K} = \mathbf{U\Lambda V}^T = \sum_i \lambda_i \mathbf{u}_i \mathbf{v}_i^T \quad (A.44)$$
$$\mathbf{K}^T = \mathbf{V\Lambda U}^T \quad (A.45)$$
$$\mathbf{V}^T \mathbf{V} = \mathbf{U}^T \mathbf{U} = \mathbf{I}_p \quad (A.46)$$
$$\mathbf{KK}^T \mathbf{U} = \mathbf{U\Lambda}^2 \quad (A.47)$$
$$\mathbf{K}^T \mathbf{KV} = \mathbf{V\Lambda}^2 \quad (A.48)$$

For $\mathbf{VV}^T$ and $\mathbf{UU}^T$ to yield unit matrices, they must be extended to be square using the remaining zero eigenvectors of $\mathbf{K}^T\mathbf{K}$ or $\mathbf{KK}^T$. Because $\mathbf{U}^T\mathbf{KV} = \mathbf{\Lambda}$ we describe **U** as left singular vectors, and **V** as right singular vectors.

Note that a square matrix has both singular vectors and values, and eigenvectors and values. In the case of symmetric matrices they will be the same, but not in the case of unsymmetric matrices.

The right vectors $\mathbf{V}$ form an orthonormal basis (coordinate system) in the row space and the left vectors form a basis in the column space. The matrix $\mathbf{K}$ maps the row space basis vector $\mathbf{v}$ into a corresponding column space basis vector $\mathbf{u}$ (apart from a scale change given by λ), and $\mathbf{K}^T$ maps $\mathbf{u}$ back into $\mathbf{v}$. Thus $\mathbf{U}$ and $\mathbf{V}$ are a natural pair of coordinate systems for the two spaces.

A.5 Determinant and Trace

The determinant and the trace are both important quantities in quantifying information content. The following elementary properties are useful:

$$\operatorname{tr}(\mathbf{A}) = \sum_i A_{ii} \tag{A.49}$$

$$\operatorname{tr}(\mathbf{A}\mathbf{A}^T) = \sum_{ij} A_{ij}^2 \tag{A.50}$$

$$\operatorname{tr}(k\mathbf{A}) = k \operatorname{tr}(\mathbf{A}) \tag{A.51}$$

$$\operatorname{tr}(\mathbf{A}+\mathbf{B}) = \operatorname{tr}(\mathbf{A}) + \operatorname{tr}(\mathbf{B}) \tag{A.52}$$

$$\operatorname{tr}(\mathbf{C}\mathbf{D}) = \operatorname{tr}(\mathbf{D}\mathbf{C}) \tag{A.53}$$

$$\operatorname{tr}(\mathbf{a}\mathbf{b}^T) = \mathbf{b}^T \mathbf{a} \tag{A.54}$$

$$\operatorname{tr}(\mathbf{B}^{-1}\mathbf{A}\mathbf{B}) = \operatorname{tr}(\mathbf{A}) \tag{A.55}$$

$$\operatorname{tr}(\mathbf{A}) = \sum_i \lambda_i \tag{A.56}$$

$$\operatorname{tr}(\mathbf{A}^{-1}) = \sum_i \lambda_i^{-1} \tag{A.57}$$

$$|\mathbf{A}^T| = |\mathbf{A}| \tag{A.58}$$

$$|k\mathbf{A}| = k^n |\mathbf{A}| \tag{A.59}$$

$$|\mathbf{A}\mathbf{B}| = |\mathbf{B}\mathbf{A}| = |\mathbf{A}||\mathbf{B}| \tag{A.60}$$

$$|\mathbf{A}^{-1}| = |\mathbf{A}|^{-1} \tag{A.61}$$

$$|\mathbf{B}^{-1}\mathbf{A}\mathbf{B}| = |\mathbf{A}| \tag{A.62}$$

$$|\mathbf{A}| = \prod_i \lambda_i \tag{A.63}$$

$$|\mathbf{I}+\mathbf{A}| = \prod_i (1+\lambda_i) \tag{A.64}$$

$$|\mathbf{I}+\mathbf{a}\mathbf{b}^T| = 1 + \mathbf{b}^T \mathbf{a} \tag{A.65}$$

where $\mathbf{A}$ and $\mathbf{B}$ are square matrices of the same order, λ_i is the i-th eigenvalue of $\mathbf{A}$, n is the order of $\mathbf{A}$, $\mathbf{C}$ and $\mathbf{D}$ are rectangular matrices such that $\mathbf{C}^T$ and $\mathbf{D}$ are the same size and shape, $\mathbf{a}$ and $\mathbf{b}$ are vectors of the same order, and k is a scalar.

A.6 Calculus with Matrices and Vectors

The differential d**x** is often used to indicate an infinitesimal vector. Throughout this book it is used quite differently. In integral expressions the following meaning is intended:

$$\int f(\mathbf{x})\,\mathrm{d}\mathbf{x} = \int \ldots \int f(x_1, x_2, \ldots, x_n)\,\mathrm{d}x_1\,\mathrm{d}x_2 \ldots \mathrm{d}x_n, \qquad (A.66)$$

i.e. d**x** is an element of volume of state space. In derivatives, the notation $\partial \mathbf{y}/\partial \mathbf{x}$ means the Fréchet derivative, whose value is a matrix:

$$\left[\frac{\partial \mathbf{y}}{\partial \mathbf{x}}\right]_{ij} = \left(\frac{\partial y_i}{\partial x_j}\right)_{x_k, k \neq j} \qquad (A.67)$$

where the order of the subscripts is such that a Taylor expansion 'looks right'. The convention is to associate the row subscript with **y** and the column subscript with **x**, so that we can write a Taylor expansion in a familiar form:

$$\mathbf{y} = \mathbf{y}_0 + \frac{\partial \mathbf{y}}{\partial \mathbf{x}}(\mathbf{x} - \mathbf{x}_0) + \ldots \qquad (A.68)$$

We often need to find the minimum or maximum of some matrix expression, very often with respect to a vector. For example what value of **x** minimises $\mathbf{x}^T \mathbf{A} \mathbf{x} + \lambda \mathbf{b}^T \mathbf{x}$? It is convenient to have a set of rules corresponding to the familiar rules of calculus with scalars so that we do not have to carry out such manipulations in components each time.

The derivative of a scalar valued expression with respect to a vector yields a vector, because it represents the set of derivatives of the scalar with respect to each element of the vector. The vector may be expressed as a row or a column, at your convenience. In components, the i-th element of the above example gives:

$$\left[\frac{\partial}{\partial \mathbf{x}}(\mathbf{x}^T \mathbf{A} \mathbf{x} + \lambda \mathbf{b}^T \mathbf{x})\right]_i = \frac{\partial}{\partial x_i}\left(\sum_{jk} x_j A_{jk} x_k + \lambda \sum_j b_j x_j\right) \qquad (A.69)$$

$$= \sum_j x_j A_{ji} + \sum_k A_{ik} x_k + \lambda b_i \qquad (A.70)$$

There is a slight subtlety in returning this to a matrix notation, because the first term would produce a row vector and the other two terms would produce column vectors if done blindly. The result can be expressed either as a row or a column:

$$\frac{\partial}{\partial \mathbf{x}}[\mathbf{x}^T \mathbf{A} \mathbf{x} + \lambda \mathbf{b}^T \mathbf{x}] = \mathbf{A}^T \mathbf{x} + \mathbf{A} \mathbf{x} + \lambda \mathbf{b} \qquad (A.71)$$

or

$$\frac{\partial}{\partial \mathbf{x}}[\mathbf{x}^T \mathbf{A} \mathbf{x} + \lambda \mathbf{b}^T \mathbf{x}] = \mathbf{x}^T \mathbf{A} + \mathbf{x}^T \mathbf{A}^T + \lambda \mathbf{b}^T \qquad (A.72)$$

Confirming that:
$$\frac{\partial}{\partial \mathbf{x}}(\mathbf{A}\mathbf{x}) = \mathbf{A} \text{ and } \frac{\partial}{\partial \mathbf{x}}(\mathbf{x}^T \mathbf{A}) = \mathbf{A}^T \tag{A.73}$$
is left as an exercise to the reader.

Note that the symbol $\nabla_\mathbf{x}$ may also be used as equivalent of $\frac{\partial}{\partial \mathbf{x}}$.

Appendix B

Answers to Exercises

Exercise 2.1

Construct a simple example of a set of linear equations which is simultaneously over- and under-determined.

See section A.1.

Exercise 2.2

For the multivariate Gaussian distribution, solve $dP/d\mathbf{x} = 0$ for the maximum probability value, and integrate $\int \mathbf{x} P(\mathbf{x}) \, d\mathbf{x}$ to obtain $\mathcal{E}\{\mathbf{x}\}$. Confirm that both are equal to $\bar{\mathbf{x}}$

At the maximum $dP/d\mathbf{x} = 0$ and $d \ln P / d\mathbf{x} = 0$, i.e.

$$\frac{\partial}{\partial \mathbf{x}}(\mathbf{x} - \bar{\mathbf{x}})^T \mathbf{S}_x^{-1}(\mathbf{x} - \bar{\mathbf{x}}) = 2\mathbf{S}_x^{-1}(\mathbf{x} - \bar{\mathbf{x}}) = 0 \tag{B.1}$$

Therefore $\mathbf{x} = \bar{\mathbf{x}}$ provided $\mathbf{S}_x^{-1}$ is not singular. Note that the '0' is a vector.

To evaluate the integral, move the origin to $\bar{\mathbf{x}}$ and rotate the coordinates to the eigenvectors of $\mathbf{S}_x$, as in Eq. (2.14). Then

$$\int \mathbf{x} P(\mathbf{x}) \, d\mathbf{x} = \prod_i \int_{-\infty}^{\infty} \frac{z_i}{(2\pi\lambda_i)^{\frac{1}{2}}} \exp\left\{-\frac{z_i^2}{2\lambda_i}\right\} dz_i \tag{B.2}$$

Each integral in the product is zero by symmetry, so the expected value is at the origin, $\bar{\mathbf{x}}$.

Exercise 2.3

Why does it not matter which square root we use for prewhitening?

Prewhitening transforms the covariance matrix to a unit matrix, or the corresponding ellipsoid (Fig. 2.4) to a sphere. The different square roots correspond to different rotations of the coordinate axes of the sphere.

Exercise 2.4

Derive the following expressions for degrees of freedom for noise:

$$d_n = \text{tr}(\mathbf{S}_\epsilon[\mathbf{KS}_a\mathbf{K}^T + \mathbf{S}_\epsilon]^{-1}) = \text{tr}([\mathbf{K}^T\mathbf{S}_\epsilon^{-1}\mathbf{K} + \mathbf{S}_a^{-1}]^{-1}\mathbf{S}_a^{-1}) \qquad (B.3)$$

From Eq. (2.47)

$$\begin{aligned} d_n &= \mathcal{E}\{\hat{\boldsymbol{\epsilon}}^T \mathbf{S}_\epsilon^{-1} \hat{\boldsymbol{\epsilon}}\} = \mathcal{E}\{\text{tr}(\mathbf{S}_\epsilon^{-1} \hat{\boldsymbol{\epsilon}} \hat{\boldsymbol{\epsilon}}^T)\} \\ &= \mathcal{E}\{\text{tr}(\mathbf{S}_\epsilon^{-1}(\mathbf{y} - \mathbf{K}\hat{\mathbf{x}})(\mathbf{y} - \mathbf{K}\hat{\mathbf{x}})^T)\} \end{aligned} \qquad (B.4)$$

From Eq. (2.31)

$$\begin{aligned} \mathbf{y} - \mathbf{K}\hat{\mathbf{x}} &= \mathbf{y} - \mathbf{K}\mathbf{x}_a - \mathbf{KS}_a\mathbf{K}^T(\mathbf{KS}_a\mathbf{K}^T + \mathbf{S}_\epsilon)^{-1}(\mathbf{y} - \mathbf{K}\mathbf{x}_a) \\ &= \mathbf{S}_\epsilon(\mathbf{KS}_a\mathbf{K}^T + \mathbf{S}_\epsilon)^{-1}(\mathbf{y} - \mathbf{K}\mathbf{x}_a) \\ &= \mathbf{S}_\epsilon(\mathbf{KS}_a\mathbf{K}^T + \mathbf{S}_\epsilon)^{-1}[\mathbf{K}(\mathbf{x} - \mathbf{x}_a) + \boldsymbol{\epsilon}] \end{aligned} \qquad (B.5)$$

so that

$$\mathcal{E}\{(\mathbf{y} - \mathbf{K}\hat{\mathbf{x}})(\mathbf{y} - \mathbf{K}\hat{\mathbf{x}})^T\} = \mathbf{S}_\epsilon(\mathbf{KS}_a\mathbf{K}^T + \mathbf{S}_\epsilon)^{-1}\mathbf{S}_\epsilon. \qquad (B.6)$$

Substituting in (B.4) gives

$$d_n = \text{tr}([\mathbf{KS}_a\mathbf{K}^T + \mathbf{S}_\epsilon]^{-1}\mathbf{S}_\epsilon) \qquad (B.7)$$

which is equivalent to the first expression. The other expression follows from using this with Eq. (2.45), giving

$$\begin{aligned} d_n &= \text{tr}(\mathbf{I}_m - \mathbf{KS}_a\mathbf{K}^T[\mathbf{KS}_a\mathbf{K}^T + \mathbf{S}_\epsilon]^{-1}) \\ &= m - \text{tr}(\mathbf{KG}) \\ &= m - \text{tr}(\mathbf{K}[\mathbf{K}^T\mathbf{S}_\epsilon^{-1}\mathbf{K} + \mathbf{S}_a^{-1}]^{-1}\mathbf{K}^T\mathbf{S}_\epsilon^{-1}) \\ &= m - n + \text{tr}(\mathbf{I}_n - [\mathbf{K}^T\mathbf{S}_\epsilon^{-1}\mathbf{K} + \mathbf{S}_a^{-1}]^{-1}\mathbf{K}^T\mathbf{S}_\epsilon^{-1}\mathbf{K}) \\ &= \text{tr}([\mathbf{K}^T\mathbf{S}_\epsilon^{-1}\mathbf{K} + \mathbf{S}_a^{-1}]^{-1}\mathbf{S}_a^{-1}) + m - n \end{aligned} \qquad (B.8)$$

These are more easily obtained by anticipating that $d_s + d_n$ should be equal to m.

Exercise 2.5

Relate the eigenvectors and values of $\mathbf{A}$ to the singular vectors and values of $\tilde{\mathbf{K}}$. Remember $\mathbf{A}$ may not be symmetric.

$$\begin{aligned} \mathbf{A} = \mathbf{GK} &= \mathbf{S}_a\mathbf{K}^T(\mathbf{KS}_a\mathbf{K}^T + \mathbf{S}_\epsilon)^{-1}\mathbf{K} \\ &= \mathbf{S}_a^{\frac{1}{2}}\tilde{\mathbf{K}}^T(\tilde{\mathbf{K}}\tilde{\mathbf{K}}^T + \mathbf{I}_m)^{-1}\tilde{\mathbf{K}}\mathbf{S}_a^{-\frac{1}{2}} \end{aligned}$$

$$
\begin{aligned}
&= \mathbf{S}_a^{\frac{1}{2}} \mathbf{V} \mathbf{\Lambda} \mathbf{U}^T (\mathbf{U} \mathbf{\Lambda}^2 \mathbf{U}^T + \mathbf{I}_m)^{-1} \mathbf{U} \mathbf{\Lambda} \mathbf{V}^T \mathbf{S}_a^{-\frac{1}{2}} \\
&= \mathbf{S}_a^{\frac{1}{2}} \mathbf{V} \mathbf{\Lambda} (\mathbf{\Lambda}^2 + \mathbf{I}_m)^{-1} \mathbf{\Lambda} \mathbf{V}^T \mathbf{S}_a^{-\frac{1}{2}}
\end{aligned}
\tag{B.9}
$$

Thus the eigenvectors of $\mathbf{A}$ are $\mathbf{R} = \mathbf{S}_a^{\frac{1}{2}} \mathbf{V}$ and $\mathbf{L} = \mathbf{S}_a^{-\frac{1}{2}} \mathbf{V}$, and its eigenvalues are $\lambda^2/(1+\lambda^2)$. If the Cholesky square root is used, then $\mathbf{S}_a^{\frac{1}{2}}$ and $\mathbf{S}_a^{-\frac{1}{2}}$ are replaced by $\mathbf{T}_a^T$ and $\mathbf{T}_a^{-T}$ respectively.

Exercise 2.6

Integrate (2.70) to obtain $S = \ln[\sigma(2\pi e)^{1/2}]$.

Change the variable to $y = (x - \bar{x})/2^{\frac{1}{2}}\sigma$, giving

$$
S = \frac{1}{\pi^{\frac{1}{2}}} \int_{-\infty}^{\infty} e^{-y^2} \left(\ln[(2\pi)^{\frac{1}{2}}\sigma] + y^2 \right) dy \tag{B.10}
$$

Using $\int \exp(-y^2)\,dy = \pi^{\frac{1}{2}}$ and $\int y^2 \exp(-y^2)\,dy = \pi^{\frac{1}{2}}/2$ we obtain

$$
\begin{aligned}
S &= \frac{1}{\pi^{\frac{1}{2}}} (\pi^{\frac{1}{2}} \ln[(2\pi)^{\frac{1}{2}}\sigma] + \pi/2^{\frac{1}{2}}/2) \\
&= \ln[\sigma(2\pi e)^{1/2}]
\end{aligned}
\tag{B.11}
$$

Exercise 2.7

Prove that $S[P(x)P(y)] = S[P(x)] + S[P(y)]$ if x and y are independent.

$$
\begin{aligned}
S[P(x)P(y)] &= \int_{-\infty}^{\infty} P(x)P(y) \ln[P(x)P(y)]\,dx\,dy \\
&= \int_{-\infty}^{\infty} P(x)P(y) \ln P(x)\,dx\,dy + \int_{-\infty}^{\infty} P(x)P(y) \ln P(y)\,dx\,dy \\
&= \int_{-\infty}^{\infty} P(x) \ln P(x)\,dx + \int_{-\infty}^{\infty} P(y) \ln P(y)\,dy \\
&= S[P(x)] + S[P(y)]
\end{aligned}
\tag{B.12}
$$

Exercise 2.8

With the aid of Bayes theorem, show that for general pdf's the information content is the same whether computed in measurement space or in state space.

Consider the entropy of the combination of the state and the measurement, i.e. the entropy of the joint *pdf* $P(\mathbf{x}, \mathbf{y})$ in the product space:

$$
S_{xy} = \int P(\mathbf{x}, \mathbf{y}) \ln[P(\mathbf{x}, \mathbf{y})]\,d\mathbf{x}d\mathbf{y} \tag{B.13}
$$

and substitute $P(\mathbf{x}, \mathbf{y}) = P(\mathbf{x}|\mathbf{y})P(\mathbf{y})$:

$$\begin{aligned} S_{xy} &= \int P(\mathbf{x}|\mathbf{y})P(\mathbf{y})\ln[P(\mathbf{x}|\mathbf{y})P(\mathbf{y})]\,d\mathbf{x}d\mathbf{y} \\ &= \int P(\mathbf{x}|\mathbf{y})P(\mathbf{y})\ln[P(\mathbf{x}|\mathbf{y})]\,d\mathbf{x}d\mathbf{y} + \int P(\mathbf{x}|\mathbf{y})P(\mathbf{y})\ln[P(\mathbf{y})]\,d\mathbf{x}d\mathbf{y} \end{aligned} \quad (B.14)$$

where the subscript on S indicates which space the entropy is evaluated in. Carry out both of the integrals over $\mathbf{x}$:

$$S_{xy} = \int S_x[P(\mathbf{x}|\mathbf{y})]P(\mathbf{y})\,d\mathbf{y} + \int P(\mathbf{y})\ln[P(\mathbf{y})]\,d\mathbf{y} \quad (B.15)$$

If the entropy of $P(\mathbf{x}|\mathbf{y})$ is independent of $\mathbf{y}$, as it is for the Gaussian linear case discussed in section 2.3.2.2, then this simplifies to:

$$S_{xy}[P(\mathbf{x}, \mathbf{y})] = S_x[P(\mathbf{x}|\mathbf{y})] + S_y[P(\mathbf{y})] \quad (B.16)$$

Likewise we can obtain:

$$S_{xy}[P(\mathbf{x}, \mathbf{y})] = S_y[P(\mathbf{y}|\mathbf{x})] + S_x[P(\mathbf{x})] \quad (B.17)$$

requiring the assumption that $S_y[P(\mathbf{y}|\mathbf{x})]$ is independent of $\mathbf{x}$, as it is in the Gaussian linear case. Equating these two forms and rearranging gives

$$S_x[P(\mathbf{x})] - S_x[P(\mathbf{x}|\mathbf{y})] = S_y[P(\mathbf{y})] - S_y[P(\mathbf{y}|\mathbf{x})]. \quad (B.18)$$

Hence $H_x = H_y$, the information content of the measurement is the same as the information content of the retrieval:

Exercise 2.9

Show that, the maximum entropy pdf of a random scalar x, when only the mean $\bar{x}$ and variance σ^2 are known, is the normal distribution.

Maximise the entropy of $P(x)$ with the given constraints:

$$\frac{d}{dP}\left[\int P(x)\ln P(x)\,dx + \lambda\int P(x)\,dx + \mu\int xP(x)\,dx + \nu\int(x-\bar{x})^2 P(x)\,dx\right] = 0 \quad (B.19)$$

Differentiate with respect to P:

$$\ln P + 1 + \lambda + \mu x + \nu(x-\bar{x})^2 = 0 \quad (B.20)$$

Hence $\ln P$ is quadratic in x. In order to satisfy the mean, variance and unit area constraints P must be a normal distribution with the given values.

Exercise 3.1

Using an eigenvector expansion of $\mathbf{A}$, give an interpretation of Eq. (3.12) in which the retrieval is expressed using scalar rather than matrix weights.

The retrieved state can be written as a linear combination of the true state and the *a priori* state, according to Eq. (3.12). This is a concept that might be difficult to visualise. Scalar weights are easy to understand, matrix weights are less so. One way to understand how matrix weights affect the solution is to use an eigenvector decomposition of $\mathbf{A}$. Remember in this case that $\mathbf{A}$ is not symmetric, so that its eigenvectors are not orthogonal (see exercise 2.5). If the matrix of right eigenvectors of $\mathbf{A}$ is $\mathbf{R}$, so that $\mathbf{AR} = \mathbf{R}\boldsymbol{\Lambda}$ and the matrix of left eigenvectors is $\mathbf{L}$ such that $\mathbf{L}^T\mathbf{A} = \boldsymbol{\Lambda}\mathbf{L}^T$, then premultiplying Eq. (3.12) by $\mathbf{L}^T$, and ignoring the terms involving random error, gives

$$\mathbf{L}^T\hat{\mathbf{x}} = \mathbf{L}^T\mathbf{A}\mathbf{x} + \mathbf{L}^T(\mathbf{I}_n - \mathbf{A})\mathbf{x}_a = \boldsymbol{\Lambda}\mathbf{L}^T\mathbf{x} + (\mathbf{I}_n - \boldsymbol{\Lambda})\mathbf{L}^T\mathbf{x}_a \qquad (B.21)$$

Take the columns of $\mathbf{R}$ to be a representation of $\mathbf{x}$, i.e., $\mathbf{x} = \mathbf{R}\mathbf{z}$, where the vector $\mathbf{z}$ forms the representation coefficients. Thus $\mathbf{z} = \mathbf{R}^{-1}\mathbf{x} = \mathbf{L}^T\mathbf{x}$, and the equation simplifies to

$$\hat{\mathbf{z}} = \boldsymbol{\Lambda}\mathbf{z} + (\mathbf{I}_n - \boldsymbol{\Lambda})\mathbf{z}_a \qquad (B.22)$$

Because $\boldsymbol{\Lambda}$ is diagonal this can be separated into components, showing that the elements of $\hat{\mathbf{z}}$ are scalar weighted means of the corresponding elements of $\mathbf{z}$ and $\mathbf{z}_a$:

$$\hat{z}_i = \lambda_i z_i + (1 - \lambda_i)z_{ai} \qquad (B.23)$$

Thus we can consider the *a priori* and true state vectors as decomposed into components, the right eigenvectors, some of which (those with $\lambda_i \sim 1$) will be well reproduced by the measurement system, and others (with $\lambda_i \sim 0$) that will come mainly from the a priori vector.

This can also be obtained by considering the primed coordinate system described in section 2.4.1, in which the forward model is transformed to $\mathbf{y}' = \boldsymbol{\Lambda}\mathbf{x}' + \boldsymbol{\epsilon}'$, and in which each element is measured independently. Exercise 2.5 shows how the vectors and values of $\mathbf{A}$ are related to this system.

Exercise 3.2

Show that the determinant of a covariance matrix of any order with given diagonal elements is greatest if the off-diagonal elements are all zero.

Let $\mathbf{S}$ be a general covariance matrix of order n, and $\mathbf{D}$ be a diagonal matrix constructed from its diagonal elements. We wish to show that $|\mathbf{S}| \leq |\mathbf{D}|$. Construct the correlation matrix $\mathbf{E}$, having all diagonal elements equal to unity and satisfying $\mathbf{S} = \mathbf{D}^{\frac{1}{2}}\mathbf{E}\mathbf{D}^{\frac{1}{2}}$. Taking the determinant of both sides gives $|\mathbf{S}| = |\mathbf{D}||\mathbf{E}|$. We can show that $|\mathbf{E}| \leq 1$ by considering its eigenvalues. The mean of the eigenvalues is equal to $\text{tr}(\mathbf{E})/n$, i.e. unity. The determinant is the product of the eigenvalues, and as the geometric mean of a set of numbers is always less than or equal to their arithmetic mean, then the determinant must be less than or equal to unity.

Exercise 4.1

Let $\mathbf{x}$ be represented by a linear combination of singular vectors of $\tilde{\mathbf{K}}$ (section 2.4.1). Show that in this representation the MAP method combines the coefficients of an exact solution and the a priori independently as scalars.

In the primed coordinate system, we have $\mathbf{x}' = \mathbf{V}^T \tilde{\mathbf{x}} = \mathbf{V}^T \mathbf{S}_a^{-\frac{1}{2}} \mathbf{x}$ and $\mathbf{y}' = \mathbf{U}^T \tilde{\mathbf{y}} = \mathbf{U}^T \mathbf{S}_\epsilon^{-\frac{1}{2}} \mathbf{y}$ with the forward model $\mathbf{y}' = \mathbf{\Lambda} \mathbf{x}' + \boldsymbol{\epsilon}'$, both covariances being unit matrices. Substituting $\mathbf{\Lambda}$ for $\mathbf{K}$ and $\mathbf{I}$ for both $\mathbf{S}_a$ and $\mathbf{S}_\epsilon$ in e.g. Eq. (4.3), we obtain for the retrieval

$$\hat{\mathbf{x}}' = (\mathbf{\Lambda}^2 + \mathbf{I})^{-1}(\mathbf{\Lambda}\mathbf{y}' + \mathbf{x}'_a) \tag{B.24}$$

or, in components, and with a little further manipulation,

$$\hat{x}'_i = (\lambda_i^2 + 1)^{-1} \left[\lambda_i^2 \left(\frac{y'_i}{\lambda_i} \right) + x'_{ai} \right], \tag{B.25}$$

showing that $\hat{x}'_i$ is a weighted mean of the exact solution y'_i/λ_i and the *a priori*.

Exercise 4.2

Show that the minimum variance solution jointly minimises every element of $\hat{\mathbf{S}}$.

The retrieval error covariance is, for a given gain matrix $\mathbf{G}$, from Eq. (4.44),

$$\hat{\mathbf{S}}(\mathbf{G}) = (\mathbf{I}_n - \mathbf{G}\mathbf{K})\mathbf{S}_a(\mathbf{I}_n - \mathbf{G}\mathbf{K})^T + \mathbf{G}\mathbf{S}_\epsilon \mathbf{G}^T. \tag{B.26}$$

For $\mathbf{G}$ to minimise every element of $\hat{\mathbf{S}}$, an arbitrary small perturbation $\delta\mathbf{G}$ should leave $\hat{\mathbf{S}}$ unchanged to first order. Consider

$$\begin{aligned} \mathbf{O} &= \hat{\mathbf{S}}(\mathbf{G} + \delta\mathbf{G}) - \hat{\mathbf{S}}(\mathbf{G}) \\ &= -(\mathbf{I}_n - \mathbf{G}\mathbf{K})\mathbf{S}_a \mathbf{K}^T \delta\mathbf{G}^T + \mathbf{G}\mathbf{S}_\epsilon \delta\mathbf{G}^T \\ &\quad -\delta\mathbf{G}\mathbf{K}\mathbf{S}_a(\mathbf{I}_n - \mathbf{G}\mathbf{K})^T + \delta\mathbf{G}\mathbf{S}_\epsilon \mathbf{G}^T, \end{aligned} \tag{B.27}$$

This will be satisfied if $\mathbf{G}\mathbf{S}_\epsilon = (\mathbf{I}_n - \mathbf{G}\mathbf{K})\mathbf{S}_a \mathbf{K}^T$, which can be rearranged to give

$$\mathbf{G} = \mathbf{S}_a \mathbf{K}^T (\mathbf{K}\mathbf{S}_a \mathbf{K}^T + \mathbf{S}_\epsilon)^{-1} \tag{B.28}$$

Exercise 4.3

Derive the following expression for the vector of contribution functions for the Backus-Gilbert retrieval:

$$\mathbf{g}(z) = \frac{\mathbf{Q}^{-1}(z)\mathbf{k}}{\mathbf{k}^T\mathbf{Q}^{-1}(z)\mathbf{k}} \qquad (B.29)$$

The derivative of Eq. (4.63) gives:

$$2\mathbf{Q}(z)\mathbf{g}(z) + \lambda(z)\mathbf{k} = 0 \qquad (B.30)$$

hence

$$\mathbf{g}(z) = -\frac{1}{2}\lambda(z)\mathbf{Q}(z)^{-1}\mathbf{k} \qquad (B.31)$$

substitute in the unit area constraint:

$$\mathbf{k}^T\mathbf{g}(z) = -\frac{1}{2}\lambda(z)\mathbf{k}^T\mathbf{Q}(z)^{-1}\mathbf{k} = 1 \qquad (B.32)$$

Solving for $\lambda(z)$ and substituting in Eq. (B.31) gives the required answer.

Exercise 5.1

Derive the alternative expression for the linearised solution:

$$\hat{\mathbf{x}} = \mathbf{x}_a + (\mathbf{K}_l^T \mathbf{S}_\epsilon^{-1} \mathbf{K}_l + \mathbf{S}_a^{-1})^{-1} \mathbf{K}_l^T \mathbf{S}_\epsilon^{-1} [\mathbf{y} - \mathbf{y}_l + \mathbf{K}_l(\mathbf{x}_l - \mathbf{x}_a)] \qquad (B.33)$$

Substitute Eq. (5.11) for $\mathbf{F}(\mathbf{x})$ in Eq. (5.5) to give

$$-\hat{\mathbf{K}}^T(\hat{\mathbf{x}})\mathbf{S}_\epsilon^{-1}[\mathbf{y} - \mathbf{y}_l - \mathbf{K}_l(\hat{\mathbf{x}} - \mathbf{x}_l)] + \mathbf{S}_a^{-1}(\hat{\mathbf{x}} - \mathbf{x}_a) = 0. \qquad (B.34)$$

Substitute $\mathbf{K}^T(\hat{\mathbf{x}}) = \mathbf{K}_l$ from the linearisation and rearrange to give

$$\mathbf{S}_a^{-1}(\hat{\mathbf{x}} - \mathbf{x}_a) + \hat{\mathbf{K}}_l \mathbf{S}_\epsilon^{-1} \mathbf{K}_l(\hat{\mathbf{x}} - \mathbf{x}_a) = \hat{\mathbf{K}}_l \mathbf{S}_\epsilon^{-1}[\mathbf{y} - \mathbf{y}_l + \mathbf{K}_l(\mathbf{x}_l - \mathbf{x}_a)], \qquad (B.35)$$

and hence the required result for $\hat{\mathbf{x}}$.

Exercise 5.2

Show that the covariance of $\delta\hat{\mathbf{y}} = \mathbf{y} - \mathbf{F}(\hat{\mathbf{x}})$ is

$$\mathbf{S}_{\delta\hat{y}} = \mathbf{S}_\epsilon (\hat{\mathbf{K}} \mathbf{S}_a \hat{\mathbf{K}}^T + \mathbf{S}_\epsilon)^{-1} \mathbf{S}_\epsilon \qquad (B.36)$$

for an optimal estimator and, by considering an eigenvector or singular vector expansion, show how it differs from $\mathbf{S}_\epsilon$.

Put $\mathbf{y} = \mathbf{F}(\mathbf{x}) + \boldsymbol{\epsilon}$ and expand $\mathbf{F}(\mathbf{x}) - \mathbf{F}(\hat{\mathbf{x}})$ to first order to give

$$\delta\hat{\mathbf{y}} = \mathbf{y} - \mathbf{F}(\hat{\mathbf{x}}) = \mathbf{K}(\mathbf{x} - \hat{\mathbf{x}}) + \boldsymbol{\epsilon} \qquad (B.37)$$

We must take care in evaluating $\mathbf{S}_{\delta\hat{y}} = \mathcal{E}\{[\mathbf{K}(\mathbf{x}-\hat{\mathbf{x}})+\boldsymbol{\epsilon}][\mathbf{K}(\mathbf{x}-\hat{\mathbf{x}})+\boldsymbol{\epsilon}]^T\}$ because $\hat{\mathbf{x}}$ is correlated with $\boldsymbol{\epsilon}$. Substituting $\hat{\mathbf{x}} - \mathbf{x} = (\mathbf{GK} - \mathbf{I})(\mathbf{x} - \mathbf{x}_a) + \mathbf{G}\boldsymbol{\epsilon}$ from Eq. (3.16) we obtain

$$\begin{aligned} \delta\hat{\mathbf{y}} &= \mathbf{K}[(\mathbf{I} - \mathbf{GK})(\mathbf{x} - \mathbf{x}_a) - \mathbf{G}\boldsymbol{\epsilon}] + \boldsymbol{\epsilon} \\ &= (\mathbf{I} - \mathbf{KG})[\mathbf{K}(\mathbf{x} - \mathbf{x}_a) + \boldsymbol{\epsilon}] \end{aligned} \quad (B.38)$$

Thus its covariance is

$$\mathbf{S}_{\delta\hat{y}} = (\mathbf{I} - \mathbf{KG})(\mathbf{K}\mathbf{S}_a\mathbf{K}^T + \mathbf{S}_\epsilon)(\mathbf{I} - \mathbf{KG})^T \quad (B.39)$$

In the case of an optimal retrieval, we can substitute for $\mathbf{G}$ from Eq. (4.41) and obtain:

$$\mathbf{S}_{\delta\hat{y}} = \mathbf{S}_\epsilon(\mathbf{K}\mathbf{S}_a\mathbf{K}^T + \mathbf{S}_\epsilon)^{-1}\mathbf{S}_\epsilon \quad (B.40)$$

See also exercise 2.4. Now express $\hat{\mathbf{K}}$ in terms of $\tilde{\mathbf{K}}$, and use $\tilde{\mathbf{K}} = \mathbf{U}\boldsymbol{\Lambda}\mathbf{V}^T$:

$$\begin{aligned} \mathbf{S}_{\hat{y}} &= \mathbf{S}_\epsilon^{\frac{1}{2}}(\tilde{\mathbf{K}}\tilde{\mathbf{K}}^T + \mathbf{I}_m)^{-1}\mathbf{S}_\epsilon^{\frac{1}{2}} \\ &= \mathbf{S}_\epsilon^{\frac{1}{2}}\mathbf{U}(\boldsymbol{\Lambda}^2 + \mathbf{I}_m)^{-1}\mathbf{U}^T\mathbf{S}_\epsilon^{\frac{1}{2}} \end{aligned} \quad (B.41)$$

To interpret this: Transform $\mathbf{S}_\epsilon^{\frac{1}{2}}$ with $\mathbf{U}$, the left singular vectors of $\tilde{\mathbf{K}}$, and reduce the components by a factor of $(1 + \lambda_i^2)^{\frac{1}{2}}$. Thus components of $\mathbf{S}_\epsilon^{\frac{1}{2}}$ corresponding to large λ_i are reduced (aliassed into errors in the retrieved profile) and those corresponding to small λ_i remain as misfit between $\mathbf{y}$ and $\mathbf{F}(\hat{\mathbf{x}})$.

Exercise 5.3

By considering how the elements of $\mathbf{B}$ are related to those of $\mathbf{T}$ one at a time in a suitable order, construct a straightforward algorithm for Cholesky decomposition. Ignore pivotting.

The problem is to solve equations of the form

$$\begin{pmatrix} B_{11} & B_{12} & B_{13} & B_{14} \\ B_{21} & B_{22} & B_{23} & B_{24} \\ B_{31} & B_{32} & B_{33} & B_{34} \\ B_{41} & B_{42} & B_{43} & B_{44} \end{pmatrix} = \begin{pmatrix} T_{11} & 0 & 0 & 0 \\ T_{21} & T_{22} & 0 & 0 \\ T_{31} & T_{32} & T_{33} & 0 \\ T_{41} & T_{42} & T_{43} & T_{44} \end{pmatrix} \begin{pmatrix} T_{11} & T_{21} & T_{31} & T_{41} \\ 0 & T_{22} & T_{32} & T_{42} \\ 0 & 0 & T_{33} & T_{43} \\ 0 & 0 & 0 & T_{44} \end{pmatrix} \quad (B.42)$$

for the elements of $\mathbf{T}$. We simply write equations for the elements of the upper triangle of $\mathbf{B}$ in turn, working across the rows. Doing this gives equations with only one unknown at every stage, namely the corresponding element of $\mathbf{T}^T$. For example

$B_{11} = T_{11}^2$, then $B_{12} = T_{11}T_{21}$, etc. The algorithm is

$$\begin{cases} T_{ii} := (B_{ii} - \sum_{k=1}^{i-1} T_{ik}^2)^{\frac{1}{2}} \\ \text{for } j := i+1 \text{ to } n \text{ do } T_{ji} := (B_{ij} - \sum_{k=1}^{i-1} T_{ik}T_{jk})/T_{ii} \end{cases} \quad (B.43)$$

Exercise 5.4

Evaluate the number of operations required, and determine when a sequential update is faster than updating with a vector of measurements

The matrix times vector $\mathbf{Sk}$ takes n^2 operations, then the vector products $\mathbf{k}^T\mathbf{x}$ and $\mathbf{k}^T(\mathbf{Sk})$ take n operations each. The symmetric vector outer product $(\mathbf{Sk})(\mathbf{Sk})^T$ takes $\frac{1}{2}n^2$ operations. Thus the total is of order $m(\frac{3}{2}n^2 + 2n)$. This is 50% slower than the n-form updates, but faster than the m-form update with a vector if $m > n$.

Exercise 5.5

Show that the covariance of an ensemble of retrievals corresponding to the a priori *is, in the nearly-linear case:*

$$\hat{\mathbf{S}}_a = \mathbf{S}_a\mathbf{K}^T(\mathbf{K}\mathbf{S}_a\mathbf{K}^T + \mathbf{S}_\epsilon)^{-1}\mathbf{K}\mathbf{S}_a \quad (B.44)$$

and relate its eigenvectors to the singular vectors of $\tilde{\mathbf{K}}$.

From Eq. (4.6)

$$\hat{\mathbf{x}} - \mathbf{x}_a = \mathbf{G}[\mathbf{K}(\mathbf{x} - \mathbf{x}_a) + \boldsymbol{\epsilon}] \quad (B.45)$$

so that

$$\begin{aligned} \hat{\mathbf{S}}_a &= \mathcal{E}\{(\hat{\mathbf{x}} - \mathbf{x}_a)(\hat{\mathbf{x}} - \mathbf{x}_a)^T\} \\ &= \mathbf{G}\mathcal{E}\{[\mathbf{K}(\mathbf{x} - \mathbf{x}_a) + \boldsymbol{\epsilon}][\mathbf{K}(\mathbf{x} - \mathbf{x}_a) + \boldsymbol{\epsilon}]^T\}\mathbf{G}^T \\ &= \mathbf{G}(\mathbf{K}\mathbf{S}_a\mathbf{K}^T + \mathbf{S}_\epsilon)\mathbf{G}^T. \end{aligned} \quad (B.46)$$

Substituting $\mathbf{G} = \mathbf{S}_a\mathbf{K}^T(\mathbf{K}\mathbf{S}_a\mathbf{K}^T + \mathbf{S}_\epsilon)^{-1}$ from Eq. (4.6), we obtain the required result. Now express $\mathbf{K}$ in terms of $\tilde{\mathbf{K}}$, and use $\tilde{\mathbf{K}} = \mathbf{U}\boldsymbol{\Lambda}\mathbf{V}^T$:

$$\begin{aligned} \hat{\mathbf{S}}_a &= \mathbf{S}_a^{\frac{1}{2}}\tilde{\mathbf{K}}^T(\tilde{\mathbf{K}}\tilde{\mathbf{K}}^T + \mathbf{I})^{-1}\tilde{\mathbf{K}}\mathbf{S}_a^{\frac{1}{2}} \\ &= \mathbf{S}_a^{\frac{1}{2}}\mathbf{V}\boldsymbol{\Lambda}(\boldsymbol{\Lambda}^2 + \mathbf{I})^{-1}\boldsymbol{\Lambda}\mathbf{V}^T\mathbf{S}_a^{\frac{1}{2}} \end{aligned} \quad (B.47)$$

Exercise 6.1

Explain why the averaging kernel $\mathbf{A} = \mathbf{W}(\mathbf{KW})^{-1}\mathbf{K}$ *is likely to be well behaved, apart from possible problems due to numerical rounding errors.*

Problems are caused primarily by the small singular values of $\mathbf{K}$. ($\mathbf{W}$ should be chosen so it does not have any small singular values.) Postmultiplying the inverse by $\mathbf{K}$ effectively eliminates the problem as follows. Putting $\mathbf{K} = \mathbf{U}\boldsymbol{\Lambda}\mathbf{V}^T$ gives

$$(\mathbf{KW})^{-1}\mathbf{K} = (\mathbf{U}\boldsymbol{\Lambda}\mathbf{V}^T\mathbf{W})^{-1}\mathbf{U}\boldsymbol{\Lambda}\mathbf{V}^T. \tag{B.48}$$

$\mathbf{U}\boldsymbol{\Lambda}$ is square, $m \times m$ and invertible, so we can write

$$(\mathbf{KW})^{-1}\mathbf{K} = (\mathbf{V}^T\mathbf{W})^{-1}\mathbf{V}^T. \tag{B.49}$$

and cancel the $\boldsymbol{\Lambda}$ even if there are small singular values. As long as $\mathbf{W}$ reasonably spans the row space of $\mathbf{K}$, $\mathbf{V}^T\mathbf{W}$ will not be poorly conditioned.

Exercise 6.2

Show that the choice of the weighting functions as the least noise-sensitive representation is independent of the measurement error covariance

Transform the measurement equation to $\mathbf{y}' = \mathbf{S}_\epsilon^{-\frac{1}{2}}\mathbf{y} = \mathbf{S}_\epsilon^{-\frac{1}{2}}\mathbf{Kx} + \mathbf{S}_\epsilon^{-\frac{1}{2}}\boldsymbol{\epsilon} = \mathbf{K}'\mathbf{x} + \boldsymbol{\epsilon}'$ so that the error covariance is $\mathbf{I}_m$. The previous analysis now applies, and we obtain

$$\hat{\mathbf{x}} = \mathbf{K}'^T(\mathbf{K}'\mathbf{K}'^T)^{-1}\mathbf{y}' = \mathbf{K}^T\mathbf{S}_\epsilon^{-\frac{1}{2}}(\mathbf{S}_\epsilon^{-\frac{1}{2}}\mathbf{K}\mathbf{K}^T\mathbf{S}_\epsilon^{-\frac{1}{2}})^{-1}\mathbf{S}_\epsilon^{-\frac{1}{2}}\mathbf{y} = \mathbf{K}^T(\mathbf{K}\mathbf{K}^T)^{-1}\mathbf{y} \tag{B.50}$$

as before.

Exercise 6.3

Show that the least squares solution to the linear problem is given by:

$$\hat{\mathbf{x}} = (\mathbf{K}^T\mathbf{K})^{-1}\mathbf{K}^T\mathbf{y} \tag{B.51}$$

We wish to minimise the sum of the squares of $\mathbf{y} - \mathbf{Kx}$:

$$\frac{\partial}{\partial \mathbf{x}}(\mathbf{y} - \mathbf{Kx})^T(\mathbf{y} - \mathbf{Kx}) = 2\mathbf{K}^T(\mathbf{y} - \mathbf{Kx}) = 0 \tag{B.52}$$

hence $\mathbf{x} = (\mathbf{K}^T\mathbf{K})^{-1}\mathbf{K}^T\mathbf{y}$.

Exercise 6.4

Derive expressions for the error resulting from ignoring errors in forward model parameters.

In the linear case, the error in $\hat{\mathbf{x}}$ due to ignoring $\mathbf{S}_b$ is the difference between Eqs. (4.33) and (4.6):

$$\delta\hat{\mathbf{x}} = \mathbf{S}_a\mathbf{K}^T(\mathbf{KS}_a\mathbf{K}^T+\mathbf{K}_b\mathbf{S}_b\mathbf{K}_b^T+\mathbf{S}_\epsilon)^{-1}(\mathbf{y}-\mathbf{Kx}_a) - \mathbf{S}_a\mathbf{K}^T(\mathbf{KS}_a\mathbf{K}^T+\mathbf{S}_\epsilon)^{-1}(\mathbf{y}-\mathbf{Kx}_a) \tag{B.53}$$

This doesn't simplify in any useful way, but it can be rearranged as

$$\begin{aligned}\delta\hat{\mathbf{x}} &= \mathbf{S}_a\mathbf{K}^T[(\mathbf{KS}_a\mathbf{K}^T+\mathbf{K}_b\mathbf{S}_b\mathbf{K}_b^T+\mathbf{S}_\epsilon)^{-1} - (\mathbf{KS}_a\mathbf{K}^T+\mathbf{S}_\epsilon)^{-1}](\mathbf{y}-\mathbf{Kx}_a) \\ &= -\mathbf{S}_a\mathbf{K}^T(\mathbf{KS}_a\mathbf{K}^T+\mathbf{K}_b\mathbf{S}_b\mathbf{K}_b^T+\mathbf{S}_\epsilon)^{-1} \\ &\qquad (\mathbf{K}_b\mathbf{S}_b\mathbf{K}_b^T)(\mathbf{KS}_a\mathbf{K}^T+\mathbf{S}_\epsilon)^{-1}(\mathbf{y}-\mathbf{Kx}_a)\end{aligned} \tag{B.54}$$

using $\mathbf{A}^{-1}-\mathbf{B}^{-1} = \mathbf{A}^{-1}(\mathbf{B}-\mathbf{A})\mathbf{B}^{-1}$, showing that the error is equivalent to an error in $\mathbf{y}$ of

$$\delta\mathbf{y} = -\mathbf{K}_b\mathbf{S}_b\mathbf{K}_b^T(\mathbf{KS}_a\mathbf{K}^T+\mathbf{S}_\epsilon)^{-1}](\mathbf{y}-\mathbf{Kx}_a) \tag{B.55}$$

Exercise 7.1

In the case where σ_ξ^2 and σ_ϵ^2 are both constant, examine the behaviour of $\hat{\sigma}_t^2$ at large times.

At large times we expect $\hat{\sigma}_t^2$ to tend to a constant. Therefore, from Eq. (7.15),

$$\begin{aligned}\hat{\sigma}_t^2(\hat{\sigma}_t^2+\sigma_\xi^2+\sigma_\epsilon^2) &= \sigma_\epsilon^2(\hat{\sigma}_t^2+\sigma_\xi^2) \\ \hat{\sigma}_t^4+\hat{\sigma}_t^2\sigma_\xi^2-\sigma_\epsilon^2\sigma_\xi^2 &= 0 \\ \hat{\sigma}_t^2 &= [(\sigma_\xi^4+4\sigma_\epsilon^2\sigma_\xi^2)^{\frac{1}{2}}-\sigma_\xi^2]/2\end{aligned} \tag{B.56}$$

Exercise 7.2

Find an expression for the backward model $\mathbf{M}''$, in terms of the forward model $\mathbf{M}$, assuming that it is independent of t.

The forward and backward processes are

$$\begin{aligned}\mathbf{x}_t &= \mathbf{Mx}_{t-1}+\boldsymbol{\xi}_t \\ \mathbf{x}_{t-1} &= \mathbf{M}''\mathbf{x}_t+\boldsymbol{\xi}''_{t-1}\end{aligned} \tag{B.57}$$

where $\boldsymbol{\xi}_t$ is uncorrelated with $\mathbf{x}_{t-1}$ and $\boldsymbol{\xi}''_{t-1}$ is uncorrelated with $\mathbf{x}_t$. The lag covariance $\mathbf{L} = \mathcal{E}\{\mathbf{x}_t\mathbf{x}_{t-1}^T\}$ can be written two ways, from the two equations

$$\begin{aligned}\mathbf{L} = \mathcal{E}\{\mathbf{x}_t\mathbf{x}_{t-1}^T\} &= \mathcal{E}\{\mathbf{Mx}_{t-1}\mathbf{x}_{t-1}+\boldsymbol{\xi}_t\mathbf{x}_{t-1}\} = \mathbf{MS}_x \\ \mathbf{L}^T = \mathcal{E}\{\mathbf{x}_{t-1}\mathbf{x}_t^T\} &= \mathcal{E}\{\mathbf{M}''\mathbf{x}_t\mathbf{x}_t+\boldsymbol{\xi}''_{t-1}\mathbf{x}_t\} = \mathbf{M}''\mathbf{S}_x.\end{aligned} \tag{B.58}$$

Hence $\mathbf{M}'' = \mathbf{S}_x\mathbf{M}^T\mathbf{S}_x^{-1}$.

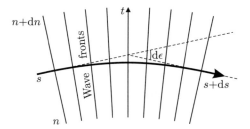

Fig. B.1 Propagation of a ray through a medium with varying refractive index.

Exercise 8.1

Follow through the suggested algebra and find an expression for the final transformed linearised instrument model.

We start with

$$\mathbf{y}'' = \mathbf{\Lambda}\mathbf{V}^T\mathbf{x} + \boldsymbol{\epsilon}_r'' + \boldsymbol{\epsilon}_s'', \tag{B.59}$$

where $\mathbf{S}_{\epsilon_s''} = \mathbf{I} - \mathbf{S}_{\epsilon_r''}$. Using the eigenvector decomposition $\mathbf{L}^T\mathbf{S}_{\epsilon_r''}\mathbf{L} = \mathbf{\Lambda}_{\epsilon_r''}$, multiply by $\mathbf{L}^T$:

$$\mathbf{L}^T\mathbf{y}'' = \mathbf{L}^T\mathbf{\Lambda}\mathbf{V}^T\mathbf{x} + \mathbf{L}^T\boldsymbol{\epsilon}_r'' + \mathbf{L}^T\boldsymbol{\epsilon}_s'', \tag{B.60}$$

which is of the form

$$\mathbf{y}''' = \mathbf{K}'''\mathbf{x} + \boldsymbol{\epsilon}_r''' + \boldsymbol{\epsilon}_s''', \tag{B.61}$$

where the error covariances of $\boldsymbol{\epsilon}_r'''$ and $\boldsymbol{\epsilon}_s'''$ are $\mathbf{\Lambda}_{\epsilon_r''}$ and $\mathbf{I} - \mathbf{\Lambda}_{\epsilon_r''}$ respectively. The description of the measurement now consists of $\mathbf{y}'''$, $\mathbf{K}'''$ and the diagonal of $\mathbf{\Lambda}_{\epsilon_r''}$.

Exercise 9.1

Derive the third form of the equation of radiative transfer:

$$L(z) = (L(0) - B(0))\tau(0, z) + B(z) - \int_0^z \tau(z', z)\frac{\mathrm{d}B(z')}{\mathrm{d}z'}\,\mathrm{d}z' \tag{B.62}$$

from the first:

$$L(z) = L(0)\tau(0, z) + \int_0^z B(z')\frac{\mathrm{d}\tau(z', z)}{\mathrm{d}z'}\,\mathrm{d}z'. \tag{B.63}$$

This follows immediately on integrating the first form by parts.

Exercise 9.2

Explain why the curvature of a ray is proportional to the component of the gradient of refractive index at right angles to the ray:

$$\frac{d\epsilon}{ds} = \frac{1}{n}\left(\frac{\partial n}{\partial t}\right)_s \tag{B.64}$$

where ϵ is the direction of propagation, s, t are coordinates along and perpendicular to the ray respectively.

Consider the propagation of the ray shown in Fig. B.1. The wavefronts are separated by a wavelength in air, which is λ_{vac}/n. Thus the difference in the distance propagated between the top and the bottom of the figure is $n\,ds/(n+dn) - ds = -(1/n)\,ds\,dn$. Note that $dn < 0$ in the diagram. This distance is also $-dt\,d\epsilon$, the negative sign because ϵ is measured upwards, so the $d\epsilon$ in the figure is negative. Equating these two gives

$$\frac{d\epsilon}{ds} = \frac{1}{n}\left(\frac{dn}{dt}\right) \tag{B.65}$$

where dn/dt is measured perpendicular to the ray, i.e. is equivalent to $(\partial n/\partial t)_s$.

Exercise 9.3

Show that the error in this approximation is:

$$\delta L = \iint f(\nu)(B(\nu, T) - \bar{B}(T))\frac{d}{dz}(\tau(\nu, z) - \bar{\tau}(z))\,d\nu\,dz \tag{B.66}$$

Expand the expression given:

$$\begin{aligned}\delta L &= \iint f(\nu)B(\nu,T)\frac{d}{dz}\tau(\nu,z)\,d\nu\,dz + \iint f(\nu)\bar{B}(T)\frac{d}{dz}\bar{\tau}(z)\,d\nu\,dz \\ &- \iint f(\nu)\bar{B}(T)\frac{d}{dz}\tau(\nu,z)\,d\nu\,dz - \iint f(\nu)B(\nu,T)\frac{d}{dz}\bar{\tau}(z)\,d\nu\,dz\end{aligned} \tag{B.67}$$

The two negative terms can both be integrated to give the same quantity as the second term. Hence

$$\delta L = \iint f(\nu)B(\nu,T)\frac{d}{dz}\tau(\nu,z)\,d\nu\,dz - \iint f(\nu)\bar{B}(\nu,T)\frac{d}{dz}\bar{\tau}(z)\,d\nu\,dz \tag{B.68}$$

which is the error in the approximation.

Exercise 10.1

How should smoothness constraints be transformed when comparing retrievals on different grid spacings?

Transforming a prior covariance matrix from **x** to **z** is carried out, from section 10.3.1.3, by

$$\mathbf{S}_{za} = \mathbf{W}^* \mathbf{S}_{xa} \mathbf{W}^{*T} \qquad (B.69)$$

A smoothness constraint **H** is of the same nature as an inverse covariance, but if it is singular we cannot evaluate the obvious transformation:

$$\mathbf{H}_z = (\mathbf{W}^* \mathbf{H}_x^{-1} \mathbf{W}^{*T})^{-1}. \qquad (B.70)$$

However this can be done in principle if the inverse is restricted to the subspace spanned by the constraint, i.e. if a pseudo inverse is used. Not all smoothness constraints can be satisfactorily transformed. The squared first and second differences can disappear if they are defined on a fine grid, and it is desired to transform them to a coarser grid - consider the meaning of doubling the grid spacing for a first difference constraint!

Exercise 10.2

What is the form of the inverse of the Markov covariance matrix?

Putting $S_{ij} = \sigma_a^2 \alpha^{|i-j|}$, where $\alpha = \exp(-\delta z/h)$, the inverse matrix is

$$\mathbf{S}^{-1} = \frac{\sigma_a^{-2}}{1-\alpha^2} \begin{pmatrix} 1 & -\alpha & 0 & \cdots & 0 & 0 \\ -\alpha & 1+\alpha^2 & -\alpha & \cdots & 0 & 0 \\ 0 & -\alpha & 1+\alpha^2 & \cdots & 0 & 0 \\ \vdots & \vdots & \vdots & \vdots & \vdots & \vdots \\ 0 & 0 & \cdots & 1+\alpha^2 & -\alpha \\ 0 & 0 & \cdots & -\alpha & 1 \end{pmatrix}. \qquad (B.71)$$

Apart from the two end elements of the diagonal, $\sigma_a^2 \mathbf{S}^{-1}$ can be interpreted as a unit matrix plus $\alpha/(1-\alpha^2)$ times a squared first derivative matrix, thus constraining that combination of the squared departure of the state from the prior profile and the squared first derivative.

Exercise 10.3

Derive expressions for the mean signal, mean retrieval and the covariance of an ensemble of retrievals, when the ensemble covariance and mean differ from the a priori covariance and mean.

Start with an instrument model which includes random and systematic errors separately, and is linearised about $\mathbf{x}_a$:

$$\mathbf{y} = \mathbf{F}(\mathbf{x}_a) + \mathbf{K}(\mathbf{x} - \mathbf{x}_a) + \boldsymbol{\epsilon}_r + \boldsymbol{\epsilon}_s \tag{B.72}$$

The ensemble mean measured signal is

$$\bar{\mathbf{y}} = \mathbf{F}(\mathbf{x}_a) + \mathbf{K}(\mathbf{x}_e - \mathbf{x}_a) + \boldsymbol{\epsilon}_s \tag{B.73}$$

in the limit of large ensemble when the random error becomes negligible. An individual retrieval characterisation is

$$\hat{\mathbf{x}} = \mathbf{x}_a + \mathbf{A}(\mathbf{x} - \mathbf{x}_a) + \mathbf{G}(\boldsymbol{\epsilon}_r + \boldsymbol{\epsilon}_s) \tag{B.74}$$

so the ensemble mean retrieval is

$$\bar{\hat{\mathbf{x}}} = \mathbf{x}_a + \mathbf{A}(\mathbf{x}_e - \mathbf{x}_a) + \mathbf{G}\boldsymbol{\epsilon}_s \tag{B.75}$$

The covariance of $\hat{\mathbf{x}}$ about this mean is

$$\begin{aligned}
\mathbf{S}_{\hat{\mathbf{x}}} &= \mathcal{E}\{(\hat{\mathbf{x}} - \bar{\hat{\mathbf{x}}})(\hat{\mathbf{x}} - \bar{\hat{\mathbf{x}}})^T\} \\
&= \mathcal{E}\{(\mathbf{A}(\mathbf{x} - \mathbf{x}_e) + \mathbf{G}\boldsymbol{\epsilon}_r)(\mathbf{A}(\mathbf{x} - \mathbf{x}_e) + \mathbf{G}\boldsymbol{\epsilon}_r)^T\} \\
&= \mathbf{A}\mathbf{S}_e\mathbf{A}^T + \mathbf{G}\mathbf{S}_{\epsilon_r}\mathbf{G}^T
\end{aligned} \tag{B.76}$$

Exercise 11.1

Investigate the relationships between the different ways of specifying the mass distribution, and determine which ones need a reference height or pressure.

The basic equations are the hydrostatic equation and the gas equation. I will use the perfect gas equation for this illustration, but it is not essential.

$$dp = -\rho g\, dz \tag{B.77}$$

$$p = \frac{\rho RT}{M} \tag{B.78}$$

leading to

$$\frac{dp}{p} = -\frac{Mg}{RT}\, dz \tag{B.79}$$

Remember that both M and g vary with altitude. I assume that numerical integration and differentiation can be carried out. Provided an interpolation rule is specified for the data, this is possible in principle.

(1) Starting with temperature as a function of absolute height, $T(z)$, integrate Eq. (B.79) to give

$$\ln p(z) = \ln p(z_0) - \int_{z_0}^{z} \frac{Mg}{RT}\, dz, \tag{B.80}$$

which requires knowledge of $p(z_0)$ at some reference level z_0. From T and p we can then obtain $\rho(z)$ from the gas equation.

(2) Starting with density as a function of absolute height, $\rho(z)$, integrate the hydrostatic equation (B.77) to give

$$p(z) = p(z_0) - \int_{z_0}^{z} \rho g \, dz. \tag{B.81}$$

This requires $p(z_0)$, which should be at some high altitude rather than the surface, or there can be numerical problems of cancellation. From ρ and p we can then obtain $T(z)$ from the gas equation.

(3) Starting with $p(z)$, we get

$$\rho(z) = -\frac{1}{g}\frac{dp}{dz} \tag{B.82}$$

from the hydrostatic equation. This does not need a reference level. $T(z)$ can then be obtained from the gas equation.

(4) Starting with $T(p)$ or $T(\ln p)$, obtain $\rho(p)$ from the gas equation. We could also start with $\rho(p)$ and obtain $T(p)$. Then the hydrostatic equation gives

$$z_g(p) - z_g(p_0) = \frac{1}{g(p_0)} \int_{z(p_0)}^{z(p)} g(z)\,dz = -\frac{1}{g(p_0)} \int_{p_0}^{p} \frac{RT(p)}{M} d\ln p \tag{B.83}$$

giving the geopotential height $z_g(p)$, which must be related to geometric height separately. This requires the height of a reference pressure level to obtain absolute height. With a little care in the numerical method, it is possible to use

$$z(p) - z(p_0) = -\int_{p_0}^{p} \frac{RT(p)}{Mg(p)} d\ln p \tag{B.84}$$

if the $z(p)$ is used to evaluate the $g[z(p)]$ inside the integral as it is calculated, as the effect is fairly small.

(5) Starting with $z(p)$, the hydrostatic equation gives

$$T(p) = -\frac{Mg}{R}\frac{dz}{d\ln p} \tag{B.85}$$

from which we can obtain $\rho(p)$ from the gas equation. This does not need a reference level.

Exercise 12.1

Show in detail how χ^2 should be calculated in this case.

We can express $\mathbf{S}_{\hat{\mathbf{x}}}$ in terms of the singular vector decomposition of $\tilde{\mathbf{K}}$ as follows:

$$\begin{aligned}\mathbf{S}_{\hat{\mathbf{x}}} &= \mathbf{S}_a\mathbf{K}^T(\mathbf{K}\mathbf{S}_a\mathbf{K}^T + \mathbf{S}_\epsilon)^{-1}\mathbf{K}\mathbf{S}_a \\ &= \mathbf{S}_a^{\frac{1}{2}}\tilde{\mathbf{K}}^T(\tilde{\mathbf{K}}\tilde{\mathbf{K}}^T + \mathbf{I}_m)^{-1}\tilde{\mathbf{K}}\mathbf{S}_a^{\frac{1}{2}} \\ &= \mathbf{S}_a^{\frac{1}{2}}\mathbf{V}\mathbf{\Lambda}(\mathbf{\Lambda}^2 + \mathbf{I}_p)^{-1}\mathbf{\Lambda}\mathbf{V}^T\mathbf{S}_a^{\frac{1}{2}}\end{aligned} \qquad (B.86)$$

Define $\mathbf{z} = \mathbf{V}^T\mathbf{S}_a^{-\frac{1}{2}}(\hat{\mathbf{x}} - \mathbf{x}_a)$. The covariance of $\mathbf{z}$ should be the diagonal matrix $\mathbf{\Lambda}(\mathbf{\Lambda}^2 + \mathbf{I}_p)^{-1}\mathbf{\Lambda}$. Thus z_i should have variance $\lambda_i^2/(1 + \lambda_i^2)$. The elements of $\mathbf{z}$ corresponding to zero variance should themselves be zero (within rounding error), and the remainder can be tested individually.

Appendix C
Terminology and Notation

C.1 Summary of Terminology

The error analysis in particular contains a number of terms and concepts that are easily confused, some of which which do not have an established terminology in the literature. The usage in this book is therefore summarised here.

Forward function. The actual relationship between the state and the measured signal. May not be known in detail.

Forward model. The relationship between the state and the measured signal as modelled numerically.

Forward model error. The error in a simulated measured signal due to using a forward model.

Forward model parameters. Parameters of the forward model which are not targets for retrieval.

Forward model parameter error. Difference between the true values of these parameters and our best estimates.

Forward function parameters. Notional equivalent of the forward model errors for the forward function. May not be known.

Full state vector. A state vector which includes all of the parameters required by the forward model, e.g. the forward model parameters as well as the atmospheric parameters at a fine grid spacing.

A priori. Any information about the state other than the measured signal being analysed.

Transfer function. Relationship between a retrieval and the true state.

Weighting function. Fréchet derivative of the forward model with respect to the state vector.

Gain matrix or *Contribution function.* Fréchet derivative of the retrieval with respect to the measured signal.

Averaging kernel matrix. Fréchet derivative of the transfer function with respect to the state.

Averaging kernel. A row of the averaging kernel matrix. Indicates which part of the true state is averaged to produce the retrieval at a single element of the retrieved state.

δ-function response. A column of the averaging kernel matrix. Indicates which part of the retrieved state is related to a single element of the true state.

(Total) measurement error. Total error in the measured signal. Difference between signal and the forward function.

Measurement noise. Random component of the measurement error.

Exact retrieval. Not one which reproduces the unknown profile exactly, but one which reproduces the measured signal exactly.

Total retrieval error. The total error in a retrieval due to all sources.

Bias. Error in a simulated retrieval of the *a priori* state using error-free simulated measured signals.

Smoothing error. Component of error due to smoothing by the averaging kernels.

Retrieval error. Error in a retrieval due to measurement error.

Retrieval noise. Error in a retrieval due to measurement noise.

Null-space error. Deprecated. An incorrect term for smoothing error.

Random error. Any error which is uncorrelated between repeated measurements. Same as 'precision'.

Systematic error. Any error which is related between repeated measurements, including both constant errors and errors which are a function of the state or the measured signal.

Gain error. An error which is a linear function of the state or the measured signal.

Precision. Same as 'random error'.

Accuracy. The total error, including random and systematic components.

Error pattern. An eigenvector of an error covariance matrix, scaled by the square root of the eigenvalue. Multi-dimensional equivalent of an 'error bar'.

C.2 List of Symbols Used

The main symbols used widely throughout the book are defined here. Others which are defined and used only within a few pages may not be listed. Subscripts and other diacritical marks are used to distinguish various kinds of e.g. covariance matrices and state & measurement vectors; not all variants of these are listed, but the meanings of the subscripts, etc., should be clear from context.

$\nabla_{\mathbf{x}}$	Gradient with respect to $\mathbf{x}$
α	Azimuth angle
$A(z, z')$	Averaging kernel function
$\mathbf{A}$	Averaging kernel matrix
$\mathbf{b}$	Vector of forward model parameters
$B(\nu, T)$	Planck Function
$c(z)$	Centre of an averaging kernel
$\mathbf{c}$	Vector of retrieval method parameters
χ	Optical depth
χ^2	The statistical quantity
d_n	Degrees of freedom for noise
d_s	Degrees of freedom for signal
ϵ_i	Measurement noise in channel i
ϵ	Angle of refraction
$\boldsymbol{\epsilon}$	Measurement noise vector
$\mathbf{e}_i$	Error pattern
$\mathbf{E}_t$	Time evolution model or operator
$\mathcal{E}$	Expected value operator
$\mathbf{f}$	Forward function
$F_i(\mathbf{x})$	Forward model for one channel
$\mathbf{F}(\mathbf{x})$	Forward model. A vector of functions.
$\mathcal{F}$	Fisher information matrix
$\mathbf{G}, \mathbf{G}_y$	Gain, or contribution function, matrix
$\mathbf{g}$	Contribution function vector: row of $\mathbf{G}$
$G_j(z)$	Individual contribution function
$\mathbf{G}_t$	Kalman gain matrix at time step t
$\mathbf{G}_b$	Sensitivity of retrieval to $\mathbf{b}$
γ	A Lagrangian multiplier
$\boldsymbol{\Gamma}$	Matrix of Lagrangian multipliers
Γ	Gamma function
$h(\mathbf{x})$	A function of the state vector
$\mathbf{h}$	A vector representing a linear $h(\mathbf{x})$
h	A length scale
H	Information content

$\mathbf{H}$	Twomey constraint matrix
$\mathbf{I}_n$	Unit matrix of order n
J	Source function; cost function
κ	Absorption coefficient
$\mathcal{K}$	Absorption coefficient matrix
$\mathbf{k}_i$	Row of $\mathbf{K}$; an individual weighting function vector
$K_i(z)$	Individual weighting function
$\mathbf{K}$	Weighting function matrix
$\mathbf{K_b}$	Sensitivity of forward model to $\mathbf{b}$
$\tilde{\mathbf{K}}$	Weighting function matrix scaled for unit covariances
λ	Singular value or eigenvalue; Lagrangian multiplier
$\mathbf{\Lambda}$	Matrix of singular value or eigenvalues
$L(\nu), L_i$	Radiance
L	Likelihood
$\mathbf{l}$	(Left) eigenvector of a (a)symmetric matrix
$\mathbf{L}$	Matrix of (left) eigenvectors of a (a)symmetric matrix
m	Length of the measurement vector
μ	Lagrangian multiplier
$\boldsymbol{\mathcal{M}}_t$	Operator representing equations of motion
$\mathbf{M}_t$	Linearised version of the operator $\boldsymbol{\mathcal{M}}_t$
n	Length of the state vector; refractive index
ν	Wavenumber
$\mathbf{O}$	Zero matrix
Ω	Solid angle
p	Pressure; rank of a matrix
p_0	Reference pressure or surface pressure
p_i	Probability of event i
P	Probability density function
ψ	Angle in a radial coordinate system
$\mathbf{Q}, Q_{ij}$	A quadratic form arising in Backus-Gilbert theory
$\mathbf{Q}$	A matrix transform used in QR decomposition
r	Radial coordinate
$r(z)$	Resolving length of an averaging kernel
$\mathbf{r}$	Right eigenvector of an asymmetric matrix
$\mathbf{R}$	Matrix of right eigenvectors; retrieval method
ρ	Density
σ	Standard deviation of noise or error
$s(z_0)$	Spread of an averaging kernel about z_0
S	Entropy
S_{ij}	Element of covariance matrix
$\mathbf{S}$	Covariance matrix
$\mathbf{S}_\epsilon$	Covariance of measurement noise

List of Symbols Used

$\mathbf{S}_a$	Covariance of prior estimate of $\mathbf{x}$
$\mathbf{S}_e$	Covariance of an ensemble of states
$\hat{\mathbf{S}}$	Covariance of retrieved state
$\tau(\nu, z)$	Transmittance
T, $T(z)$	Temperature
$\mathbf{T}$	Upper triangular matrix. Usually with the same subscript as the corresponding $\mathbf{S}$.
θ	Zenith angle of propagation of a ray
$\mathbf{u}$	Left singular vector
$\mathbf{U}$	Matrix of left singular vectors
$\mathbf{v}$	Right singular vector
$\mathbf{V}$	Matrix of right singular vectors
$w(A)$	Width of an averaging kernel
w_j	Profile representation coefficient
$W_j(z)$	Profile representation function
$\mathbf{W}$	Representation function; working matrix
x_i	State vector element
$\mathbf{x}$	State vector
$\mathbf{x}_0$	Any reference state
$\mathbf{x}_a$	Prior estimate of the state
$\mathbf{x}_e$	Member of an ensemble of states
$\hat{\mathbf{x}}$	Retrieved state
$\hat{\mathbf{y}}$	Measurement computed from retrieved state
y_i	Measurement vector element
$\mathbf{y}$	Measurement vector
$\mathbf{y}_a$	Measurement computed from prior state
z	Altitude, generally in units of distance
$\mathbf{z}$	A linear transformation of $\mathbf{x}$ or of $\mathbf{y}$
ζ	$-\ln(p)$ used as an altitude coordinate

Bibliography

Andersson, E., Pailleux, J., Thepaut, J. N., Eyre, J. R., McNally, A. P., Kelly, G. A., Courtier, P. (1994), "Use of cloud-cleared radiances in 3-dimensional and 4-dimensional variational data assimilation", *Quart. J. Roy. Meteorol. Soc.*, **120**, 627.

Andersson, E., Haseler, J., Undén, P., Courtier, P., Kelly, G., Vasiljević, D., Branković, C., Cardinali, C., Gaffard, C., Hollingsworth, A., Jakob, C., Janssen, P., Klinker, E., Lanzinger, A., Miller, M., Rabier, F., Simmons, A., Strauss, B., Thepaut, J. N., Viterbo, P. (1998), "The ECMWF implementation of three-dimensional variational assimilation (3D-Var). III: Experimental results", *Quart. J. Roy. Meteorol. Soc.*, **124**, 1831.

Andrews, D. G., Holton, J. R. and Leovy, C. B. (1987), *Middle Atmosphere Dynamics*, Academic Press.

Arfken, G. (1995), *Mathematical methods for physicists*, 4th Ed., Academic Press, London.

Atkinson, K. E. (1989) *An Introduction to Numerical Analysis*, Wiley.

Aumann, H. H. and Pagano, R. J. (1994), "Atmospheric infrared sounder on the Earth Observing System", *Optical Eng.*, **33**, 776.

Backus, G. E. and Gilbert, J. F. (1970), "Uniqueness in the inversion of inaccurate gross earth data", *Phil. Trans. R. Soc. Lond.*, **266**, 123.

Banwell, C. N. and McCash, E.M. (1994), *Fundamentals of Molecular Spectroscopy*, 4th edition, McGraw-Hill, London.

Barnett, T. L. (1969) "Application of a nonlinear least squares method to atmospheric temperature sounding", *J. Atmos. Sci.*, **26**, 457.

Bayes, Rev. Mr. (1763), "An Eſſay towards ſolving a Problem in the Doctrine of Chances". *Phil. Trans. R. Soc. Lond.*, **53**, 370.

Bergthorsson, P. and Döös, B. (1955), "Numerical weather map analysis", *Tellus*, **5**, 329.

Birch, K. P. and Downs, M. J. (1993), "An updated Edlén equation for the refractive index of air", *Metrologia*, **30**, 155.

Birch, K. P. and Downs, M. J. (1994), "Correction to the updated Edlén equation for the refractive index of air", *Metrologia*, **31**, 315.

Brasseur, G. P. (1997), *The Stratosphere and its Role in the Climate System*, Springer-Verlag, Berlin.

Bretherton, C. S., Smith, C., and Wallace, J. M. (1992), "An intercomparison of methods for finding coupled patterns in climate data", *J. Climate*, **5**, 541.

Bovensmann, H., Burrows J. P., Buchwitz M., Frerick J., Noel S., Rozanov V. V., Chance K. V., Goede A. P. H. (1999) "SCIAMACHY: Mission objectives and measurement modes", *J. Atmos. Sci.*, **56**, 127.

Burrows J. P., Weber M., Buchwitz M., Rozanov V., Ladstatter-Weissenmayer A., Richter A., DeBeek R., Hoogen R., Bramstedt K., Eichmann K. U., Eisinger M. (1999), "The global ozone monitoring experiment (GOME): Mission concept and first scientific results", *J. Atmos. Sci.*, **56**, 151.

Chahine, M. T. (1968), "Determination of the temperature profile in an atmosphere from its outgoing radiance", *J. Opt. Soc. Am.*, **58**, 1634.

Chahine, M. T. (1970), "Inverse problems in radiative transfer, : A determination of atmospheric parameters", *J. Atmos. Sci.*, **27**, 960.

Ciddor, P. E. (1996), "Refractive index of air: new equations for the visible and near infrared", *Appl. Opt.*, **35**, 1566.

Clough, S. A., Kneizys, F. X., Shettle, E. P. and Anderson, G. P. (1985), "Atmospheric radiance and transmittance: FASCOD2", *Proc. Sixth Conference on Atmospheric Radiation*, Am. Meteorol. Soc., Boston, Mass.

Clough, S. A., Iacono, M. J. and Moncet, J.-L. (1992), "Line-by-line calculations of atmospheric fluxes and cooling rates: application to water vapor", *J. Geophys. Res.*, **97**, 15 785.

Cohn, S. E. and Parrish, D. F. (1991), "The behavior of forecast error covariances for a Kalman filter in two dimensions", *Mon. Wea. Rev.*, **119**, 1757.

Connor, B. J. and Rodgers, C. D.(1988), "A Comparison of Retrieval Methods: Optimal Estimation, Onion Peeling, and a Combination of the Two", *Advances in Remote Sensing Retrieval Methods 1988*, Ed A. Deepak. A. Deepak Publishing.

Conrath, B. J. (1972), Vertical resolution of temperature profiles obtained from remote radiation measurements, *J. Atmos. Sci.*, **29**, 1262

Conrath, B. J., Hanel, R. A., Kunde, V. G. and Prabhakara, C. (1970), "The infrared radiometer experiment on Nimbus 3", *J. Geophys. Res.*, **75**, 5831.

Courtier, P., Andersson, E., Heckley, W., Pailleux, J., Vasiljević, D., Hamrud, M., Hollingsworth, A., Rabier, E., Fisher, M. (1998), "The ECMWF implementation of three-dimensional variational assimilation (3D-Var). I: Formulation", *Quart. J. Roy. Meteorol. Soc.*, **124**, 1783.

Cowling, T. G. (1950), "Atmospheric absorption of heat radiation by water vapour", *Phil. Mag.*, **41**, 109.

Cressman, G. P. (1959), "An operational objective analysis scheme", *Mon. Wea. Rev.*, **87**, 367.

Curtis, A. R. (1953), Discussion of "A statistical model for water vapour absorption" by R. M. Goody, *Quart. J. Roy. Meteorol. Soc.*, **78**, 638.

Daley, R. (1991), *Atmospheric Data Analysis*, Cambridge University Press, 457 pp.

Derber, J. C. and Wu, W.S. (1998), "The use of TOVS cloud-cleared radiances in the NCEP SSI analysis system", *Mon. Wea. Rev.*, **126**, 2287.

Deutsch, R. (1965), *Estimation Theory*, Prentice Hall, Englewood Cliffs, N.J.

Dobson, G. M. B. (1968), "Forty years' research on atmospheric ozone at Oxford", *Appl. Opt.*, **7**, 387.

Edwards, D. P. (1992), "GENLN2: A general line-by-line atmospheric transmittance and radiance model. Version 3.0 description and users guide", Rep. NCAR/TN-367+STR, National Center for Atmospheric Research, Boulder, Colorado.

English, S. J., Eyre, J. R. and Smith, J. A. (1999), "A cloud-detection scheme for use with satellite sounding radiances in the context of data assimilation for numerical weather prediction", *Quart. J. Roy. Meteorol. Soc.*, **125**, 2359.

Eliassen, A. (1954), "Provisional report on calculation of spatial covariance and autocorrelation of the pressure field", Report No 5, Videnskaps-Akademiets Institutt for Vaer-Og Klimaforskning, Oslo, Norway, 12 pp. Reprinted in Bengtsson, L., M. Ghil and E. Källén, (eds.), (1981), *Dynamical Meteorology. Data Assimilation Methods*, Springer Verlag, New York, U.S.A., 319.

Eyre, J. R. (1987) "On systematic errors in satellite sounding products and their climatological mean values", *Quart. J. Roy. Meteorol. Soc.*, **113**, 279.

Eyre, J. R., Kelly, G. A., McNally, A. P., Andersson, E., Persson, A. (1993), "Assimilation of TOVS radiance information through one-dimensional variational analysis", *Quart. J. Roy. Meteorol. Soc.*, **119**, 1427.

Fisher, R. A. (1921), "On the mathematical foundation of theoretical statistics", *Phil. Trans. R. Soc. Lond.*, **A222**, 309.

Fletcher, R (1971), "A modified Marquardt subroutine for nonlinear least squares fitting", Report R6799, A.E.R.E., Harwell.

Fletcher, R. (1987), *Practical Methods of Optimization*, 2nd edition, John Wiley and Sons, Chichester, New York.

Gandin, L.S. (1963), "Objective Analysis of Meteorological Fields", *Gidrometeorologischeskoe Izdatelstvo*, Leningrad, 242 pp. English translation by Israel Program for Scientific Translations, Jerusalem, 1965.

Gelb, A., ed. (1974), *Applied Optimal Estimation*, MIT Press.

Gelman, M. E., Miller A. J. and Woolf, H.M. (1972), "Regression techniques for determining temperature profiles in the upper stratosphere from satellite-measured radiances", *Mon. Wea. Rev.*, **100**, 542.

Gill, P. E., Murray, W, and Wright, M. H. (1981), *Practical Optimization*, Academic Press, London.

Gill, P. E., Murray, W, and Wright, M. H. (1990), *Numerical Linear Algebra and Optimization*, Addison-Wesley, Redwood City, California.

Gordley, L. G., Marshall, B. T. and Chu, D. A. (1994), "LINEPAK: Algorithms for modelling spectral transmittance and radiance", *J. Quant. Spectrosc. Radiat. Transfer*, **52**, 563.

Godson, W. L. (1953), "The evaluation of infrared fluxes due to atmospheric water vapour", *Quart. J. Roy. Meteorol. Soc.*, **79**, 367.

Golub, G. H. and Van Loan, C. F. (1996) *Matrix Computations*, 3rd edition, Johns Hopkins University Press.

Goody, R. M. and Yung, Y. L. (1989) *Atmospheric Radiation, Theoretical Basis*, Second edition, Oxford University Press.

Gordley, L. L. and Russell, J. M. III (1981), "Rapid Inversion of Limb Radiance Data using an Emissivity Growth Approximation", *Appl. Opt.*, **20**, 807.

Götz, F. W. P., A. R. Meetham and G. M. B. Dobson (1934), "The vertical distribution of ozone in the stratosphere", *Proc. R. Soc. Lond.* **A145**, 416.

Hanel, R. A., Schlachman, B., Clark, F. D., Prokesh, C. H. Taylor, J. B., Wilson, W. M. and Chaney, L. (1970) "The Nimbus III Michelson interfereometer", *Appl. Opt.*, **9**, 1767.

Hanel, R. A., Conrath, B. J., Jennings, D. E. and Samuelson, R. E. (1992) *Exploration of the Solar System by infrared remote sounding*, Cambridge University Press.

Hays, P. B., Abreu, V. J., Dobbs, M. E., Gell, D. A., Grassl H. J. and Skinner, W. R. (1993), " The High-Resolution Imager on the Upper-Atmosphere Research Satellite", *J. Geophys. Res.*, **98**, 10,713.

Heath, D. A., Krueger, A. J. and Park, H. (1978), "The Solar Backscatter Ultraviolet (SBUV) and Total Ozone Mapping Spectrometer (TOMS) experiment", in *The NIMBUS 7 User's Guide*, ed. C. R. Madrid, p 175. NASA Goddard Space Flight Center, Greenbelt, MD.

Houghton, J. T., Taylor, F. W. and Rodgers, C. D.(1984), *Remote Sounding of Atmo-*

spheres, Cambridge University Press.

Ide K., Courtier P., Ghil M., Lorenc A. C. (1997), "Unified notation for data assimilation: Operational, sequential and variational", *J. Met. Soc. Japan*, **75**, 181.

Jazwinski, A. H. (1970), *Stochastic Processes and Filtering Theory*, Academic Press.

Joiner, J. and da Silva, A. M. (1998), "Efficient methods to assimilate remotely sensed data based on information content", *Quart. J. Roy. Meteorol. Soc.*, **124**, 1669.

Kalman, R. E. (1960), "A new approach to linear filtering and prediction problems", *Trans. AMSE, Ser. D, J. Basic Eng.*, **82**, 35.

Kaplan, L. D. (1959), "Inference of atmospheric structure from remote radiation measurements", *J. Opt. Soc. Am.*, **49**, 1004.

Kaye, G. W. C. and Laby, T. H. (1973), *Tables of physical and chemical constants and some mathematical functions*, 14th ed., Longman, London.

Kidder, S. Q. and Vonder Haar, T. H. (1995), *Satellite Meteorology*, Academic Press.

King, J. I. L. (1956) in *Scientific Uses of Earth Satellites*, p133, J. A. Van Allen, ed., University of Michigan Press, Ann Arbor.

King, J. I. L. (1959) "Deduction of vertical thermal structure of a planetary atmosphere from a satellite", *General Electric, Tech. Info. Series* No. R59SD477.

Kneizys, F,. X., Shettle, E. P., Gallery, WE. O., Chetwynd, J. H., Abreu, L. W., Selby, J. E. A., Chough, S. A. and Fenn, R. W. (1983) *Atmospheric Transmittance/Radiance: Computer Code LOWTRAN 6*, Environmental Research Papers No 846, AFGL-TR-83-0187, Air Force Geophysics Laboratory, Hanscomb AFB, Mass. 01731, U.S.A.

Kuntz M. and Hopfner M. (1999), "Efficient line-by-line calculation of absorption coefficients", *J. Quant. Spectrosc. Radiat. Transfer*, **63**, 97.

Lanczos, C. (1961), *Linear Differential Operators*, D. Van Nostrand. Princeton, N. J.

Levenberg, K. (1944), "A method for the solution of certain nonlinear problems in least squares", *Quart. Appl. Math.*, **2**, 164.

Liou, Kuo-Nan (1992), *Radiation and Cloud Processes in the Atmosphere: Theory, Observation, and Modeling*, Oxford University Press, New York.

López-Puertas, M. and Taylor, F. W. (2000), *Non-Local Thermodynamic Equilibrium in Planetary Atmospheres*, World Scientific Publishing Co., Singapore.

Marquardt, D. W. (1963), "An algorithm for least-squares estimation of nonlinear parameters", *SIAM J. Appl. Math.*, **11**, 431.

Mateer, C. L. (1964), "A study of the information content of umkehr observations", Technical Report No 2, 04682-2-T, Department of Meteorology and Oceanography, College of Engineering, University of Michigan.

Mateer, C. L. (1965), "On the information content of umkehr observations", *J. Atmos. Sci.*, **22**, 370.

McMillin, L. M., Crone, L. J. and Kleespies, T. J. (1995), "Atmospheric transmittance of an absorbing gas. 5. Improvements to the OPTRAN approach.", *Appl. Opt.*, **34**, 8396.

Menke, W. (1989), *Geophysical data analysis : discrete inverse theory*, Revised edition, Academic Press.

Moré, J. J. (1978), "The Levenberg-Marquardt algorithm: implementation and theory", in *Proceedings of the 1977 Dundee conference on numerical analysis, Lecture notes in mathematics,*, **630**, 105. Ed. G. A. Watson. Springer Verlag.

Moré, J. J. and Wright, S. J. (1993), *Optimization Software Guide*, SIAM, Philadelphia.

Noble, B. and Daniel, J, W. (1988), *Applied Linear Algebra*, Prentice-Hall International.

Popescu A. and Ingmann P. (1993), "ENVISAT Global Ozone Monitoring By Occultations Of Stars Instrument – GOMOS", *ESA Bulletin*, **76**, 36, European Space Agency.

Parker, R. L. (1994), *Geophysical Inverse Theory*, Princeton University Press.

Press, W. H., Teukolksy, S., Vetterling, W. T. and Flannery, B. (1995), *Numerical recipes: the art of scientific computing*, Second edition, Cambridge University Press.

Purser, R. J. and Huang, H.-L. (1993), "Estimating effective data density in a satellite retrieval or an objective analysis", *J. Appl. Met.*, **32**, 1092.

Rabier, F., McNally, A., Andersson, E., Courtier, P., Undén, P., Eyre, J. R., Hollingsworth, A. and Bouttier, F. (1998), "The ECMWF implementation of three-dimensional variational assimilation (3D-Var). II: Structure functions", *Quart. J. Roy. Meteorol. Soc.*, **124**, 1809.

Rizzi, R. and Matricardi, M. (1998), "The use of TOVS clear radiances for numerical weather prediction using an updated forward model", *Quart. J. Roy. Meteorol. Soc.*, **124**, 1293.

Rodgers, C. D.(1976), "Retrieval of Atmospheric Temperature and Composition From Remote Measurements of Thermal Radiation", *Rev. Geophys. and Space Phys.*, **14**, 609.

Rodgers, C. D.(1990), "The Characterization and Error Analysis of Profiles Retrieved from Remote Sounding Measurements", *J. Geophys. Res.*, **95**, 5587.

Rosencrantz, P. W. (1975), "Shape of the 5 mm oxygen band in the atmosphere", *IEEE Trans. Antennas Propag.*, **AP-23**, 498.

Rothman L. S., Rinsland C. P., Goldman A., Massie S. T., Edwards D. P., Flaud J. M., Perrin A., Camy-Peyret C., Dana V., Mandin J. Y., Schroeder J., McCann A., Gamache R. R., Wattson R. B., Yoshino K., Chance K. V., Jucks K. W., Brown L. R., Nemtchinov V., Varanasi P. (1998), "The HITRAN molecular spectroscopic database and HAWKS (HITRAN Atmospheric Workstation): 1996 edition", *J. Quant. Spectrosc. Radiat. Transfer*,**60**, 665.

Russell J. M., Gordley L. L., Park J. H., Drayson S. R., Hesketh W. D., Cicerone R. J., Tuck A. F., Frederick J. E., Harries J. E. and Crutzen P. J. (1993), The Halogen Occultation Experiment", *J. Geophys. Res.*, **98**, 10,777.

Russell, J. M. and Drayson, S. R. (1972), "The inference of atmospheric ozone using satellite horizon measurements in the 1042 cm^{-1} band", *J. Atmos. Sci.* **29**, 376.

Scales, L. E. (1985), *Introduction to Non-linear Optimization*, Macmillan, London.

Shannon, C. E. and Weaver, W. (1949), *The Mathematical Theory of Communication*, Paperback edition, University of Illinois Press, Urbana, 1962.

Shepherd G. G., Thuillier G., Gault W. A., Solheim B. H., Hersom C., Alunni J. M., Brun J. F., Brune S., Charlot P., Cogger L. L., Desaulniers D. L., Evans W. F. J., Gattinger R. L., Girod F., Harvie D., Hum R. H., Kendall D. J. W., Llewellyn E. J., Lowe R. P., Ohrt J., Pasternak F., Peillet O., Powell I., Rochon Y., Ward W. E., Wiens R. H. and Wimperis J. (1993), "WINDII, The Wind Imaging Interferometer on the Upper-Atmosphere Research Satellite", *J. Geophys. Res.*, **98**, 10,725.

Simeoni D., Singer C., Chalon G. (1997), "Infrared atmospheric sounding interferometer", *Acta Astronautica*, **40**, 113.

Smith, W. L., Woolf, H. M. and Jacob, W. J. (1970), "A regression method for obtaining real-time temperature and geopotential height profiles from satellite spectrometer measurements, and its applications to NIMBUS-3 SIRS observations", *Mon. Wea. Rev.*, **98**, 582.

Sparks, L. (1997), "Efficient line-by-line calculation of absorption coefficients to high numerical accuracy", *J. Quant. Spectrosc. Radiat. Transfer*, **57**, 631.

Stephens, G. L. (1994), *Remote Sensing of the Lower Atmosphere*, Oxford University Press.

Strand, O. N. (1974), "Coefficient errors caused by using the wrong covariance matrix in the general linear model", *Ann. Statistics*, **2**, 935.

Strow, L. L. (1988), "Line mixing in infrared atmospheric spectra", *SPIE Vol 928 Modeling* of the Atmosphere, 194.

Strow L. L., Motteler H. E., Benson R. G., Hannon S. E. and De Souza-Machado S. (1998), "Fast computation of monochromatic infrared atmospheric transmittances using compressed look-up tables", *J. Quant. Spectrosc. Radiat. Transfer*, **59**, 481.

Stuart, A., Ord, J. K. and Arnold, S. (1999), *Kendall's Advanced Theory of Statistics, Volume 2A*, 6th edition, Arnold and Oxford University Press.

Tarantola, A. (1987), *Inverse Problem Theory : Methods for Data Fitting and Model Parameter Estimation*, Elsevier, Amsterdam.

Thomas, G. E. and Stamnes, K. (1999), *Radiative Transfer in the Atmosphere and Ocean*, Cambridge University Press, Cambridge.

Tikhonov, A.-N. (1963), "On the solution of incorrectly stated problems and a method of regularization", *Dokl. Acad. Nauk SSSR*, **151**, 501.

Todling, R. and Cohn, S. E. (1994), "Suboptimal schemes for atmospheric data assimilation based on the Kalman filter", *Mon. Wea. Rev.*, **122**, 2530.

Twomey, S. (1963), "On the numerical solution of Fredholm integral equation of the first kind by the inversion of the linear system produced by quadrature", *J. Ass. Comput. Mach.*, **10**, 97.

Twomey, S., Herman, B. and Rabinoff, R. (1977), "An extension to the Chahine method of inverting the radiative transfer equation", *J. Atmos. Sci.*, **34**, 1085.

Twomey, S. (1996), *Introduction to the Mathematics of Inversion in Remote Sensing and Indirect Measurements*, Dover Publications Inc, Mineola, New York.

Ware R., Exner M., Feng D., Gorbunov M., Hardy K., Herman B., Kuo Y., Meehan T., Melbourne W., Rocken C., Schreiner W., Sokolovskiy S., Solheim F., Zou X., Anthes R., Businger S. and Trenberth K. (1996), "GPS sounding of the atmosphere from low earth orbit: Preliminary results", *Bull. Amer. Met. Soc*, **77**, 19.

Wark, D. Q. (1961), "On indirect temperature soundings of the stratosphere from satellites", *J. Geophys. Res.*, **66**, 77.

Wark, D. Q. and Hilleary, D. T. (1969), 'Atmospheric Temperature: Successful test of remote probing', *Science*, **165**, 1256.

Wark, D. Q. (1970), 'SIRS, An experiment to measure the free air temperature from a satellite', *Appl. Opt.*, **9**, 1761

Weinreb, M. P. and Neuendorffer, A. C. (1973), "Method to Apply Homogeneous-path Transmittance Models to Inhomogeneous Atmospheres", *J. Atmos. Sci.*, **30**, 662.

Wilkinson, J. H. (1965), *The Algebraic Eigenvalue Problem*, Oxford University Press.

Wiscombe, W. J. (1976), "Extension of the doubling method to inhomogeneous sources", *J. Quant. Spectrosc. Radiat. Transfer*, **16**, 477.

Yamamoto, G. (1961), "Numerical method for estimating the stratospheric temperature distribution from satellite measurements in the CO_2 band", *J. Met.*, **18**, 581.

Index

a priori, 25, **159–173**
   approximate, 110
   effect on retrieval, 161
   interpretation, 46
accuracy and precision, 50
adjoint, 15
adjoint equation, 134
AIRS, 136
analytic Jacobians, 145
approximate
   a priori, 110
   measurement error covariance, 111
   weighting functions, 111
assimilation, **129–140**
   adjoint methods, 132
   interface with sounders, 137
   optimal interpolation, 131
   radiances, 135
   successive correction, 130
   systematic errors, 138
averaging kernel
   area, 47
   function, 75
   Kalman filter, 126
   matrix, 31, 37, 47
   nonlinear case, 86
   standard example, 56
   width, 54

Backus–Gilbert, 55, **74–79**
band modelling, 154
   multiple absorbers, 157
basis, 17
Bayes' theorem, 22
Bayesian approach, 20–26

bias, 46

Chahine's method, 116
characterisation, **43–48**, 186
   nonlinear case, 86
χ^2 test, 187
Cholesky decomposition, 95, 200
climatology, 88
column amount retrieval, 73
column space, 17
condition number, 102
configurational entropy, 118
constrained exact solution, 101
contribution function, 9, 30
convergence
   correct, 90
   rate, 87
   slow, 91
   tests, 89
covariance, 20
covariance matrix, 20
   due to forward model parameter error, 50
   of MAP solution, 67
   of minimum variance solution, 72
   of retrieval due to measurement noise, 50
   of smoothing error, 49
   representing, 51
Curtis–Godson approximation, 155

data density, 54, 61, 165
degrees of freedom, 27
   for noise, 30
   for signal, 29, 54

from averaging kernel, 31, 37
δ-function response, 47
design of observing systems, **175–184**
   forward model, 176
   instruments, 175
   optimisation, 178
   retrieval and diagnostics, 177
   retrieval method, 179
diagnostics, 177
Dobson, 2

effective rank, 27
effective row space, 27
eigenvectors and values, 199
emissivity growth approximation, 156
empirical orthogonal function, 99
entropy, 33
   configurational, 118
   of a Gaussian distribution, 34
error analysis, **43**, 48–52, 186
   nonlinear case, 86
   onion peeling, 120
error patterns, 52, 55, 59, 165
error, measurement, 14
exact retrieval, 9

first guess, 88
Fisher, 32
   information matrix, 32, 69, 86
forward function, 14, 44
   parameters, 44
forward model, 13, 14, 43, **141**
   error, 48, 50
   parameter error, 48, 49
Fréchet derivative, **204**, 225
Fredholm integral equation, 8
full state vector, 69

gain
   Kalman, 122
   matrix, 30
gain error, 51
Gauss–Newton method, 85
Gaussian distribution function, 20
   multivariate, 21
grid spacing, 162
grossly nonlinear, 81

Hessian, 85

IASI, 136
impact parameter, 150
information, **13–41**
   content, 32
      linear Gaussian case, 36
   matrix, 32
information content
   from averaging kernel, 37
intercomparison of instruments, 192
internal consistency, 188
inverse Hessian method, 85
inverse method, 44

Jacobian, 15, 45, 98, 141
   analytic, 145

Kalman filter, **121–128**, 129, 134
   averaging kernel, 126
   error analysis, 126
   extended, 125
   linear, 122
Kalman gain, 122
Kalman smoother, 124
Kaplan, 2
kernel, 15

least squares solution, **105**
   overconstrained, 105
   underconstrained, 106
Levenberg–Marquardt, 92
likelihood, 32
limb sounders, 3
linear Gaussian problem
   information, 24
linear in χ, 145
linear methods, **65–79**
linear problems
   no error, 17
   with error, 20
linear relaxation, 114
linearity
   classification, 81
   degree of, 82
local thermodynamic equilibrium, 142
LTE, 142

m-form, 26, 67, 97
MAP, 66, 84
MaxEnt method, 118
maximum *a posteriori*, 66, 84

algebraic forms for linear problem, 66
algebraic forms for nonlinear problem, 85
maximum entropy, 40, 118
maximum likelihood, 32
   incorrect usage, 66
McMillin–Fleming method, 156
measure function, 34, 41
measurement error, 44
measurement noise, 14, 44
measurement space, 16
measurement vector, 13
MEM, 118
minimum variance, 71
mixed-determined, 18
ML, 66
model parameter error, 48
model resolution function, 47
model resolution matrix, 31
modelling error, 60
moderately nonlinear, 81
multiple regression, 113

n-form, 26, 67, 94
nadir sounders, 3
nearly linear, 81
Newton's method, 85
noise
   measurement, 14
   retrieval, 48, 50
nonlinear problem, **81–100**
   formulation, 83
nonlinear relaxation, 116
nonlinearity
   classification, 81
   degree of, 82
normal distribution function, 20
normal equations, 106
null space error, 49
numerical efficiency, 93

onion peeling, 119
   error analysis, 120
optimal, 43
optimal methods
   approximations, 110
   linear, **65–79**
   nonlinear, **81–100**
OPTRAN, 156
over-determined, 18

point spread function, 47
popular mistake, 88
precision and accuracy, 50
prewhitening, 28
prior information, 22
pseudo inverse, 108, 162

QR decomposition, 96

radiative transfer equation, 141
random error, 14, 50
range, 17
rank of a matrix, 17
ray tracing, 147
Rayleigh, 54
reference ellipsoid, 148
relaxation
   linear, 114
   nonlinear, 116
representations
   optimising, 99, 163
representing covariances, 51
resolution, 52–55, 61
resolving length, 55, 77
retrieval
   error, 46
   grid spacing, 162
   method, 44
   noise, 48, 50
   of column amount, 73
row space, 17

sequential updating, 97
Shannon, 32
signal-to-noise ratio, 33
single scattering albedo, 142
singular vector decomposition, 18, **201**
   truncated, 107
smoothing, 46
smoothing error, 48, 164
spread, 55, 62, 75
square root of a matrix, 200
standard example, 9
   averaging kernel, 56
   Backus–Gilbert, 77
   convergence of linear relaxation, 115
   effective rank, 29
   error analysis for MAP retrieval, 55
   exact solution
      condition number, 102

information content and degrees of
      freedom, 37
  least squares solution, 105
  nonlinearity, 83
  singular vectors of the weighting
      functions, 19
state space, 15
state vector, 13
state vector choice, 180
systematic error, 50, 138, 172

tangent linear model, 15, 134
trade-off
  parameter, 78
  plot, 53
transfer function, 45
transmittance modelling, 152
  band transmittance, 154
  line-by-line, 153
triangulation and back substitution, 95
truncated singular vector decomposition,
    107
Twomey–Tikhonov, **108**, 159

Umkehr, 2
under-determined, 17

validation, **185**
vector space, 15, 197
vertical grid choice
  coordinate, 181
  spacing, 162
virtual measurement, 22

weighting function, 8, 15, 45
  approximate, 111